Programming 32-bit
Microcontrollers in C

Exploring the PIC32

Programming 32-bit Microcontrollers in C

Exploring the PIC32

Lucio Di Jasio

ELSEVIER

AMSTERDAM • BOSTON • HEIDELBERG • LONDON
NEW YORK • OXFORD • PARIS • SAN DIEGO
SAN FRANCISCO • SINGAPORE • SYDNEY • TOKYO
Newnes is an imprint of Elsevier

Newnes

Newnes is an imprint of Elsevier
30 Corporate Drive, Suite 400, Burlington, MA 01803, USA
Linacre House, Jordan Hill, Oxford OX2 8DP, UK

Recognizing the importance of preserving what has been written, Elsevier prints its
books on acid-free paper whenever possible.

Library of Congress Cataloging-in-Publication Data
Application submitted

British Library Cataloguing-in-Publication Data
A catalogue record for this book is available from the British Library.

ISBN: 978-0-7506-8709-6

For information on all Newnes publications
visit our Web site at www.books.elsevier.com

08 09 10 11 12 10 9 8 7 6 5 4 3 2 1

Typeset by Charon Tec Ltd (A Macmillan Company), Chennai, India
www.charontec.com

Printed in the United States of America

Working together to grow
libraries in developing countries

www.elsevier.com | www.bookaid.org | www.sabre.org

ELSEVIER   BOOK AID International   Sabre Foundation

Dedicated to my son, Luca.

Acknowledgments

Once more this project would have never been possible if I did not have 110% support from my wife Sara, who understands my passion(s) and constantly encourages me to pursue them. Special thanks go to Steve Bowling and to Garry Champ. Their passion and experience in embedded control application caused them to volunteer for reviewing the technical content of this book. While Garry did not know what he was signing up to, Steve should have known better having been my primary technical resource for the previous book. I owe big thanks also to Patrick Johnson, who enthusiastically supported this book idea from the very beginning and pulled all the stops to make sure that I would be able to work in direct contact with his most advanced design and application teams working on the PIC32 project. Thanks to Joe Triece, "the architect", for being always available to me and always curious about my experiences and impressions. Thanks to Joe Drzewiecky for assembling such a complex tool suite, always working hard to make MPLAB© IDE a better tool. Special thanks also go to the entire PIC32 application team headed by Nilesh Rajbharti and a special mention to Adrian Aur, Dennis Lehman, Larry Gass and Chris Smith for addressing quickly all my questions and offering so much help and insight into the inner workings of the microcontroller, the peripherals and its libraries. But I would like to extend my gratitude to all my friends, the colleagues at Microchip Technology and the many embedded control engineers I have been honored to work with over the years. You have so profoundly influenced my work and shaped my experience in the fantastic world of embedded control.

Finally, since the publication of my previous book on Programming 16-bit microcontrollers in C, I have received so much feedback and so many readers have written to me to congratulate but also to point out errors and issues. This has been a very humbling but also rewarding experience and I want to thank you all. I tried to incorporate as many of your suggestions as possible in this new work but I am still looking for your continued support and advice.

Contents

Introduction .... *xix*

Part 1: Exploring **1**

Day 1: The Adventure Begins **3**

The Plan..3
Preparation 3
The Adventure Begins...6
Compiling and Linking 8
The Linker Script..10
Building the First Project 11
Using the Simulator..12
Finding a Direction...14
The JTAG Port...16
Testing PORTB...17
Mission Debriefing..19
Notes for the Assembly Experts 20
Notes for the PIC MCU Experts..22
Notes for the C Experts 22
Tips & Tricks 22
Exercises..23
Books..24
Links...24

Day 2: Walking in Circles **25**

The Plan..25
Preparation 25
The Exploration..27
While Loops 28

An Animated Simulation...31
Using the Logic Analyzer..35
Debriefing..37
Notes for the Assembly Experts...38
Notes for the 8-Bit PIC Microcontroller Experts.......................38
Notes for the 16-Bit PIC Microcontroller Experts.....................38
Notes for the C Experts...39
Notes for the MIPS Experts...39
Tips & Tricks...39
Notes on Using the Peripheral Libraries....................................40
Exercises..42
Books...42
Links...42

Day 3: Message in a Bottle...**43**
The Plan...43
Preparation..43
The Exploration...43
Do Loops..44
Variable Declarations..45
for Loops..47
More Loop Examples...48
Arrays...49
Sending a Message..50
Testing with the Logic Analyzer...53
Testing with the Explorer 16 Demonstration Board...................54
Testing with the PIC32 Starter Kit..55
Debriefing..57
Notes for the Assembly Experts...57
Notes for the PIC Microcontroller Experts.................................58
Notes for the C Experts...58
Tips & Tricks...59
Exercises..60
Books...60
Links...60

Day 4: NUMB3RS...**61**
The Plan...61
Preparation..61
The Exploration...61
On Optimizations (or Lack Thereof)...64
Testing...64

Going long long 65
Integer Divisions 67
Floating Point 69
Measuring Performance 70
Debriefing...73
Notes for the Assembly Experts 73
Notes for the 8-Bit PIC® Microcontroller Experts 75
Notes for the 16-Bit PIC and dsPIC® Microcontroller Experts 76
Tips & Tricks 77
Exercises..78
Books...79
Links 79

Day 5: Interrupts.. 81
The Plan...81
Preparation...81
The Exploration...81
Interrupts and Exceptions 82
Sources of Interrupt...84
Interrupt Priorities 85
Interrupt Handlers Declaration 88
The Interrupt Management Library 90
Single Vector Interrupt Management 90
Managing Multiple Interrupts 95
Multivectored Interrupt Management...................................98
A Simple Application 103
The Secondary Oscillator 108
The Real-Time Clock Calendar (RTCC) 109
Debriefing..111
Notes for the PIC Microcontroller Experts 111
Tips & Tricks 112
Exercises..113
Books...113
Links 113

Day 6: Memory 115
The Plan...115
Preparation...115
The Exploration...116
Memory Space Allocation..118
Looking at the MAP 123
Pointers 127

The Heap 128
The PIC32MX Bus..129
PIC32MX Memory Mapping 130
The Embedded-Control Memory Map 134
Debriefing...135
Notes for the C Experts 135
Notes for the Assembly Experts.......................................136
Notes for the PIC Microcontroller Experts 136
Tips & Tricks 137
Exercises..137
Books 138
Links..138

Part 2: Experimenting...**139**
Day 7: Running **141**
The Plan...141
Preparation..141
The Exploration 142
Performance vs. Power Consumption 144
The Primary Oscillator Clock Chain 146
The Peripheral Bus Clock..147
Initial Device Configuration..148
Setting Configuration Bits in Code 150
Heavy Stuff...152
Ready, Set, Go!..158
Fine-Tuning the PIC32: Configuring Flash Wait States.....................160
Fine-Tuning the PIC32: Enabling the Instruction and Data Cache..........163
Fine-Tuning the PIC32: Enabling the Instruction Pre-Fetch...............164
Fine-Tuning the PIC32: Final Notes 165
Debriefing..167
Notes for the Assembly Experts.......................................167
Notes for the PIC® Microcontroller Experts 167
Tips & Tricks 168
Exercises..171
Books 171
Links..171

Day 8: Communication **173**
The Plan...173
Preparation..173
The Exploration 174

Synchronous Serial Interfaces 174
Asynchronous Serial Interfaces 176
Parallel Interfaces 177
Synchronous Communication Using the SPI Modules 178
Testing the Read Status Register Command 182
Writing Data to the EEPROM 186
Reading the Memory Contents...187
A 32-bit Serial EEPROM Library 187
Testing the New SEE Library..191
Debriefing...193
Notes for the C Experts 193
Notes for the Explorer 16 Experts 193
Notes for the PIC24 Experts 194
Tips & Tricks..194
Exercises...195
Books 196
Links 196

Day 9: Asynchronous Communication 197
The Plan..197
Preparation...197
The Exploration 197
UART Configuration 200
Sending and Receiving Data 202
Testing the Serial Communication Routines 204
Building a Simple Console Library 206
Testing a VT100 Terminal...209
The Serial Port as a Debugging Tool...211
The Matrix Project...211
Debriefing...214
Notes for the C Experts 214
Notes for the PIC® Microcontroller Experts 215
Tips & Tricks...215
Exercises...216
Books 216
Links 217

Day 10: Glass = Bliss.. 219
The Plan..219
Preparation...219
The Exploration 219
HD44780 Controller Compatibility..221

The Parallel Master Port..223
Configuring the PMP for LCD Module Control 224
A Small Library of Functions to Access an LCD Display 225
Building an LCD Library and Using the PMP Library 231
Creating the *include* and *lib* Directories..237
Advanced LCD Control..240
Progress Bar Project 241
Debriefing..245
Notes for the PIC24 Experts...245
Tips & Tricks..246
Exercises...246
Books...247
Links...247

Day 11: It's an Analog World **249**
The Plan..249
Preparation...249
The Exploration..249
The First Conversion 253
Automating Sampling Timing..254
Developing a Demo 255
Creating Our Own Mini ADC Library 257
Fun and Games..258
Sensing Temperature 261
Debriefing..266
Notes for the PIC24 Experts...266
Tips & Tricks..267
Exercises...267
Books...268
Links...268

Part 3: Expansion... **269**
Day 12: Capturing User Inputs **271**
The Plan..271
Preparation...271
Buttons and Mechanical Switches...272
Button Input Packing..275
Button Inputs Debouncing..277
Rotary Encoders 280
Interrupt-Driven Rotary Encoder Input 283
Keyboards..288
PS/2 Physical Interface...288

The PS/2 Communication Protocol 289
Interfacing the PIC32 to the PS/2 290
Input Capture 290
Testing Using a Stimulus Scripts 296
The Simulator Profiler 301
Change Notification 302
Evaluating Cost 308
I/O Polling 309
Testing the I/O Polling Method 314
Cost and Efficiency Considerations 317
Keyboard Buffering 319
Key Code Decoding 324
Debriefing 328
Notes for the PIC24 Experts 329
Tips & Tricks 329
Exercises 330
Books 330
Links 331

Day 13: UTube 333
The Plan 333
Preparation 333
The Exploration 334
Generating the Composite Video Signal 337
The Output Compare Modules 342
Image Buffers 345
Serialization, DMA, and Synchronization 346
Completing a Video Library 353
Testing the Composite Video 357
Measuring Performance 360
Seeing the Dark Screen 360
Test Pattern 362
Plotting 364
A Starry Night 366
Line Drawing 368
Bresenham Algorithm 370
Plotting Math Functions 373
Two-Dimensional Function Visualization 376
Fractals 381
Text 389
Printing Text on Video 391
Text Test 394

The Matrix Reloaded..395
Debriefing...398
Notes for the PIC24 Experts...399
Tips & Tricks...399
Exercises...401
Books..402
Links...402

Day 14: Mass Storage... **403**
The Plan..403
Preparation..403
The Exploration..404
The Physical Interface..405
Interfacing to the Explorer 16 Board..406
Starting a New Project..407
Selecting the SPI Mode of Operation...408
Sending Commands in SPI Mode...408
Completing the SD Card Initialization..411
Reading Data from an SD/MMC Card...413
Writing Data to an SD/MMC Card..416
Testing the SD/MMC Interface..419
Debriefing...424
Tips & Tricks...425
Exercises...426
Books..426
Links...426

Day 15: File I/O... **427**
The Plan..427
Preparation..427
The Exploration..428
Sectors and Clusters..428
The File Allocation Table..429
The Root Directory...430
The Treasure Hunt...433
Opening a File...444
Reading Data from a File..454
Closing a File..459
The Fileio Module..460
Testing fopenM() and freadM()..463
Writing Data to a File...465
Closing a File, Take Two...471

Accessory Functions..473
Testing the Complete Fileio Module 476
Code Size..480
Debriefing..481
Tips & Tricks..481
Exercises..482
Books...482
Links..483

Day 16: Musica, Maestro!.. **485**
The Plan...485
Preparation..485
The Exploration 486
OC PWM Mode..488
Testing the PWM as a D/A Converter 490
Producing Analog Waveforms...492
Reproducing Voice Messages 497
A Media Player...498
The WAVE File Format 500
The `Play()` Function 501
The Audio Routines..510
A Simple WAVE File Player 513
Debriefing..515
Tips & Tricks..516
Exercises..516
Books...516
Links..517
Disclaimer 517
Final Note for the Experts 517

Index **519**

Introduction

The first step in almost every rehabilitation program is A- Acknowledge . . . your limitations. So this is how I need to start this book, I will admit it: I am an *8-bitter*!

I have been programming 8-bit microcontrollers since I was in high school and for most of my professional career. And there is worse, while I am relatively fluent in several high level programming languages, I truly love assembly programming!

There, I said it! I love that kick that I get when I know I used every single machine cycle in every microsecond my embedded applications run. I am also obsessed with control: I like to know of every configuration bit in every peripheral I use. As a consequence, in general, I don't trust compilers or other people's libraries unless I really cannot live without them or I have them completely disassembled.

So why would I write a book about 32-bit programming in C?

In fact I started what I should call my "rehabilitation program" a couple of years ago by approaching the programming of 16-bit microcontrollers first. The introduction of the PIC24 family of microcontrollers gave me the motivation to try and migrate to C programming with a new and exciting architecture. As a result of my experience, I wrote the first book: "Programming 16-bit microcontrollers in C. Learning to fly the PIC24". But by the time the book was published, rumors circulated in Microchip that a new 32-bit chip had just come out of the "ovens" and I had to have one!

I'll spare you the details of how I got my hands around one of the very first test chips, but what you need to know is that in a matter of days I had most of the code, originally developed for the PIC24 book, ported and running on the PIC32 plugged in my old Explorer16 board.

Microchip marketing folks will tell you that the PIC32 architecture was specifically designed so to make the "migration" from 8-bit and 16-bit PIC architectures smooth and seamless, but I had to see it with my eyes to believe it.

So who better than an assembly-loving, control-obsessed, 8-bitter can tell you about the exploration of the PIC32?

Who Should Read this Book?

The PIC32 turns out to be a remarkably easy to use device, but nonetheless, it is a truly powerful machine based on a well established 32-bit core (MIPS) and supported by a large number of tools, libraries and documentation. This book can only offer you a small glimpse into such a vast world and in fact I call it a first "exploration". It is my strong belief that learning should be fun, and I hope you will have a good time with some of the "playful" exercises and projects I present throughout each chapter in the book. However you will need quite some preparation and hard work in order to be able to digest the material I am presenting at a pace that will accelerate rapidly through the first few chapters.

This book is meant for programmers of a basic to intermediate level of experience, but not for "absolute" beginners; so don't expect me to start with the basics of the binary numbers, the hexadecimal notation or the fundamentals of programming. Although, we will briefly review the basics of C programming as it relates to the applications for the latest generation of general-purpose 32-bit microcontrollers, before moving on to more challenging projects. My assumption is that you, the reader, belong to one of the following four categories:

- Embedded Control programmer: experienced in assembly-language micro-controllers programming, but with only a basic understanding of the C language.

- PIC® microcontroller expert: with a basic understanding of the C language.

- Student or professional: with some knowledge of C (or C++) programming for PCs.

- Other SLF (superior life forms): I know programmers don't like to be classified that easily so I created this special category just for you!

Depending on your level and type of experience, you should be able to find something of interest in every chapter. I worked hard to make sure that every one of them contained both C programming techniques and new hardware peripherals details. Should you already be familiar with both, feel free to skip to the experts section at the end of the chapter, or consider the additional exercises, book references and links for further research/reading.

A special note is reserved for those of you who have already read my previous book on programming 16-bit microcontrollers in C. First of all let me thank you, then let me explain why you will get a certain sensation of deja vu. No, I did not try to cheat my way through the old 16-bit material to produce a new book, but I have re-produced most of the projects to demonstrate practically the main claims of the PIC32 architecture and toolset: its seamless migration from 8 and 16-bit PIC applications, the vastly increased performance and nonetheless the great ease of use. For you, at the end of every chapter, I have included a special section where I detail the differences encountered, the enhancements and other information that will help you port your applications faster and with greater confidence.

These are some of the things you will learn:

- The structure of an embedded-control C program: loops, loops and more loops

- Basic timing and I/O operations

- Basic embedded control multitasking in C, using the PIC32 interrupts

- New PIC32 peripherals, in no specific order:

  1. Input Capture

  2. Output Compare

  3. Change Notification

  4. Parallel Master Port

  5. Asynchronous Serial Communication

  6. Synchronous Serial Communication

  7. Analog-to-Digital conversion

- How to control LCD displays

- How to generate video signals

- How to generate audio signals

- How to access mass-storage media

- How to share files on a mass-storage device with a PC

Structure of the Book

Each chapter of the book is offered as a day of exploration in the 32-bit embedded programming world. There are three parts. The first part contains six small chapters of increasing levels of complexity. In each chapter, we will review one basic hardware peripheral of the PIC32MX family of microcontrollers and one aspect of the C language, using the MPLAB C32 compiler (Student Version included in the CD-ROM). In each chapter, we will develop at least one demonstration project. Initially, such projects will require exclusive use of the MPLAB SIM software simulator (a part of the MPLAB toolsuite included in the CD-ROM), and no actual hardware will be necessary; although, an Explorer 16 demonstration board or a PIC32 Starter kit might be used.

In the second part of the book, titled "Experimenting" and containing five more chapters, an Explorer 16 demonstration board (or third-party equivalent) will become more critical, as some of the peripherals used will require real hardware to be properly tested.

In the third part of the book, titled "Expansion", there are five larger chapters. Each one of them builds on the lessons learned in multiple previous chapters while adding new peripherals to develop projects of greater complexity. The projects in the third part of the book require the use of the Explorer 16 demonstration board and basic prototyping skills too (yes, you might need to use a soldering iron). If you don't want to or you don't have access to basic PCB prototyping tools, an ad hoc expansion board (AV32) containing all the circuitry and components necessary to complete all the demonstration projects will be made available on the companion web site: *http://www.exploringpic32.com*

All the source code developed in each chapter is also available for immediate use on the companion CD-ROM.

What this Book is Not

This book is not a replacement for the PIC32 datasheet, reference manual and programmer's manual published by Microchip Technology. It is also not a replacement for the MPLAB C32 compiler user's guide, and all the libraries and related software tools offered by Microchip. Copies are available on the companion CD-ROM, but I expect you to download the most recent versions of all those documents and tools from Microchip's Web site (*http://www.microchip.com*). Familiarize yourself with them and keep them handy. I will often refer to them throughout the book, and I might present small block diagrams and other excerpts here and there as necessary. But, my narration cannot replace the information presented in the official manuals. Should you notice a conflict between my narration and the official documentation, ALWAYS refer to the latter. However please send me an email if a conflict arises, I will appreciate your help and I will publish any correction and useful hint I will receive on the companion web site: *http://www. exploringpic32.com*

This book is also not a primer on the C language. Although a review of the language is performed throughout the first few chapters, the reader will find in the references several suggestions on more complete introductory courses and books on the subject.

Checklists

Although this book is not directly making references to aviation and flight training as my previous book was, I decided to maintain some important elements introduced there.

The use of checklists to perform every single procedure before and during each project is one of them. Pilots don't use checklists because the procedures are too long to be memorized or because they suffer from short memory problems. They use checklists because it is proven that the human memory can fail, and tends to do so more often when stress is involved. Pilots can perhaps afford less mistakes than other categories, and they value safety above their pride. There is nothing really dangerous that you, as a programmer can do or forget to do, while developing code for the PIC32. Nonetheless, I have prepared a number of simple checklists to help you perform the most common programming and debugging tasks. Hopefully, they will help you in the early stages, when learning to use the new PIC32 toolset or later if you are, like most of us, alternating between several projects and development environments from different vendors.

The Pilot Checklist – MPLAB® IDE Quick Start Guide

New Project Setup

Project>Project Wizard	Start
Step 1: Device	PIC32MX360F512L
Step 2: ToolSuite	MPLAB C32 C Compiler
Step 3: NewProject dialog box	Select BROWSE
Folder	Select or create new
Project name	Type new name here
Step 4: Copy files	Only if necessary
Step 5: Complete wizard	Click on Finish

Manual Device Configuration (if not using pragmas)

Configure>Configuration Bits	Open window
Configuration bits set in ocde	Unchecked
ICE/ICD Comm channel select	ICE EMUC2/EMUD2 share with PGCD2
Boot Flash Write Protect	Boot Flash is writable
Code Protect	Protection Disabled
Oscillator Selection bits	Primary OSC with PLL (XT, HS, EC)
Secondary Oscillator Enable	Enabled
Internal External Switchover	Disabled
Primary Oscillator Configuration	XT osc mode
CLKO output signal active on OSCO	Disabled
Peripheral Clock Divisor	PB clock is Sys clock/2
Clock Switching and Monitor	Disabled and clock monitor disabled
Watchdog Timer Postscaler	Any
Watchdog Timer Enable	Disabled
PLL Input Divider	2× Divider
PLL Multiplier	18× Multiplier
System PLL output clock divider	PLL Divide by 1

Create New File and Add to Project

Project>AddNewProjectFile	Assign name (.c or .h)
File>Open	Select "\c32\include\Template.c"
if main source file & using pragmas	Select "\c32\include\Template wPragmas.c"
Header/comments	Copy
Add code	As needed
File>Save	Select
Project>SaveProject	Select

MPLAB SIM Debugger Setup

Debugger>Select Tool	Select MPLAB SIM
Debugger>Settings	Select
1. Osc/Trace Tab	Select
1.1 Processor Frequency	72 MHz
1.2 Trace Options	Trace All
2. Animation/Real Time Updates	Select Tab
2.1 Animate Step	Slow 500 ms/Fast 10 ms
3. Apply/OK	Select

PIC32MX Family Characteristics

Vdd range	2.0V to 3.6V
Digital input pins	5V tolerant
Analog input pins	0V to 3.6V max

MPLAB ICD2 In Circuit Debugger Setup

Target Board	Power Up
ICD2 to Target	Connect
ICD2 to PC	Connect (wait for triple ding-dong)
Debugger>SelectTool	Select MPLAB ICD2
Debugger>Settings	Select
1. Status Tab	Select
1.1 Automatically Connect	Verify NOT Checked
2. Power Tab	Select
2.1 Power target from ICD2	Verify NOT Checked
3. Program Tab	Select
3.1 Allow ICD2 to select ranges	Verify Checked
3.2 Program after successful build	Select if desired (not recommended)
3.3 Run after successful program	Select if desired (not recommended)
OK button	Click
Debugger>Connect	Select

Emergency: USB Drivers Re-start (Debugger fails to connect)

Debugger>SelecTool	Select None
Project>Close	Save Project and close
File>Exit	Terminate MPLAB
USB cable	Disconnect
Target	Cycle Power
MPLAB	Launch
USB cable	Connect (wait for enumeration)
Debugger>SelecTool	Select Debugger model
Debugger>Connect	Select (not required for REAL ICE)

Emergency: Breakpoint Cannot Be Set (debugging)

1. Verify the C source code line is not commented
2. Verify you have not used more than six breakpoints (see breakpoints list F2)
3. Verify the C source line does not contain only a variable declaration
4. Verify the C source file is part of the Project Files list
5. Verify the project has been Built before placing a breakpoint

Explorer16 Demonstration Board

Power Supply	9 V to 15V (reversed polarity protected)
Main oscillator	8 MHz crystal (use 4× PLL to obtain 32MHz)
Secondary oscillator	32,768 Hz (connected to TMR1 oscillator)

The Pilot Checklists – MPLAB® IDE Quick Start Guide (Debugginh and Emergencies)

Project Build

1.	Project>Build Configuration	Select "Debug"
2.	Project>BuildOptions>Project	Open Dialog box
2.1	Directories Tab	Select
2.2	Show Directories for:	Select "Include Search path"
2.3	"New" Button	Press
2.4	" … " Button	Press and select "\C32\include" directory
3.	MPLAB PIC32 C Compiler Tab	Select
3.1	Categories	Select "General"
3.2	Generate debugging information	Checked
3.3	Categories	Select Optimization
3.4	Optimization Level	Select 0 during debugging
3.5	All other optimization options	Unchecked during debugging
4.	MPLAB PIC32 Linker Tab	Select
4.1	Categories	Select "General"
4.2	Heap Size	Assign generously if malloc() used
5.	OK button	Click

Add all (.c) (.h) and (.o) required — Use "Add Files to a Project" checklists (A, B or C)
Project>BuildAll — Select (CTRL+F10)
or Project>Make — Select (F10) if only a few modules modified

Adding Files to a Project

View>Project	
Project>AddFilesToProject	
1. Select directory	Checked
2. Select files of type	Select
3. Select File name	If required (.c), (.h) or (.o)
Project>SaveProject	Select

Method A

Checked
Select
If required
(.c), (.h) or (.o)
Select

Adding Files to a Project

File>Open	
With cursor inside Editor	
Editor pop up menu	
Project>SaveProject	

Method B (text files only)

Open existing file
Right Click
Select AddToProject
Select

Adding Files to a Project

View>Project	
With cursor on File folder	
Project pop up menu	
Project>SaveProject	

Method C (from Project window)

Checked
Right Click
Select Add Files…
Select

Simulator Logic Analyzer Setup

View>SimulatorLogicAnalyzer	Select
Debugger>Settings>Osc/Trace	Select
TraceOptions>TraceAll	Verify Checked
Channels button	Click
Available Signals	Select all required
Signals Order	Move Up/Down
OK button	Click

PIC32MX360F512L Characteristics

Maximum operating speed	72 MHz
General Purpose RAM available	32,768 bytes
FLASH Program memory	512k bytes

MPLAB REAL ICE In Circuit Debugger Setup

Target Board	Power Up
ICD2 to Target	Connect
ICD2 to PC	Connect (wait for enumeration)
Debugger>SelectTool	Select MPLAB REAL ICE

PIC32 Starter Kit In Circuit Debugger Setup

PIC32 Starter Kit to Target	Connect
Target Board	Power Up
PIC32 Starter Kit to PC	Connect (Wait for enumeration)
Debugger>SelectTool	Select PIC32MX Starter Kit

Emergency: Lost Cursor while Single Stepping/Animate

Program Counter value — Check in MPLAB status bar (bottom)
1. Place cursor on first line of main() Execute Run To Cursor
2. Continue single stepping until the cursor reappears in the main program
3. Search for the PC in the Memory Window

Else — Most likely you Stepped IN a library function

1. Place the cursor on the next C statement execute Run To Cursor
2. If you have one or more breakpoints already set, execute Run

IF all else seems to fail
Send RESET command and start again

Emergency: After Pressing Halt, MPLAB Freeze (ICD2 debugging)

Wait!
1. MPLAB could be uploading the content of a large variable/array in the Watch window
2. MPLAB could be refreshing the Special Function Registers window (if open)
3. MPLAB could be updating the Disassembly window (if open)
4. MPLAB could be updating the Local Variables window (if open and contains a large object)

After regaining control, close any data window or remove any large object before continuing

Exploring

The Adventure Begins

The Plan

This will be our first experience with the PIC32 32-bit microcontroller and, for some of you, the first project with the MPLAB® IDE Integrated Development Environment and the MPLAB C32 language suite. Even if you have never heard of the C language, you might have heard of the famous "Hello World!" programming example. If not, let me tell you about it.

Since the very first book on the C language, written by Kernighan and Ritchie several decades ago, every decent C language book has featured an example program containing a single statement to display the words "Hello World" on the computer screen. Hundreds, if not thousands, of books have respected this tradition, and I don't want my books to be the exception. However, our example will be just a little different. Let's be realistic—we are talking about programming microcontrollers because we want to design *embedded-*control applications. Though the availability of a monitor screen is a perfectly safe assumption for any personal computer or workstation, this is definitely not the case in the embedded-control world. For our first embedded application we'd better stick to a more basic type of output: a digital I/O pin. In a later and more advanced chapter, we will be able to interface to an LCD display and/or a terminal connected to a serial port. But by then we will have better things to do than writing "Hello World!"

Preparation

Whether you are planning a small outdoor trip or a major expedition to the Arctic, you'd better make sure you have the right equipment with you. Our exploration of the PIC32 architecture is definitely not going to be a matter of life or death, but you will appreciate the convenience of following the few simple steps outlined here before getting your foot out the door . . . ahem, I mean before starting to type the first few lines of code.

So, let's start by verifying that we have all the necessary pieces of equipment ready and installed (from the attached CD-ROM and/or the latest version available for download from Microchip's PIC32 Web site at *www.microchip.com/PIC32*). You will need the following:

- MPLAB IDE, free Integrated Development Environment (v8.xx or later)

- MPLAB SIM, free software simulator (included in MPLAB)

- MPLAB C32, C compiler (free Student Edition)

Now let's use the New Project Setup checklist to create a new project with the MPLAB IDE. From the **Project** menu, select the **Project Wizard**. This will bring up a short but useful sequence of little dialog boxes that will guide us through the few steps required to create a new project in an orderly and clean way:

1. The first dialog box will ask you to choose a specific device model. Select the **PIC32MX360F512L** device and click **Next**. Although we will use only the simulator, and for the purpose of this project we could use pretty much any PIC32 model, we will stick to this particular part number throughout our exploration.

2. In the second dialog box, select the **PIC32 C-Compiler Tool Suite** and click **Next**. Many other tool suites are available for all the other PIC© architectures, and at least one other tool suite is already available for development on the PIC32 in assembly; don't mix them up!

3. In the third dialog box, you are asked to assign a name to the new project file. Instead click the **Browse** button and create a new folder. Name the new folder **Hello,** and inside it create the project file **Hello World**, then click **Next**.

4. In the fourth dialog box, simply click **Next** to proceed to the following dialog box since there is no need to copy any source files from any previous projects or directories.

5. Click **Finish** to complete the project setup.

    Since this is our first time, let's continue with the following additional steps:

6. Open a new editor window by selecting **File | New**, typing the **Ctrl + N** keyboard shortcut or by clicking the corresponding ⬜ (**New File**) button in the MPLAB standard toolbar.

7. Type the following three comment lines:

```
/*
**Hello Embedded World!
*/
```

8. Select **File | Save As** to save the file as Hello.c.

9. Now right-click with your mouse on the editor window to bring up the editor's context menu and select the **Add To Project** item. This will tell MPLAB that the newly created file is an integral part of the project.

10. Select **Project | Save Project** to save the project.

Note

You will notice that, after saving the file, the color of the three lines of text in the editor window changes to green. This is because the MPLAB Editor has been able to recognize your file as a C language source file (the .c extension tipped it off) and is now applying the default context-sensitive color rules. According to theses rules, green is the color assigned to comments, blue is the color assigned to language keywords, and black is used for all the remaining code.

Once you are finished, your project window should look like the one in Figure 1.1. If you cannot see the project window, select **View | Project**. A small check mark should appear next to the item in the View menu. Also make sure that the Files tab is selected. We will review the use of the other tab (Symbols) in a later chapter.

Figure 1.1: The "Hello World" Project window.

Depending on your personal preferences, you might now want to "dock" this window to assign it a specific place on your workspace rather than keeping it floating. You can do so by right-clicking with your mouse on the title bar of the small window to access the context menu and selecting the **Dockable** option. You can then drag it to the desired edge of the screen, where it will stick and split the available space with the editor.

The Adventure Begins

It is time to start writing some code. I can sense your trepidation, especially if you have never written any C code for an embedded-control application before. Our first line of code is:

```
#include <p32xxxx.h>
```

This is not yet a proper C statement but an instruction for the preprocessor (which feeds the compiler) with the request to include the content of a device-specific file before proceeding any further. The pic32xxxx.h file, in its turn, contains more `#include` instructions designed so that the file relative to the device currently selected in the project is included. That file in our case is p32mx360f512l.h. We could have used its name directly, but we chose not to in order to make the code more independent and hopefully easier to port, in the future, to new projects using different models.

If you decide to further inspect the contents of the p32mx360f512l.h file (it is a simple text file that you can open with the MPLAB editor), you will see that it contains an incredibly long list of definitions for all the names of the internal special-function registers (often referred to in the documentation as the *SFRs*) of the chosen PIC32 model. If the include file is accurate, those names reflect exactly those being used in the device datasheet and the PIC32 reference manual.

Here is a segment of the p32mx360f512l.h file in which the special-function register that controls the watchdog module (WDTCON) and each of its individual bits are assigned their conventional names:

```
...
extern volatile unsigned int    WDTCON__attribute__
((section("sfrs")));
typedef union {
  struct {
    unsigned WDTCLR:1;
```

```
    unsigned WDTWEN:1;
    unsigned SWDTPS0:1;
    unsigned SWDTPS1:1;
    unsigned SWDTPS2:1;
    unsigned SWDTPS3:1;
    unsigned SWDTPS4:1;
    unsigned :7;
    unsigned FRZ:1;
    unsigned ON:1;

  };

...
```

Back to our Hello.c source file; let's add a couple more lines that will introduce you to the `main()` function:

```
main()
{

}
```

What we have now is already a complete, although still empty and pretty useless, C language program. In between those two curly brackets is where we will soon put the first few instructions of our embedded-control application.

Independently of this function position in the file, whether in the first lines on top or the last few lines in a million-lines file, the `main()` function is the place where the microcontroller will go first at power-up or after each subsequent reset. This is actually an oversimplification. After a reset or at power-up, but before entering the `main()` function, the microcontroller will execute a short initialization code segment automatically inserted by the MPLAB C32 linker. This is known as the *Startup* code or *crt0* code (or simply c0 in the traditional C language literature). The Startup code will perform basic housekeeping chores, including the all important initialization of the stack, among many other things.

Our mission is to activate for the first time one or more of the output pins of the PIC32. For historical reasons, and to maintain the greatest compatibility possible with the many previous generations of PIC microcontrollers, the input/output (I/O) pins of the PIC32 are grouped in modules or ports, each comprising up to 16 pins, named in alphabetical order from A to H. We will start logically from the first group known as PortA. Each port has several special-function registers assigned to control its operations; the main one, and the easiest to use, carries traditionally the same name as the module (PORTA).

Notice how, to distinguish the control register name from the module name in the following, we will use a different notation for the two: PORTA (all uppercase) will be used to indicate one of the control registers; PortA will refer to the entire peripheral module.

According to the PIC32 datasheet, assigning a value of 1 to a bit in the PORTA register turns the corresponding output pin to a logic high level (3.3 V). Vice versa, assigning a value of 0 to the same bit will produce a logic level low on the output pin (0 V).

Assignments are easy in C language—we can insert a first *assignment statement* in our project as in the following example:

```
#include <p32xxxx.h>

main()
{
   PORTA = 0xff;
}
```

First, notice how statements in C must be terminated with a semicolon. Then notice how they resemble mathematical equations—they are not!

An assignment statement has a right side, which is computed first. A resulting value is obtained (in this case it was simply a constant expressed in hexadecimal notation) and it is then transferred to the left side, which acts as a receiving container. In this case it was the special-function PORTA register of the microcontroller.

> **Note**
>
> In C language, by prefixing the literal value with *0x* (zero x), we indicate the use of the hexadecimal radix. For historical reasons a single *0* (zero) prefix is used for the octal notation (does anybody use octal anymore?). Otherwise the compiler assumes the default decimal radix.

Compiling and Linking

Now that we have completed the main() and only function of our first C program, how do we transform the source into a binary executable?

Using the MPLAB Integrated Development Environment (IDE), it's very easy! It's a matter of a single click of your mouse in an operation called a *Project Build*. The sequence of events is actually pretty long and complex, but it is mainly composed of two steps:

1. *Compiling.* The MPLAB C32 compiler is invoked and an object code file (.o) is generated. This file is not yet a complete executable. Though most of the code generation is complete, all the addresses of functions and variables are still undefined. In fact this is also called a *relocatable code object.* If there are multiple source files, this step is repeated for each one of them.

2. *Linking.* The linker is invoked and a proper position in the memory space is found for each function and each variable. Also, any number of precompiler object code files and standard library functions may be added at this time, as required. Among the several output files produced by the linker is the actual binary executable file (.hex).

All this is performed in a very rapid sequence as soon as you ask MPLAB to *build* your project. Each group of files, as presented in the project window (refer back to Figure 1.1), will be used during the project build to assist in the compiling or linking phase:

- Every source code (.c) file in the *Source Files* list will be compiled to produce relocatable object files.

- Each additional object file in the *Object Files* list will then be linked together with the previous object files.

- The *Library Files* list will also be used during the linking process to search for and extract library modules that contain functions, if any have been used in the project.

- Finally, the *Linker Script* section might contain an additional file that can be used to provide additional instructions to the linker to change the order and priority of each data and code section as they are assembled in the final binary executable file. The MPLAB C32 tool suite offers a *default linker script* mechanism that is sufficient for most general applications and certainly for all the applications we will review in this book. As a consequence, for the rest of this book we will safely leave this section of the project window empty, accepting the default setting provided.

The last two sections of the project window are treated differently:

- The *Header Files* section is designed to contain the names of the include files (.h) used. However, they don't get processed directly by the compiler. They are listed here only to document the project dependencies and for your convenience; if you double-click them they will open immediately in the editor window.

- The *Other Files* section is designed to contain the names of any additional file, not included in any of the previous categories but used in the project. Once more this section serves a documentation purpose more than anything else.

The Linker Script

Just like the p32xxxx.h include file tells the compiler about the names (and sizes) of device-specific SFRs, the (default) linker script informs the linker about the SFRs' predefined position in memory (according to the selected device datasheet). It also provides other essential services such as:

- Listing the total amount of FLASH memory available

- Listing the total amount of RAM memory available

- Listing their respective address ranges

- Listing the position of critical entry points such as the reset and exception vectors

- Listing the position of the interrupt vectors and the vectors table

- Listing the position of the device configuration words

- Including additional processor-specific object files

- Determining the position and size of the software stack and the heap (via parameters passed from MPLAB project files, as we will see in the next chapters)

Now, if you are curious like me, you might want to take a look inside. The linker script file, it turns out, is a simple text file, although with the .ld extension. It can be opened and inspected using the MPLAB editor. Assuming you accepted the default values when you installed MPLAB on your hard drive, you will find the default linker script for the PIC32MX360F512L microcontroller by opening the procdefs.ld file found in the following directory:

```
C:\Program Files\Microchip\PIC32-Tools\pic32-libs\proc\
32MX360F512L
```

Wow, I know, my head is spinning, too! It took me half an hour to find my way through the labyrinth of subdirectories created during the MPLAB installation. But the reality is that the linker will find it and use it automatically, and you will hardly ever have to see or worry about it again. Here is a segment of the script where the address of the reset vector, the general exception vector, and a few other critical entry points are defined:

```
. . .
/***********************************************************
 *Memory Address Equates
 ***********************************************************
_RESET_ADDR      = 0xBFC00000;
_BEV_EXCPT_ADDR  = 0xBFC00380;
_DBG_EXCPT_ADDR  = 0xBFC00480;
_DBG_CODE_ADDR   = 0xBFC02000;
_GEN_EXCPT_ADDR  = _ebase_address + 0x180;
. . .
```

Note

Don't try to open the procdefs.ld from Windows Explorer or using the default Windows Notepad application; it won't look pretty. This file was generated in a Unix environment and does not contain the standard end-of-line sequence used by Windows programs. Instead use the MPLAB Editor as I suggested.

Building the First Project

Select the option **Build All** from the **Project** menu or click the corresponding 📇 (Build All) button in the project toolbar. MPLAB will open a new window; the content of yours should be very similar to what I obtained, shown in Figure 1.2.

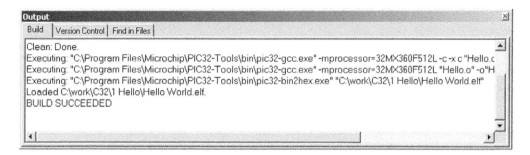

Figure 1.2: The content of the Output Window Build tab after a successful build.

Should you prefer a command-line interface, you will be pleased to learn that there are alternative methods to invoke the compiler and the linker and achieve the same results without using the MPAB IDE, although you will have to refer to the MPLAB C32 compiler user guide for instructions. In this book, we will stick with the MPLAB IDE interface and will use the appropriate checklists to make it even easier.

Using the Simulator

Select **Debugger | Select Tool | MPLAB SIM** to choose and activate the software simulator as the main debugging tool for this project. I recommend that you get in the habit of using the **MPLAB SIM debugger setup** checklist to configure a number of parameters that will improve your simulation experiences, although we won't need it during this first simulation. Let's perform instead another and all-important general configuration step of MPLAB itself.

Select the **Configure | Settings** item from the MPLAB menu and, inside the large and complex dialog box that will pop up, select the **Debugger** tab.

As illustrated in Figure 1.3, I recommend that you check three of the options available to instruct MPLAB to automatically perform a few useful tasks:

- Save all the files you changed in the Editor window before running the code.

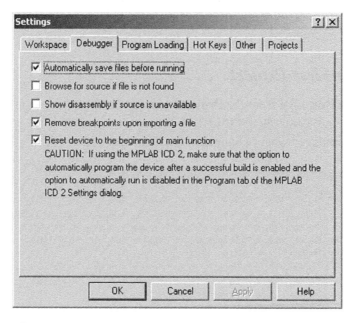

Figure 1.3: MPLAB Settings dialog box Debugger tab.

- Remove existing breakpoints before importing a new executable.

- After any device reset, position the debugger cursor at the beginning of the main function.

The last task, in particular, might seem redundant, but it is not. If you remember, as was briefly mentioned at the beginning of this chapter, there is a small segment of code (*crt0* or Startup code) that the linker places automatically for us between the actual reset vector and our code. If we do not instruct MPLAB otherwise, the simulator will attempt to step through it, and since there is no C source code to show for it, it would have to happen in the *disassembly* window. Not that there would be anything wrong with that; actually, I invite you to try that sometime to inspect this mysterious (but so useful) segment of code. The fact is that we are just not ready for it yet and, after all, our focus in this exploration is 100 percent on the C language programming of the PIC32 rather than the underlying MIPS assembly.

If all is well, before trying to execute the code let's also open a Watch window and add the PORTA special-function register to it:

1. Select **View | Watch** from the main menu to access the Watch window (see Figure 1.4).

2. Type or select **PORTA** in the SFR selection list (top left).

3. Click the **Add SFR** button.

4. Press the simulator reset button 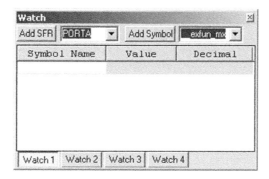 (Reset) in the Debug toolbar or select **Debugger | Reset**.

Figure 1.4: MPLAB IDE Watch window.

5. Observe the contents of the PORTA register; it should be cleared (all zeroes) at reset.

6. Also notice that a large green arrow has appeared right next to the first opening curly bracket of the main function. It points at the part of our code that is going to be executed next.

7. Now, since we need to learn to walk before we "run," let's use the ⟦**0⁺**⟧ (Step Over) or the ⟦**{}**⟧ (Step In) buttons in the Debugger toolbox, or the **Debugger | Step In** and **Debugger | Step Over** commands from the main menu, to execute the one and only statement in our first program.

8. Observe how the content of PORTA changes in the watch window. Or, I should say, notice how nothing happens. Surprise!

Finding a Direction

It is time to hit the books, specifically the PIC32MX datasheet (Chapter 13 focuses on the I/O ports detail). PortA is a pretty complex, 12-pin-wide port. Each one of the pins is controlled by a small block of logic, represented in Figure 1.5.

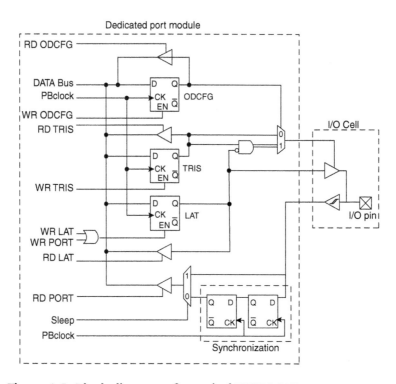

Figure 1.5: Block diagram of a typical PIC32 I/O port structure.

Although completely understanding the diagram in Figure 1.5 is beyond the scope of our explorations today, we can start by making a few simple observations. There are only three signals that eventually reach the I/O cell. They are the data output, the data input, and the tristate control signals. The latter is essential to decide whether the pin is to be used as an input or an output, which is often referred to as the *direction* of the pin.

From the datasheet, again, we can determine the default direction for each pin—that is, in fact, configured as an input after each reset or power up event. This is a safety feature and a standard for all PIC microcontrollers. The PIC32 makes no exception.

The *TRISA* special-function register allows us to change the direction of each individual pin on PortA. The rule is simple to remember:

- Clear a bit to **0** for an **O**utput pin.
- Set a bit to **1** for an **I**nput pin.

So, we need to add at least one more assignment to our program if we want to change the direction of all the pins of PortA to output and see their status change. Here is how our simple project looks after the addition:

```
#include <p32xxxx.h>

main()
{
  // configure all PORTA pins as output
  TRISA = 0;
  PORTA = 0xff;
}
```

We can now retest the code by repeating the following few steps:

1. Rebuild the project (select **Project | Build All**, use **Ctrl + F10,** or click the **Build All** button in the project toolbox).

2. Execute a couple of single-steps and . . . you have it (see Figure 1.6)!

If all went well, you should see the content of PORTA change to 0xFF, highlighted in the Watch window in red. Hello Embedded World!

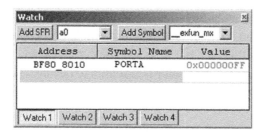

Figure 1.6: The Watch window after PortA content has changed!

The JTAG Port

Our first choice of PortA was dictated partially by the alphabetical order and partially by the fact that on the Explorer16 demonstration boards, PortA pins, RA0 through RA7, are conveniently connected to 8 LEDs. So, if you try and execute this example code on the actual demo board using an in-circuit debugger, you will have the satisfaction of seeing all the LEDs turn on, nice and bright . . . or perhaps not?

There is one more important detail affecting the operation of a few PortA pins that you need to be aware of. Where previous generations of PIC microcontrollers used a two-wire protocol to connect to an in-circuit programmer and/or debugger, known as the *ICSP/ICD interface*, the PIC32 offers an additional interface, widely adopted among 32-bit architectures, known as the *JTAG interface*.

Note

The PIC24 experts will not fail to point out that several 16-bit large pin-count devices were already offering JTAG to support *boundary scan* features. With the PIC32 architecture, the JTAG functionality is extended to include all programming and debugging features.

In fact, for all debugging and programming purposes, the JTAG and the ICSP/ICD interface are now equivalent and the choice between the two will be dictated more by personal preference, the availability and cost of (Microchip own and third-party) tools, and/or the number of pins required. In this last respect, the ICSP/ICD interface has a small advantage over the JTAG interface since it requires only half the microcontroller I/Os. On the other side, if the boundary scan functionality is required, the JTAG interface is the one and only option.

As a consequence of the decision to offer both interfaces, the designers of the PIC32 had to make sure that both debugging options were available by default upon reset or power-up of the device. The JTAG port pins are multiplexed with PortA pins RA0, RA1, RA4, and RA5, over which they take priority.

The PIC32 Starter Kit is an example of a programming and debugging tool that uses the JTAG port. The MPLAB REAL ICE and the MPLAB ICD2 instead use the traditional ICSP/ICD port.

If you intend to test the code developed so far on the Explorer 16 board using the MPLAB REAL ICE or the MPLAB ICD2 in circuit debuggers, you will have to remember to disable the JTAG port to gain access to all the pins of PortA and therefore all the LEDs. Here is all it takes:

```
// disable the JTAG port
   DDPCONbits.JTAGEN = 0;
```

After all, only one more assignment statement needs to be added at the top of the main function. Instead of assigning a new value to the entire DDPCON register (in charge of the configuration of the Debug Data Ports), we used the special C language notation to access individual bits (or groups of bits) within a word. We will expand on these subjects in the next few chapters.

If you intend to test the code on the Explorer 16 board using the PIC32 Starter Kit and a 100-pin PIM adapter, you must *not* disable the JTAG port. You will still have control on the remaining pins of PortA: RA2, RA3, RA6, and RA7. Don't be envious; you have three more LEDs that you can control on the Starter Kit board itself, connected to PortD instead: RD0, RD1, and RD2. In fact, even if you don't have an Explorer 16 board but just a PIC32 Starter Kit, you could change the code in the previous examples, replacing all references to PortA registers with the PortD equivalents: TRISD and PORTD. Perhaps it will be less spectacular but equally instructive!

Testing PORTB

To complete our day of exploration, we will now investigate the use of one more I/O port, PortB. It is simple to edit the program and replace the two PortA control registers assignments with TRISB and PORTB.

Rebuild the project and follow the same steps we did in the previous exercise and you'll get a new surprise: The same code that worked for PortA does *not* work for PortB!

Don't panic—I did it on purpose. I wanted you to experience a little PIC32 migration pain. It will help you learn and grow stronger.

It is time to go back to the datasheet and study in more detail the PIC32 pin-out diagrams. There are two fundamental differences between the 8-bit PIC microcontroller architectures and the new 16- and 32-bit architectures:

- Most PortB pins are multiplexed with the analog inputs of the Analog-to-Digital Converter (ADC) peripheral. The 8-bit architecture reserved PortA pins primarily for this purpose; the roles of the two ports have been swapped!

- If a peripheral module input/output signal is multiplexed on an I/O pin, as soon as the module is enabled, it takes complete control of the I/O pin—independently of the direction (TRISx) control register content. In the 8-bit architectures it was up to the user to assign the correct direction to each pin, even when a peripheral module required its use.

By default, pins multiplexed with "analog" inputs are disconnected from their "digital" input ports. This explains what was happening during our last attempt. All PortB pins of the PIC32 are, by default at power-up, assigned an analog input function; therefore, reading the PORTB register returns all 0s. Notice, though, that the output latch of PortB has been correctly set, although we cannot see it through the PORTB register. To verify it, check the contents of the LATB register instead.

To reconnect the PortB input pins to the digital inputs, we have to act on the ADC module configuration. From the datasheet, we learn that the SFR AD1PCFG controls the analog/digital assignment of each pin (see Figure 1.7).

Assigning a 1 to each bit in the AD1PCGF SFR will accomplish the task and convert the pin into a digital input. Our new and complete program example is now:

```
#include <p32xxxx.h>

main()
{
    // configure all PORTB pins as output
    TRISB=0,                 // all PORTB as output
    AD1PCFG=0xffff;          // all PORTB as digital
    PORTB=0xff;

}
```

r-0	r-0	r-0	r-0	r-0	r-0	r-0	r-0
—	—	—	—	—	—	—	—
bit 31							bit 24

r-0	r-0	r-0	r-0	r-0	r-0	r-0	r-0
—	—	—	—	—	—	—	—
bit 23							bit 16

R/W-0	R/W-0	R/W-0	R/W-0	R/W-0	R/W-0	R/W-0	R/W-0
PCFG15	PCFG14	PCFG13	PCFG12	PCFG11	PCFG10	PCFG9	PCFG8
bit 15							bit 8

R/W-0	R/W-0	R/W-0	R/W-0	R/W-0	R/W-0	R/W-0	R/W-0
PCFG7	PCFG6	PCFG5	PCFG4	PCFG3	PCFG2	PCFG1	PCFG0
bit 7							bit 0

Legend:
R = Readable bit   W = Writable bit   P = Programmable bit   r = Reserved bit
U = Unimplemented bit   -n = Bit Value at POR: ('0', '1', x = Unknown)

bit 31-16   **Reserved:** Reserved for future use, maintain as '0'

bit 15-0   **PCFG<15:0>:** Anlog Input Pin Configuration Control bits

> 1 = Anlog input pin in Digital mode, port read input enabled, ADC input multiplexer input for this analog input connected to AVss
> 0 = Anlog input pin in Analog mode, digital port read will return as a '1' without regard to the voltage on the pin, ADC samples pin voltage

Note: The AD1PCFG register functionality will vary depending on the number of ADC inputs available on the seleced device. Please refer to the specific device data sheet for additional details on this register.

Figure 1.7: AD1PCFG: ADC port configuration register.

This time, compiling and single-stepping through it will give us the desired results (see Figure 1.8).

Mission Debriefing

After each expedition, there should be a brief review. Sitting on a comfortable chair in front of a cool glass of . . . water, it's time to reflect on what we have learned from this first experience.

Writing a C program for a PIC32 microcontroller can be very simple, or at least no more complicated than an assembly or 8-bit equivalent project. Two or three instructions, depending on which port we plan to use, can give us direct control over the most basic tool available to the microcontroller for communication with the rest of the world: the I/O pins.

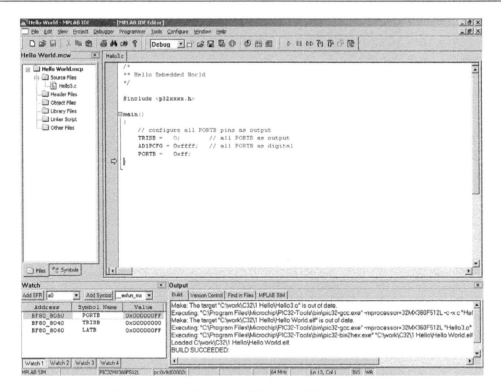

Figure 1.8: Hello Embedded World using PortB.

Also, there is nothing the MPLAB C32 compiler can do to read our minds. Just as in assembly, we are responsible for setting the correct direction of the I/O pins. We are still required to study the datasheet and learn about the small differences between the 8-bit and 16-bit PIC microcontrollers we might be familiar with and the new 32-bit breed.

As high level as the C programming language is thought to be, writing code for embedded-control devices still requires us to be intimately familiar with the finest details of the hardware we use.

Notes for the Assembly Experts

If you have difficulties blindly accepting the validity of the code generated by the MPLAB C32 compiler, you might find comfort in knowing that, at any given point in time, you can decide to switch to the *Disassembly Listing* view (see Figure 1.9). You can quickly inspect the code generated by the compiler, since each C source line is shown in a comment that precedes the segment of code it generated.

```
Disassembly Listing                                              _□×
--- C:\work\C32\1 Hello\Hello3.c ----------------------------------
1:              /*
2:              ** Hello Embedded World
3:              */
4:
5:              #include <p32xxxx.h>
6:
7:              main()
8:              {
9D000000  27BDFFF8  addiu     sp,sp,-8
9D000004  AFBE0000  sw        s8,0(sp)
9D000008  03A0F021  addu      s8,sp,zero
9:                  // configure all PORTB pins as output
10:                 TRISB =   0;      // all PORTB as output
9D00000C  3C02BF81  lui       v0,0xbf81
9D000010  AC408040  sw        zero,-32704(v0)
11:                 AD1PCFG = 0xffff;  // all PORTB as digital
9D000014  3C03BF81  lui       v1,0xbf81
9D000018  3402FFFF  ori       v0,zero,0xffff
9D00001C  AC629060  sw        v0,-28576(v1)
12:                 PORTB =   0xff;
9D000020  3C03BF81  lui       v1,0xbf81
9D000024  240200FF  addiu     v0,zero,255
9D000028  AC628050  sw        v0,-32688(v1)
13:             }
```

Figure 1.9: Disassembly Listing window.

You can even single-step through the code and do all the debugging from this view, although I strongly encourage you *not* to do so or limit the exercise to a few exploratory sessions as we progress through the first chapters of this book. Satisfy your curiosity, but gradually learn to trust the compiler. Eventually, use of the C language will give a boost to your productivity and increase the readability and maintainability of your code.

As a final exercise, I would encourage you to open the *Memory Usage Gauge* window by selecting **View | Memory Usage Gauge** (see Figure 1.10).

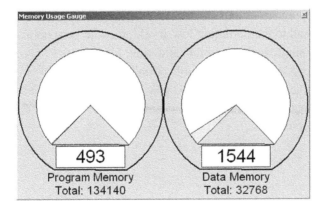

Figure 1.10: MPLAB IDE Memory Usage Gauge window.

Don't be alarmed, even though we wrote only three lines of code in our first example and the amount of program memory used appears to be already up to 490 or more words. This is not an indication of any inherent inefficiency of the C language. There is a minimum block of code that is always generated (for our convenience) by the MPLAB C32 compiler. This is the Startup code (*crt0*) that we mentioned briefly before. We will return to it, in more detail, in the following chapters as we will discuss variable initialization, memory allocation, and interrupts.

Notes for the PIC MCU Experts

Those of you who are familiar with the PIC16, PIC18, and even the PIC24 architecture will find it interesting that *all* PIC32 SFRs are now 32-bit wide. But in particular, if you are familiar with the PIC24 and dsPIC architecture, it might come to you as a surprise that the ports did *not* scale up! Even if PORTA and TRISA are now 32-bit wide registers, the PortA module still groups fewer than 16 pins, just like in the PIC24. You will realize in the following chapters how this has several positive implications for easy code migration up from the 16-bit architectures while granting optimal performance to the 32-bit core.

Whether you are coming from the 8-bit or the 16-bit PIC/dsPIC world, with the PIC32 peripheral set you will feel at home in no time!

Notes for the C Experts

Certainly we could have used the printf() function from the standard C libraries. In fact they are readily available with the MPLAB C32 compiler. But we are targeting embedded-control applications and we are not writing code for multigigabyte workstations. Get used to manipulating low-level hardware peripherals inside the PIC32 microcontrollers. A single call to a library function, like printf(), could have added several kilobytes of code to your executable. Don't assume a serial port and a terminal or a text display will always be available to you. Instead develop a sensibility for the "weight" of each function and library you use in light of the limited resources available in the embedded design world.

Tips & Tricks

The PIC32MX family of microcontrollers is based on a 3 V CMOS process with a 2.0 V to 3.6 V operating range. As a consequence, a 3.3 V power supply (Vdd) is used on most applications and demonstration boards; this limits the output voltage of each I/O pin when producing a logic high output. Interfacing to 5 V legacy devices and applications, though, is really simple:

- To drive a 5 V output, use the ODCx control registers (ODCA for PortA, ODCB for PortB, and so on) to set individual output pins in open-drain mode and connect external pull-up resistors to a 5 V power supply.

- Digital input pins instead are already capable of tolerating up to 5 V. They can be connected directly to 5 V input signals.

Watch out

Be careful with I/O pins that are multiplexed with analog inputs (most PortB pins, for example); they cannot tolerate voltages above 3.6 V!

Exercises

If you have the Explorer 16 board and an in-circuit debugger:

- Use the MPLAB REAL ICE Debugging or the MPLAB ICD2 Debugging checklists to help you prepare the project for debugging.

- Insert the instructions required to disable the JTAG port.

- Test the PortA example, connecting the Explorer 16 board and checking the visual output on LED0-7.

If you have the PIC32 Starter Kit:

- Use the PIC32 Starter Kit Debugging checklist to help you prepare the project for debugging.

- Modify the code to operate on PortD, but do *not* disable the JTAG port.

- Test the code by checking the visual output on LED0-2 on the PIC32 Starter Kit itself.

In both cases you can:

- Test the PortB example by connecting a voltmeter (or DMM) to pin RB0, if you can identify it on your board, and watching the needle move between 0 and 3.3 V as you single-step through the code.

Books

Kernighan, B., and Ritchie, D., *The C Programming Language* (Prentice-Hall, Englewood Cliffs, NJ). When you read or hear programmers talk about the "K&R," also known as "the white book," they mean *this* book. The C language has evolved quite a bit since the first edition was published in 1978. The second edition (1988) includes the more recent ANSI C standard definitions of the language, which are closer to the standard the MPLAB C32 compiler adheres to (ISO/IEC 9899:1990 also known as C90).

Links

http://en.wikibooks.org/wiki/C_Programming. This is a Wiki-book on C programming and as such it is a bit of a work in progress. It's convenient if you don't mind doing all your reading online. Hint: Look for the chapter called "A Taste of C" to find the omnipresent "Hello World!" example.

Walking in Circles

The Plan

It is funny how many stories of expeditions gone wrong culminate with a revealing moment where the explorers realize they got desperately lost and have been walking in circles for a while. In embedded-control programming it's the opposite: Our programs need a framework, a structure so that the flow of code can be managed, and this usually is built around one *main loop*.

Today we will review the basics of the loops syntax in C, and we'll also take the opportunity to introduce a first peripheral module: the 16-bit Timer1. Two new MPLAB© SIM features will be used for the first time: the *Animate* mode and the *Logic Analyzer* view.

Preparation

For this second lesson, we will need the same basic software components we installed (from the attached CD-ROM and/or the latest versions available for download from Microchip's Web site) and used before, including:

- MPLAB IDE (Integrated Development Environment)

- MPLAB SIM (software simulator)

- MPLAB C32 compiler (free Student Edition)

We will also reuse the New Project Setup checklist to create a new project with the MPLAB IDE.

Select the **Project Wizard** from the **Project** menu and proceed through the few steps that follow:

1. The first dialog box will ask you to choose a specific device model. Select the **PIC32MX360F512L** device and click **Next**.

2. In the second dialog box, select the **PIC32 C-Compiler Tool Suite** and click **Next**. Make sure to select the C compiler suite, not the assembly suite!

3. In the third dialog box, you are asked to assign a name to the new project file. Instead, click the **Browse** button and create a new folder. Name the new folder **Loops**, and inside it create the project file **Loops,** then click **Next**.

4. In the fourth dialog box, simply click **Next** to proceed to the following dialog box, since there is no need to copy any source files from any previous projects or directories.

5. Click **Finish** to complete the project wizard.

6. Open a new editor window by selecting **File | New**, typing the **Ctrl + N** keyboard shortcut, or clicking the corresponding (**New File**) button in MPLAB standard toolbar.

7. Type the following three comment lines:

    ```
    /*
    ** Loops
    */
    ```

8. Select **File | Save As** to save the file as Loops.c.

9. Now right-click with your mouse on the editor window to bring up the editor's context menu and select the **Add To Project** item. This will tell MPLAB that the newly created file is an integral part of the project.

10. Select **Project | Save Project** to save the project.

Soon, after you repeat these same steps a few more times, they will become automatic to you, but you will always have the option to refer to the **Create New File** and **Add to Project** checklists conveniently included in this book.

The Exploration

One of the key questions that might have come to mind after you worked through the previous lesson is, "What happens when all the code in the main() function has been executed?" Well, nothing really happens, literally!

When the main() function terminates and returns back to the startup code (crt0), a new function _exit() is called and the PIC32 remains stuck there in a tight loop from which it can escape only if a processor reset is performed. Notice that this is something that depends on the MPLAB C32 tool suite and that is not a C language proper feature. C compilers normally are designed to return control to an operating system when the main() function returns, but as you understand, there is no operating system to return to in our case.

Note

The _exit() function, just like the startup code, is not visible in the editor window (not our code) and is not visible even from the disassembly window (not a library). The only way you can find out about it is if you open the **Memory** window and you select the **Code View** pane.

The good news is that we can easily define a replacement for the _exit() function if we have a better idea of what to do with it. We could, for example, mimic what the MPLAB C30 tool suite used to do for PIC24 and dsPIC applications—that is, insert a reset instruction in there and have the entire application repeat over and over again. But what we truly want in embedded control is an application that runs continuously, from the moment the power switch has been flipped on until the moment it is turned off. So, letting the program run through entirely, reset, and execute again might seem like a convenient way to arrange the application so that it keeps repeating as long as there is "juice."

The reset option might work in a few limited cases, but what you will soon discover is that running in this "loop," you develop a "limp." Upon reaching the end of the program, executing the reset instruction takes the microcontroller back to the reset vector to again execute the startup code. As short as the startup can be, it will make the loop very unbalanced. Going through all the SFR and global variable initializations each time is probably not necessary and it will certainly slow down the application. A better option, instead, is to code a proper application *main loop* ourselves. To begin, let's review the most basic control flow mechanisms available in C language.

While Loops

In C there are at least three ways to code a loop. Here is the first: the `while` loop:

```
while ( x)
{
  // your code here...
}
```

Anything you put in between those two curly brackets { } will be repeated for as long as the *logic expression* in parenthesis (`x`) returns a true value. But what is a logic expression in C?

First of all, in C there is no distinction between logic expressions and *arithmetic expressions*. In C, the Boolean logic *true* and *false* values are represented just as integer numbers with a simple rule:

- *false* is represented by the integer `0`

- *true* is represented by *any* integer except `0`

So `1` is "true," but so are `13` and `-278`!

To evaluate logic expressions, a number of logic operators are defined, such as:

| | | the "logic OR" operator
`&&` | the "logic AND" operator
`!` | the "logic NOT" operator

These operators consider their operands as logical (Boolean) values using the rule mentioned previously, and they return a logical value. Here are some trivial examples (assume that `a = 17` and `b = 1`, or in other words they are both true):

```
( a || b)     is true
( a && b)     is true
( !a)         is false
```

There are, then, a number of operators that compare numbers (integers of any kind, and floating-point values too) and return logic values. They are:

`==` | the "equal-to" operator, notice it is composed of two equal signs to distinguish it from the "assignment" operator we used before.
`!=` | the "NOT-equal to" operator

>   the "greater-than" operator
>=   the "greater-or-equal to" operator
<   the "less-than" operator
<=   the "less-or-equal to" operator

Here are some examples (assuming `a = 10`):

```
( a > 1)      is true
(-a >= 0)     is false
( a == 17)    is false
( a != 3)     is true
```

Back to the `while` loop: We said that as long as the expression in parentheses produces a true logic value (that is, any integer value but `0`), the program execution will continue around the loop. When the expression produces a false logic value, the loop will terminate and the execution will continue from the first instruction after the closing curly bracket.

Notice that the evaluation of the expression is done first, before the curly bracket content is executed (if it ever is), and is then reevaluated each time.

Here are a few curious loop examples to consider:

```
while ( 0)
{
  // your code here...
}
```

A constant false condition means that the loop will never be executed. This is not very useful. In fact I believe we have a good candidate for the "world's most useless code" contest!

Here is another example:

```
while ( 1)
{
  // your code here...
}
```

A constant true condition means that the loop will execute forever. This is useful and is in fact what we will use for our main program loops from now on. For the sake of readability, a few purists among you will consider using a more elegant approach, defining a couple of constants:

```
#define FALSE      0
#define TRUE       !FALSE
```

And using them consistently in their code, as in:

```
While ( TRUE)
{
  // your code here...
}
```

It is time to add a few new lines of code to the loops.c source file and put the while loop to good use:

```
#include <p32xxxx.h>
main()
{
  // initialization
  DDPCONbits.JTAGEN = 0;    // disable the JTAG port
  TRISA = 0xff00;           // PORTA pin 0..7 as output

  // application main loop
  while( 1)
  {
    PORTA = 0xff;           // turn pin 0-7 on
    PORTA = 0;              // turn all pin off
  }
}
```

The structure of this example program is essentially the structure of every embedded control program written in C. There will always be two main parts:

- The *initialization*, which includes both the device peripherals initialization and variables initialization, executed only once at the beginning

- The *main loop*, which contains all the control functions that define the application behavior and is executed continuously

An Animated Simulation

Use the Project Build checklist to compile and link the loops.c program. Also use the MPLAB SIM Simulator Setup checklist to prepare the software simulator.

To test the code in this example with the simulator, I recommend you use the *Animate* mode (**Debugger | Animate**). In this mode, the simulator executes one C program line at a time, pausing shortly after each one to give us time to observe the immediate results. If you add the PORTA special-function register to the Watch window, you should be able to see its value alternating rhythmically between 0xff and 0x00.

The speed of execution in Animate mode can be controlled with the **Debug | Settings** dialog box, selecting the **Animation/Real Time Updates** tab, and modifying the **Animation Step Time** parameter, which by default is set to 500 ms. As you can imagine, the Animate mode can be a valuable and entertaining debugging tool, but it gives you quite a distorted idea of what the actual program execution timing will be. In practice, if our example code was to be executed on a real hardware target, say an Explorer16 demonstration board (where the PIC32 is running at, say, 72 MHz), the LEDs, connected to the PortA output pins, would blink too fast for our eyes to notice. In fact, each LED would be turned on and off several million times each second.

To slow things down to a point where the LEDs would blink nicely just a couple of times per second, I propose we use a timer so that in the process we learn to use one of the key peripherals integrated in all PIC® microcontrollers. For this example we will choose Timer1, the first of five modules available inside the PIC32MX360FJ512L models (see Figure 2.1). This is one of the most flexible and simple peripheral modules. All we need is to take a quick look at the PIC32 datasheet, check the block diagram and the details of the Timer1 control registers, and find the ideal initialization values.

We quickly learn that there are three SFRs that control most Timer1 functions. They are:

- TMR1, which contains the 16-bit counter value
- T1CON, which controls the activation and the operating mode of the timer
- PR1, which can be used to produce a periodic reset of the timer (not required here)

We can clear the TMR1 register to start counting from zero:

```
TMR1 = 0;
```

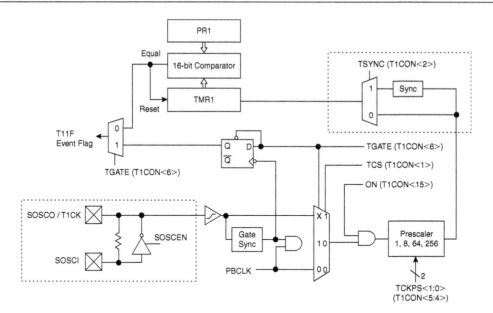

Figure 2.1: 16-bit Timer1 module block diagram.

Then we can initialize T1CON so that the timer will operate in a simple configuration, where:

- Timer1 is activated: TON = 1

- The main MCU clock serves as the source (Fpb): TCS = 0

- The prescaler is set to the maximum value (1:256): TCKPS = 11

- The input gating and synchronization functions are not required, since we use the MCU internal clock directly as the timer clock: TGATE = 0, TSYNC = 0

- We do not worry about the behavior in IDLE mode: SIDL = 0 (default)

Virtual Address	Name		Bit 31/23/15/7	Bit 30/22/14/6	Bit 29/21/13/5	Bit 28/20/12/4	Bit 27/19/11/3	Bit 28/18/10/2	Bit 25/17/9/1	Bit 24/16/8/0
BF80_0600	T1CON	31:24	—	—	—	—	—	—	—	—
		23:16	—	—	—	—	—	—	—	—
		15:8	ON	FRZ	SIDL	TMWDIS	TMWIP	—	—	—
		7:0	TGATE	—	TCKPS<1:0>		—	TSYNC	TCS	—

Figure 2.2: T1CON: Timer1 control register.

Once we assemble all the bits in a single 32-bit value, to assign to T1CON, we get:

```
T1CON = 1000 0000 0011 0000
```

or, in a more compact hexadecimal notation:

```
T1CON = 0x8030;
```

Once we are done initializing the timer, we enter a loop where we just wait for TMR1 to reach the desired value set by the constant DELAY.

```
while( TMR1 < DELAY)
{
  // wait
}
```

Assuming a 36 MHz peripheral bus clock frequency will be used, we need to assign quite a large value to DELAY to obtain a delay of about a quarter of a second. In fact, the following formula dictates the total delay time produced by the loop:

$$\text{Tdelay} = (\text{Fpb}) * 256 * \text{DELAY}$$

With Tdelay = 256 ms and resolving for DELAY, we obtain the value 36000:

```
#define DELAY 36000
```

By putting two such delay loops in front of each PORTA assignment inside the main loop, we get our latest and best code example:

```
/*
** Loops
*/
#include <p32xxxx.h>

#define DELAY 36000          // 256 ms

main()
{
  // 0. initialization
  DDPCONbits.JTAGEN = 0;   // disable JTAGport, free up PORTA
  TRISA = 0xff00;          // all PORTA as output
  T1CON = 0x8030;          // TMR1 on, prescale 1:256 PB=36 MHz
  PR1 = 0xFFFF;            // set period register to max
```

```
// 1. main loop
while( 1)
{
  //1.1 turn all LED ON
  PORTA = 0xff;
  TMR1 = 0;
  while ( TMR1 < DELAY)
  {
    // just wait here
  }

  // 1.2 turn all LED OFF
  PORTA = 0;
  TMR1 = 0;
  while ( TMR1 < DELAY)
  {
    // just wait here
  }
} // main loop
} // main
```

Note

Programming in C, the number of opening and closing curly brackets tends to increase rapidly as your code grows. After a very short while, even if you stick religiously to the best indentation rules, it can become difficult to remember which closing curly brackets belong to which opening curly brackets. By putting little reminders (comments) on the closing brackets, I try to make the code easier to follow and more readable. Also, by using the Ctrl + M shortcut in the editor window, you can quickly jump and alternate between matching brackets in your code.

It is time now to build the project and verify that it is working. If you have an Explorer 16 demonstration board available, you could try to run the code right away. The LEDs should flash at a comfortably slow pace, with a frequency of about two flashes per second.

Trying to run the same code with the MPLAB SIM simulator, though, you will discover that things are now way too slow. I don't know how fast your PC is, but on mine, MPLAB SIM cannot get anywhere close to the execution speed of a true PIC32 microcontroller.

If you use the Animate mode, things get even worse. As we saw before, the animation adds a further delay of about half a second between the execution of each individual line of code. So, for pure debugging purposes, on the simulator feel free to change the DELAY constant to a much smaller value—36, for example!

Using the Logic Analyzer

To complete this lesson and make things more entertaining, after building the project I suggest we play with a new simulation tool: the MPLAB SIM Logic Analyzer.

The Logic Analyzer gives you a graphical and extremely effective view of the recorded values for any number of the device output pins, but it requires a little care in the initial setup.

Before anything else, you should make sure that the *Tracing* function of the simulator is turned on:

1. Select the **Debug | Settings** dialog box and then choose the **Osc/Trace** tab.

2. In the Tracing options section, check the **Trace All** box.

3. Now you can open the **Analyzer** window from the **View | Simulator Logic Analyzer** menu.

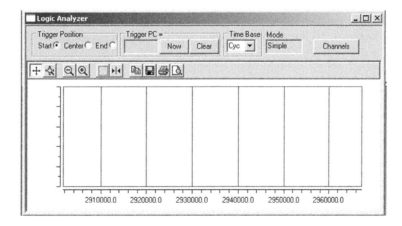

Figure 2.3: MPLAB SIM Logic Analyzer window.

4. Now click the **Channels** button, to bring up the channel selection dialog box.

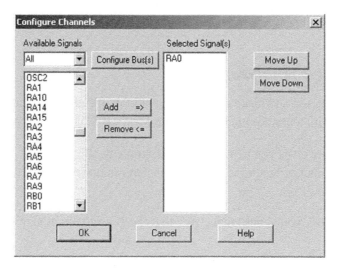

Figure 2.4: Logic Analyzer Channels Configuration dialog box.

5. From here, you can select the device output pins you would like to visualize. In our case, select **RA0** and click **Add =>**.

6. Click **OK** to close the channel selection dialog box.

For future reference, all the preceding steps are listed in the Logic Analyzer Setup checklist.

7. Run the simulation by pressing the ▷ (**Run**) button on the Debugger toolbar, selecting the **Debugger | Run** menu, or pressing the **F9** shortcut key.

8. After a short while, press the ⏸ (**Halt**) button on the Debugger toolbar, select the **Debugger | Halt** menu, or press the **F5** shortcut key.

The Logic Analyzer window should display a neat square wave plot, as shown in Figure 2.5.

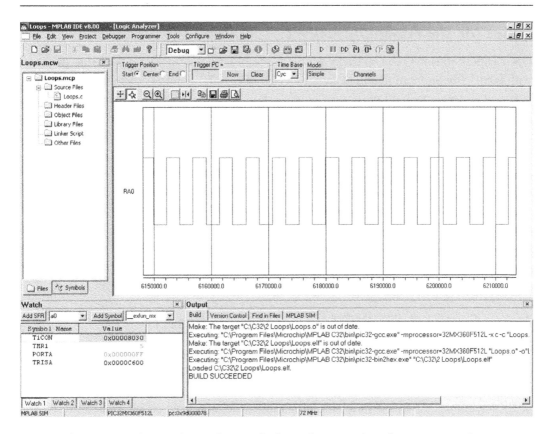

Figure 2.5: The Logic Analyzer window after running the Loops project.

Debriefing

In this brief excursion, we learned about the way the MPLAB C32 compiler deals with program termination. For the first time, we gave our little project a bit of structure—separating the main() function in an initialization section and an infinite main loop. To do so, we learned about the while loop statements, and we took the opportunity to touch briefly on the subject of logical expressions evaluation. We closed the day with a final example, where we used a timer module for the first time and we played with the Logic Analyzer window to plot the RA0 pin output.

We will return to all these elements, so don't worry if you have more doubts now than when we started; this is all part of the learning experience.

Notes for the Assembly Experts

Logic expressions in C can be tricky for the assembly programmer who is used to dealing with *binary operators* of identical names (AND, OR, NOT. . .). In C there is a set of binary operators, too, but I purposely avoided showing them in this lesson to avoid mixing things up. Binary logic operators take pairs of bits from each operand and compute the result according to the defined table of truth. Logic operators, on the other hand, look at each operand (independently of the number of bits used) as a single Boolean value.

See the following examples on byte sized operands:

```
              11110101                          11110101      (TRUE)
binary OR     00001000      logical OR          00001000      (TRUE)
              --------                          --------
gives         11111101      gives               00000001      (TRUE)
```

Notes for the 8-Bit PIC Microcontroller Experts

I am sure you noticed: Timer0 has disappeared! The good news is, you are not going to miss it. In fact, the remaining five timers of a PIC32 are so loaded with features that there is no functionality in Timer0 that you are going to feel nostalgic about. All the SFRs that control the timers have similar names to the ones used on PIC16 and PIC18 microcontrollers and are pretty much identical in structure. Still, keep an eye on the datasheet; the designers managed to cram in several new features, including:

- All timers are now 16 bits wide.

- Each timer has a 16-bit period registers.

- A new 32-bit mode timer-pairing mechanism is available for Timer2/3 and Timer4/5.

- A new external clock gating feature has been added on Timer1.

Notes for the 16-Bit PIC Microcontroller Experts

For the PIC24 and dsPIC experts among you there will be no surprises with the PIC32. The timer modules are designed to be highly compatible with the previous 16-bit generation architecture. In fact, the same is true for all the peripheral modules of the

PIC32MX family, with the PIC24 H series being the closest. Still, occasionally here and there the step up to a 32-bit bus has offered opportunities for improvements that the designers of the PIC32 could not resist.

The most dramatic difference, though, is represented by the decoupling between the core bus clock and the peripherals bus clock. This is a radical departure, for the first time in the PIC architectures history, from all previous generations' bus designs. It was a necessary step that allows the MIPS core of the PIC32 to be free from the speed limitations of the Flash memory array and of the peripheral modules, to achieve much higher performance levels without sacrificing compatibility while operating within a very low power budget. In the next chapters we will learn more about the two internal buses, the oscillator module, and their proper configuration.

Notes for the C Experts

If you are used to programming in C on a personal computer or workstation, you expect that, upon termination of the main() function, control will be returned to the operating system. Though several *real-time operating systems* (RTOSs) are available for the PIC32, a large number of applications won't need and won't use one. This is certainly true for all the simple examples in this book. By default, the MPLAB C32 compiler assumes that there is no operating system to return control to.

Notes for the MIPS Experts

The MIPS experts among you might have been looking for a mention of the *core* 32-bit timer (yes, there are truly six timers inside the PIC32) and the hardware control registers typically offered for access through the *coprocessor 0* (CP0) instructions. It was tempting to mention them, but I intentionally avoided it and decided not to use any of them for as long as possible. My purpose is to force you, the reader, to familiarize yourself with the PIC environment in which the MIPS core has been implanted. My intention is to demonstrate the use of the PIC32 and its peripherals as a true PIC microcontroller, the fastest ever designed so far, but still a true PIC machine.

Tips & Tricks

Some embedded applications are designed to run their main loops for months or years in a row without ever being turned off or receiving a reset command. But the control registers of a microcontroller are simple RAM memory cells. The probability that a power supply fluctuation (un-detected by the brown-out reset circuit), an electromagnetic pulse emitted by some noisy equipment in the proximity, or even a cosmic ray could alter their contents is a small but finite number. Given enough time (years) and depending on the application, you might see it happen. When you design applications that have to operate reliably on huge time scales, you should start seriously considering the need to provide a periodic "refresh" of the most important control registers of the essential peripherals used by the application.

Group the sequence of initialization instructions in one or more functions. Call the functions once at power-up, before entering the main loop, but also make sure that inside the main loop the initialization functions are called when idling and no other critical task is pending, so that every control register is reinitialized periodically.

Notes on Using the Peripheral Libraries

The MPLAB C32 tool suite comes with a complete set of standard C libraries and an additional set of *peripherals libraries* designed to simplify and standardize the use of all the internal resources of the PIC32. The peripheral libraries are specifically designed to provide an even higher level of compatibility with previous Microchip 16-bit architectures and in particular with the PIC24 series of microcontrollers. The following example uses the timers' library timer.h to exemplify the advantages and disadvantages of relying on libraries.

Should we need to initialize the Timer1 module using the peripheral libraries, as in the "loops" projects we developed today, in place of the direct access to the Timer1 module registers:

```
TMR1 = 0;
T1CON = 0x8030;  // or TMR1bits.ON = 1; TMR1bits.TCKPS=3;
PR1 = 0xFFFF;
```

we could use the following code:

```
WriteTimer1( 0);
OpenTimer1( T1_ON | T1_PS_1_256, 0xFFFF);
```

The clear advantage is that you don't need to add many comments to the two lines of code; they read pretty well already. This code is self-documenting. Additionally, if you misspell one of the parameter names, the compiler will promptly complain and point it out.

But it is not all roses, either. Although the function parameters are checked for spelling errors, in most cases there is no way for the compiler to tell whether you used the right parameter for the right function. For example, when configuring Timer2, the following error would go undetected:

```
OpenTimer2( T2_ON | T1_PS_1_256, 0xFFFF);
```

It seems a pretty innocent mistake, but it would probably cause you to spend a few hours scratching your head to understand why the Timer2 prescaler is configured wrong, whereas it is all fine by the compiler.

The best advantage of using the libraries, the abstraction they offer, is also another source of potential frustration. Since they hide the implementation details from us, we are not given to know if, for example, the TMR1 register is already being cleared by the OpenTimer1() function or if we need to do it ourselves before invoking it. It turns out it is not, but you can verify that only if you visually get access to the library source files or you inspect them in the disassembly listing.

Further, although the PIC32MX device datasheet defines the official names for all the control registers (T1CON) and for each bit inside them (TCKPS), the parameters defined in the peripheral libraries have different names and spelling (T1_PS_1_256), although they try to mimic them closely. The new names can be found only in a separate set of documentation. You need to either study the Peripheral Library User Guide or inspect the timer.h include file and verify where each parameter is defined.

So, my personal recommendation regarding the use of the peripheral libraries is one of cautious and deliberate choice on a case-by-case basis. For some simple peripherals such as the I/O ports and the timers, I cannot see much of an advantage in using the library. After all, to select the correct parameters, you will still need to learn about each and every bit in each control register and be familiar with their meaning and correlation. Besides, is WriteTimer1(0); really that much more readable than TMR1=0;?

When the complexity of the peripheral module is greater and the work the library functions are performing for us bring more value, such as is the case, for example, of the DMA library we will use later in the book, I recommend we take advantage of it.

In any case, throughout the rest of the book you will have several examples of both types of approaches and, as is often the case, it will be your personal programming style that will dictate when and where you will feel comfortable using the peripheral libraries, direct register access, or a mix of the two.

Exercises

1. Output a counter on the PortA pins instead of the alternating on and off patterns. Use PortD if you have a PIC32 Starter Kit.

2. Use a rotating pattern instead of alternating on and off.

3. Rewrite the loops project using exclusively peripheral library functions to control PortA pins; set, configure, and read the timer; and disable the JTAG port if necessary.

Books

Ullman, L., and Liyanage, M., *C Programming* (Peachpit Press, Berkeley, CA, 2005). This is a fast-reading and modern book, with a simple step-by-step introduction to the C programming language.

Links

http://en.wikipedia.org/wiki/Control_flow#Loops. A wide perspective on programming languages and the problems related to coding and taming loops.

http://en.wikipedia.org/wiki/Spaghetti_code. Your code gets out of control when your loops start knotting . . .

Message in a Bottle

The Plan

Yesterday we learned that there is a loop at the core of every embedded-control application, and we learned to code it in C using the `while` statement. Today we will continue exploring a variety of other techniques available to the C programmer to perform loops. Along the way, we will take the opportunity to briefly review integer variable declarations and increment and decrement operators, quickly touching on array declarations and usage. By the end of the day you will be ready for a hopefully entertaining project that will make use of all the knowledge you acquired during the day by creating a survival tool you'll find essential should you ever be stranded on a deserted island.

Preparation

In this lesson we will continue to use the MPLAB© SIM software simulator, but once more an Explorer 16 demonstration board could add to the entertainment. In preparation for the new demonstration project, you can use the New Project Setup checklist to create a new project called Message and a new source file called Message.c.

The Exploration

In a `while` loop, a block of code enclosed by two curly brackets is executed if, and for as long as, a logic expression returns a Boolean true value (not zero). The logic expression is evaluated before the loop, which means that if the expression returns false right from the beginning, the code inside the loop might never be executed.

Do Loops

If you need a type of loop that gets executed at least once but only subsequent repetitions are dependent on a logic expression, you have to look at a different type of loop.

Let me introduce you to do loop syntax:

```
do {
  // your code here...
} while ( x);
```

Don't be confused by the fact that the do loop syntax is using the while keyword again to close the loop; the behavior of the two is very different.

In a do loop, the code found between the curly brackets is always executed first; only then is the logic expression evaluated. Of course, if all we want to get is an infinite loop for our main() function, it makes no difference if we choose the do or the while:

```
main()
{
  // initialization code
  ...

  // main application loop
  do {
    ...
  } while ( 1)
} // main
```

Looking for curious cases, we might analyze the behavior of the following loop:

```
do{
  // your code segment here...
} while ( 0);
```

You will realize that the code segment inside the loop is going to be executed once and, no matter what, only once. In other words, the loop syntax around the code is, in this case, a total waste of your typing efforts and another good candidate for the "most useless piece of code in the world" contest.

Let's now look at a more useful example, where we use a `while` loop to repeatedly execute a piece of code for a predefined and exact number of times. First, we need a variable to perform the count. In other words, we need to allocate one or more RAM memory locations to store a counter value.

Note

In the previous two lessons we have been able to skip, almost entirely, the subject of variable declarations because we relied exclusively on the use of what are in fact predefined variables: the special-function registers of the PIC32.

Variable Declarations

We can declare an integer variable with the following syntax:

```
int i;
```

Since we used the keyword `int` to declare `i` as a 32-bit (signed) integer, the MPLAB C32 compiler will make arrangements for 4 bytes of memory to be used. Later, the linker will determine where those 4 bytes will be allocated in the physical RAM memory of the selected PIC32 model. As defined, the variable `i` will allow us to count from a negative minimum value $-2,147,483,648$ to a maximum positive value of $+2,147,483,647$. This is quite a large range of values—so large that most 8- and 16-bit compilers would have been so generous only for the next type up in the hierarchy of integer types, known as `long`, as in:

```
long l;
```

But this is one of the advantages of using a 32-bit microcontroller. The arithmetic and logic unit (ALU) of the PIC32 is actually performing all arithmetic operations with equal ease (same number of clock cycles) for 32-bit integers just as it would for a 16-bit or an 8-bit integer. The MPLAB C32 compiler therefore defaults immediately to 32-bit for the basic integer type (`int`) and makes `long` just a synonym for it.

This is all nice and dandy from a performance point of view, but it comes with a price in terms of memory space. The RAM memory space allocated to hold each integer variable in your program is now double what it used to be on an 8 or 16-bit PIC© microcontroller. Though it is true that we have more of it on the PIC32 models, RAM often remains one of the most precious resources in an embedded-control application.

So if you don't have a use for the huge range of values that the PIC32's int and long types can offer and you are looking for a smaller counter, and you can accept a range of values from, say, −128 to +127, you can use the char integer type instead:

```
char c;
```

The MPLAB C32 compiler will use only 8 bits (a single byte) to hold c.

If a range of values from −32768 and +32767 is more what you were looking for, the short integer type is the right type for you:

```
short s;
```

The MPLAB C32 compiler will use only 16 bits (two bytes) to hold s. All four types can further be modified by the unsigned attribute, as in:

```
unsigned char  c;      // ranges from 0..255
unsigned short s;      // ranges from 0..65,535
unsigned int   i;      // ranges from 0..4,294,967,295
unsigned long  l;      // ranges from 0..4,294,967,295
```

Now, if you really need a large range of values, nothing beats the long long type and its unsigned variant:

```
long long l;           // ranges from -2^63 to +2^63-1
unsigned long long l;  // ranges from 0 to +2^64
```

Note

The MPLAB C32 compiler will allocate 64 bits (8 bytes or RAM) for each long long variable, which can seem like a lot, but the workload you can expect from the PIC32 to crunch these numbers is not going to be much different than what it used to be for a PIC16 to work on a simple 16-bit integer.

There are then variable types defined for use in floating-point arithmetic:

```
float       f;         // defines a 32 bit floating point
long double d;         // defines a 64 bit floating point
```

But for our looping purposes, let's stick with integers for now.

for Loops

Returning to our counter example, all we need is a simple integer variable to be used as index/counter, capable of covering the range from 0 to 5. Therefore, a char integer type will do:

```
char i;                  //declare i as an 8-bit integer with sign

i = 0;                   // init the index/counter
while ( i<5)
{
  // insert your code here ...
  // it will be executed for i= 0, 1, 2, 3, 4

  i = i+1;               // increment
}
```

Whether counting up or down, this is something you are going to do a lot in your everyday programming life. In C language, there is a third type of loop that has been designed specifically to make coding this common case easy. It is called the for loop, and this is how you would have used it in the previous example:

```
for ( i=0; i<5; i=i+1)
{
  // insert your code here ...
  // it will be executed for i=0, 1, 2, 3, 4
}
```

You will agree that the for loop syntax is compact, and it is certainly easier to write. It is also easier to read and debug later. The three expressions separated by semicolons and enclosed in the brackets following the for keyword are exactly the same three expressions we used in the prior example:

- Initialize the index

- Check for termination using a logic expression

- Advance the index/counter, in this case incrementing it

You can think of the `for` loop as an abbreviated syntax of the `while` loop. In fact, the logic expression is evaluated first and, if it's false from the beginning, the code inside the loop's curly brackets may never be executed.

Perhaps this is also a good time to review another convenient shortcut available in C. There is a special notation reserved for the increment and decrement operations that uses the operators:

  `++` *increment*, as in: `i++;` is equivalent to: `i = i+1;`

  `--` *decrement*, as in: `i--;` is equivalent to: `i = i-1;`

There will be much more to say on the subject in later chapters, but this will suffice for now.

More Loop Examples

Let's see some more examples of the use of the `for` loop and the increment/decrement operators. First, a count from 0 to 4:

```
for ( i=0; i<5; i++)
{
  // insert your code here ...
  // it will be executed for i= 0, 1, 2, 3, 4
}
```

Then a count down from 4 to 0:

```
for ( i=4; i>=0; i--)
{
  // insert your code here ...
  // it will be executed for i= 4, 3, 2, 1, 0
}
```

Can we use the for loop to code an (infinite) main program loop? Sure we can! Here is an example:

```
main()
{
  // 0. initialization code
  // insert your initialization code here...
```

```
  // 1. the main application loop
  for ( ; 1; )
  {
    // insert your main loop here...
  }
} // main
```

If you like it, feel free to use this form. As for me, from now on I will stick to the `while` syntax (it is just an old habit).

Arrays

Before starting to code our next project, we need to review one last C language feature: *array variable types*. An array is just a contiguous block of memory containing a given number of identical elements of the same type. Once the array is defined, each element can be accessed via the array name and an index. Declaring an array is as simple as declaring a single variable—just add the desired number of elements in square brackets after the variable name:

```
char c[10];     // declares c as an array of 10 x 8-bit integers
short s[10];    // declares s as an array of 10 x 16-bit integers
int i[10];      // declares i as an array of 10 x 32-bit integers
```

The same squared-brackets notation is used to refer to the content or assign a value to each element of an array, as in:

```
a = c[0];          // copy the value of the 1st element of c
                      into a
c[1] = 123;        // assign the value 123 to the second element
                      of c
i[2] = 12345;      // assign the value 12,345 to the third element
                      of i
i[3] = 123* i[4];  // compute 123 x the value of the fifth element
                      of i
```

___

Note

In C language, the first element of an array has index 0, whereas the last element has index *N*-1, where *N* is the declared array size.

It is when we manipulate arrays that the `for` type of loop comes in very handy. Let's see an example where we declare an array of 10 integers and we initialize each element of the array to a constant value of 1:

```
int a[10];            // declare array of 10 integers: a[0], a[1],
                         a[2] ... a[9]
int i;                // to be used as the loop index
for ( i=0; i<10; i++)
{
  a[ i] = 1;
}
```

Sending a Message

It's time to take all the new elements of the C language we have reviewed so far and put them to use in our next project. We will try once more to communicate with the outside world, this time using an entire row of LEDs connected to PortA, as they happen to be connected on the Explorer 16 demo board, flashing in a rapid sequence so that when we move the board left and right rhythmically they will display a short text message.

How about "Hello World!" or perhaps more modestly "HELLO"? Here is the code:

```
#include <p32xxxx.h>

// 1. define timing constants
#define SHORT_DELAY      400
#define LONG_DELAY      3200
```

```
// 2. declare and initialize an array with the message bitmap
char bitmap[30] = {
  0xff,      // H
  0x08,
  0x08,
  0xff,
  0,
  0,
  0xff,      // E
  0x89,
  0x89,
  0x81,
  0,
  0,
  0xff,      // L
  0x80,
  0x80,
  0x80,
  0,
  0,
  0xff,      // L
  0x80,
  0x80,
  0x80,
  0,
  0,
  0x7e,      // O
  0x81,
  0x81,
  0x7e,
  0,
  0
  };

// 3. the main program
main()
{
  // disable JTAG port
  DDPCONbits.JTAGEN = 0;

  // 3.1 variable declarations
  int i;                    // i will serve as the index
```

```
// 3.2 initialization
TRISA = 0xff00;        // PORTA pins connected to LEDs are outputs
T1CON = 0x8030;        // TMR1 on, prescale 1:256 Tpb=36MHz
PR1 = 0xFFFF;          // max period (not used)

// 3.3 the main loop
while( 1)
{
  // 3.3.1 display loop, hand moving to the right
  for( i=0; i<30; i++)
  { // update the LEDs
    PORTA = bitmap[i];
    // short pause
    TMR1 = 0;
    while ( TMR1 < SHORT_DELAY)
    {
    }
  } // for i

  // 3.3.2 long pause, hand moving back to the left
  PORTA = 0;                 // turn LEDs off
  // long pause
  TMR1 =  0;
  while ( TMR1 < LONG_DELAY)
  {
  }
} // main loop
} // main
```

In section 1, we define a couple of timing constants so that we can control the flashing sequence speed for execution and debugging.

In section 2, we declare and initialize an 8-bit integer array of 30 elements, each containing an LED configuration in the sequence.

Hint

Convert the hex values in the array initialization to binary on a piece of paper and, using a highlighter or a red pen, mark each 1 on the page to see the message emerge.

Section 3 contains the main program, with the variable declarations (3.1) at the top, followed by the microcontroller initialization (3.2) and eventually the main loop (3.3).

The main (`while`) loop, in turn, is further divided in two sections: Section 3.3.1 contains the actual LED Flash sequence, composed of 30 steps, to be played when the board is swept from left to right. A `for` loop is used for accessing each element of the array, in order. A `while` loop is used to wait on Timer1 for the proper sequence timing. Section 3.3.2 contains a pause for the sweep back, implemented using a `while` loop with a longer delay on Timer1.

Testing with the Logic Analyzer

To test the program, we will initially use the MPLAB SIM software simulator and the Logic Analyzer window:

1. Build the project using the **Project Build** check list.

2. Open the **Logic Analyzer** window.

3. Click the **Channel** button to add, in order, all the I/O pins from RA0 to RA7 connected to the row of LEDs.

The MPLAB SIM Setup and Logic Analyzer Setup checklists will help you make sure that you don't forget any detail.

4. Then I suggest you go back to the editor window and set the cursor on the first instruction of the 3.3.2 section.

5. Right-click to select the **context** menu and choose the **Run to Cursor** command. This will let the program execute the entire portion containing the message output (3.3.1) and will stop just before the long delay.

6. As soon as the simulation halts on the cursor line, you can switch to the **Logic Analyzer** window and verify the output waveforms. They should look like Figure 3.1.

To help you visualize the output, I added a few red dots to represent the LEDs being turned on during the first few steps of the sequence. If you squeeze your eyes a bit and imagine you see an LED on wherever the corresponding pin is at the logic high level, you will be able to read the message.

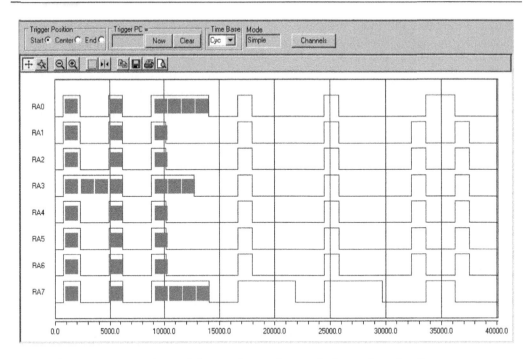

Figure 3.1: Snapshot of the Logic Analyzer window after the first sweep.

Testing with the Explorer 16 Demonstration Board

If you have an actual Explorer 16 demonstration board and an MPLAB REAL ICE programmer and debugger available, the fun can be doubled:

1. Use the **Setup** checklist for your in circuit debugger of choice.

2. Use the **Device Configuration** checklist to verify that the device configuration bits are properly set for use with the Explorer 16 demonstration board.

3. Use the **Programming** checklist to program the PIC32 in circuit.

After dimming the light a bit in the room, you should be able to see the message flashing as you "shake" the board. The experience is going to be far from perfect, though. With the Simulator and the Logic Analyzer window, we can choose which part of the sequence we want to visualize with precision and "freeze" it on the screen. On the demonstration board, you might find it quite challenging to synchronize the board's movement with the LED sequence.

Consider adjusting the timing constants to your optimal speed. After some experimentation, I found that the values 400 and 3200, respectively, for the short and long delays were ideal, but your preferences might differ.

Testing with the PIC32 Starter Kit

If you have a PIC32 Starter Kit, it will be harder but not impossible to adapt our example to use only the three available LEDs connected to the PortD pins RD0, RD1, and RD2. Unfortunately, even if you get hold of a PIM adapter board to attach the Starter Kit to an Explorer 16 board, you won't be able to see the demo in its full glory, because the Starter Kit uses the JTAG port, and that means that four out of the eight LEDs on PortA are not available.

This is not fair. In fact, I believe we need to change our strategy and find another way to send our message out to the world with the PIC32 Starter Kit. The idea is to use the old and trusty Morse code! Here is the sequence of light flashes required:

```
H        E  L       L       O
.  .  .  .   .  -  .  .   -  .  .   -  -  -
```

The rules are simple: Once chosen a basic pulse length for the *dot* (a couple tenths of a second), every other interval is required to generate a proper Morse code message based on integer multiples of it. A dash will be three times longer. The pause between dash and dots is going to be one single dot long, the pause between letters will be three dots long, and finally the pause between words will be five dots long. Once more, we can encode the entire message using an array of alternating 1s and 0s. Here is the modified code example:

```
#include <p32xxxx.h>

// 1. define timing constant
#define DOT_DELAY 18000

// 2. declare and initialize an array with the message bitmap
char bitmap[] = {
  // H ....
  1,0,1,0,1,0,1,0,0,0,
  // E .
  1,0,0,0,
  // L .-..
  1,0,1,1,1,0,1,0,1,0,0,0,
```

```
  // L .-..
  1,0,1,1,1,0,1,0,1,0,0,0,
  // ---
  1,1,1,0,1,1,1,0,1,1,1,
  // end of word
  0,0,0,0,0
  };

// 3. the main program
main()
{
  // 3.1 variable declarations
  int i;              // i will serve as the index

  // 3.2 initialization
  TRISD = 0;          // all PORTD as output
  T1CON = 0x8030;     // TMR1 on, prescale 1:256 PB=36MHz
  PR1 = 0xFFFF;       // max period (not used)

  // 3.3 the main loop
  while( 1)
  {
      // 3.3.1 display loop, spell a letter at a time
      for( i=0; i<sizeof(bitmap); i++)
      {
        PORTD = bitmap[i];

        // short pause
        TMR1 = 0;
        while ( TMR1 < DOT_DELAY)
        {
        }
      } // for i

  } // main loop
} // main
```

Notice that, to avoid having to count the dots and dashes manually to allocate the right amount of space for the bitmap array, I used a little trick. By leaving the square brackets ([]) empty in the declaration of the array, I essentially told the compiler to figure out by itself the right size based on the number of integers used in the follow list (between curly brackets { }). Of course, this would have not worked if there had been no initialization list immediately following the array declaration. A problem would have occurred later in the `for` loop if I had no other way to know how many elements had eventually been added to the array. Luckily, the `sizeof()` function came to my rescue, giving me a byte count (the size of the array in bytes), and since each array element is a `char` type integer, that coincides with the exact number of elements I was looking for.

Debriefing

In this lesson we reviewed the declaration of a few basic variable types, including integers and floating points of different sizes. Array declarations and their initialization were also used to create an original "shaking" LED display first and Morse code later, using `for` loops to send messages to the world.

Notes for the Assembly Experts

The `++` and `--` operators are actually much smarter than you might think. If the variable they are applied to is an integer, as in our trivial examples, there is little they can do to help, apart from saving you a few keystrokes. But if they are applied to a pointer (which is a variable type that contains a memory address), they actually increase the address by the exact number of bytes required to represent the quantity pointed to. For example, a pointer to 16-bit integers will increment its address by two, while a pointer to a 32-bit integer will increment its address by four, and so on.

The increment and decrement operators can also be applied inside a generic expression to operate *before* or *after* a variable content is fetched. Here are a few examples (assuming the initial conditions `a=0` and `b=1`):

```
a = b++;     // a = 1, b = 2
```

In this first case, `a` is assigned the value of `b` first, and `b` is incremented later.

```
a = ++b;     // a = 2, b = 2
```

In this second case, `b` is incremented first and then its (new) value is passed to `a`.

Use these interesting options with moderation, though. The actual convenience (as in reduction of keystrokes) is counterbalanced by an increased obfuscation of the code.

As per a potential increase in the efficiency, it is most probably negligible. In fact, whether you use the increment/decrement operators or not, the MPLAB C32 compiler optimizer, even at the lowest settings, can probably do a better job of optimizing the use of the PIC32 registers in a generic expression without you having to fiddle with these details.

Let me add one last word on loops. It can be confusing to see so many options: Should you test the condition at the beginning or the end? Should you use the `for` type or not? The fact is, in some situations the algorithm you are coding will dictate which one to use, but in many situations you will have a degree of freedom, and more than one type might do. Choose the one that makes your code more readable, and if it really doesn't matter, as in the main loop, just choose the one you like and be consistent.

Notes for the PIC Microcontroller Experts

Depending on the target microcontroller architecture and ultimately the arithmetic and logic unit (ALU), operating on bytes versus operating on word quantities can make a big difference in terms of code compactness and efficiency. In the PIC16 and PIC18 8-bit architectures there is a strong incentive to use byte-sized integers wherever possible; in the PIC32, 32-bit word-sized integers can be manipulated with the same efficiency. The only limiting factor, preventing us from always using 32-bit integers with the MPLAB C32 compiler, is the consideration of the relative preciousness of the internal resources of the microcontroller, and in this case the RAM memory.

Notes for the C Experts

Even if PIC32 microcontrollers have a relatively large RAM memory, larger than the Flash memory of most 8-bit microcontrollers, embedded-control applications will always have to contend with the reality of cost and size limitations. If you learned to program in C on a PC or a workstation, you probably never thought twice about using an `int` whenever you needed an integer. Well, this is the time to think again. Shaving one byte at a time off the requirements of your application might, in some cases, mean you're able to fit in a smaller PIC32 microcontroller, saving fractions of a dollar that when multiplied by the thousands or millions of units (depending on your production run rates) can mean real money added to the bottom line of your company. In other words, if you learn to keep the size of your variables to the strict minimum necessary, you will become a better embedded-control designer. Ultimately, this is what engineering is all about.

Tips & Tricks

Since the first day I have introduced you to the mysteries of the startup (crt0) code, that little piece of code that the linker places automatically in between the main function and the reset vector. Today you might have not realized how the crt0 code helped us once more. In this last project we declared an array called bitmap[] and we asked for it to be initialized with a specific series of values, but the array, being a data structure, resides in RAM during execution. It is one of the crt0 code responsibilities to copy the contents of the array from a table in Flash memory to RAM, immediately before the main program execution.

Another useful service performed by the crt0 code is to initialize every globally declared variable to 0. In most cases this will have the effect of making your code safer and more predictable (you always initialize your variables before use, don't you?), but it will come at a cost. If you have large arrays allocated in RAM, and even if you chose not to initialize them explicitly, it will take a small but finite amount of time to the crt0 code to fill them with zeros before your main program will be able to execute. In embedded-control applications, there can be cases when this delay is not acceptable. In some applications, a few microseconds can make the difference between blowing an expensive power MOSFET, for example, or having your application recovering fast and safe from a critical reset condition. In these special cases you can define the special function _on_reset(), as in the following example:

```
void _on_reset( void)
{
  // something urgent that needs to be done immediately
  // after a reset or at power up
    your code here ...
}
```

This function will replace an empty place holder that the crt0 code is normally calling before getting to the initialization part. Be careful, though, to make it short and not to make too many assumptions at this point. First, remember that this function will be called *every* time the PIC32 goes through a reset sequence. Second, apart from the stack, you cannot count on your program functions and global variables to be available and initialized yet!

Exercises

1. Improve the display/hand synchronization, waiting for a button to be pressed before the hand sweep is started.

2. Add a switch to sense the sweep movement reversal and play the LED sequence backward on the back sweep.

Books

Rony, P., Larsen, D., and Titus, J., *The 8080A Bugbook, Microcomputer Interfacing And Programming* (Howard W. Sams & Co., Inc, Indianapolis, IN, 1976). This is the book that introduced me to the world of microprocessors and changed my life forever. No high-level language programming here, just the basics of assembly programming and hardware interfacing. (Too bad this book is already considered museum material; see link below.)

Links

www.bugbookcomputermuseum.com/BugBook-Titles.html. A link to the "Bugbooks museum"; 30 years since the introduction of the Intel 8080 microprocessor and it is like centuries have already passed.

http://en.wikipedia.org/wiki/Morse_code. Learn about the Morse code, its history, and its applications.

NUMB3RS

The Plan

Just yesterday we learned about different types of C variables, and we stressed the importance of using the right type of variable for each application to preserve a precious resource: RAM. I don't know about you, but I am now very curious about putting those variables to work and seeing how the MPLAB© C32 compiler performs basic arithmetic on them. Knowing that the PIC32 has a set of 32 "working" registers and a 32-bit ALU, I am expecting to see some very efficient code, but I also want to compare the relative performance of the same operation performed on different data types and, in particular, floating-point types. Hopefully after today we will have a better understanding of how to balance performance and memory resources, real-time constraints, and complexity to better fit the needs of our embedded-control applications.

Preparation

This entire lesson will be performed exclusively with software tools that include the MPLAB IDE, MPLAB C32 compiler, and the MPLAB SIM simulator.

Use the New Project Setup checklist to create a new project called **NUMB3RS** and a new source file called **NUMB3RS.c**.

The Exploration

To review all the available data types, I recommend you take a look at the MPLAB C32 User Guide. You can start in Chapter 1.5, where you can find a first list of the supported integer types (see Table 4.1).

Table 4.1: MPLAB C32 integer types comparison table.

Type	Bits	Min	Max
char, signed char	8	−128	127
unsigned char	8	0	255
short, signed short	16	−32768	32767
unsigned short	16	0	65535
int, signed int, long, signed long	32	-2^{31}	$2^{31}-1$
unsigned int, unsigned long	32	0	$2^{32}-1$
long long, signed long long	64	-2^{63}	$2^{63}-1$
unsigned long long	64	0	$2^{64}-1$

As you can see, there are 10 different integer types specified in the ANSI C standard, including char, int, short, long, and long long, both in the signed (default) and unsigned variant. The table shows the number of bits allocated specifically by the MPLAB C32 compiler for each type and, for your convenience, spells out the minimum and maximum values that can be represented.

It is expected that when the type is signed, one bit must be dedicated to the sign itself. The resulting absolute value is halved, while the numerical range is centered around zero. We have also noted before (in our previous explorations) how the MPLAB C32 compiler treats int and long as synonyms by allocating 32 bits (4 bytes) for both of them. In fact, 8-, 16-, and 32-bit quantities can be processed with equal efficiency by the PIC32 ALU. Most of the arithmetic and logic operations on these integer types can be coded by the compiler using single assembly instructions that can be executed very quickly—in most cases, in a single clock cycle.

The long long integer type (added to the ANSI C extensions in 1999) offers 64-bit support and requires 8 bytes of memory. Since the PIC32 core is based on the MIPS 32-bit architecture, operations on long long integers must be encoded by the compiler using short sequences of instructions inserted inline. Knowing this, we are already expecting a small performance penalty for using long long integers; what we don't know is how large it will be.

Let's look at a first integer example; we'll start by typing the following code:

```
main ()
{
int i,j,k;
  i = 1234;    // assign an initial value to i
  j = 5678;    // assign an initial value to j
  k = i * j;   // multiply and store the result in k
}
```

After building the project (**Project | Build All** or **Ctrl+F10**), we can open the Disassembly window (**View | Disassembly Listing**) and take a look at the code generated by the compiler:

```
12:                          i = 1234;
9D00000C   240204D2   addiu      v0,zero,1234
9D000010   AFC20000   sw         v0,0(s8)
13:                          j = 5678;
9D000014   2402162E   addiu      v0,zero,5678
9D000018   AFC20004   sw         v0,4(s8)
```

Even without knowing the PIC32 (MIPS) assembly language, we can easily identify the two assignments. They are performed by loading the literal values to register v0 first and from there to the memory locations reserved for the variable i (pointed to by the s8 register), and later for variable j (pointed to by the s8 register with an offset of 4).

In the following line, the multiplication is performed by transferring the values from the locations reserved for the two integer variables i and j back to registers v0 and v1 and then performing a single 32-bit multiplication mul instruction. The result, available in v0, is stored back into the locations reserved for k (pointed to by s8 with an offset of 8)—pretty straightforward!

```
14:                          k = i*j;
9D00001C   8FC30000   lw         v1,0(s8)
9D000020   8FC20004   lw         v0,4(s8)
9D000024   70621002   mul        v0,v1,v0
9D000028   AFC20008   sw         v0,8(s8)
```

Note

It is beyond the scope of this book to analyze in detail the MIPS assembly programming interface, but I am sure you will find it interesting to note that the mul instruction, like all other arithmetic instructions of the MIPS core, has three operands—although in this case the compiler is using the same register (v0) as both one of the sources and the destination. Note how the MIPS core belongs to the so-called *load and store* class of machines, as all arithmetic operands have first to be fetched from RAM into registers (load) before arithmetic operations can be performed, and later the result has to be transferred back to RAM (store). Finally, if you are even minimally interested in the MIPS assembly, note how the compiler chose to use the addiu instruction to load more efficiently a literal word into a register. In reality this performs an addition of an immediate value with a second operand that was chosen to be the aptly named register zero.

On Optimizations (or Lack Thereof)

You will notice how the overall program, as compiled, is somewhat redundant. The value of j, for example, is still available in register v0 when it is reloaded again—just before the multiplication. Can't the compiler see that this operation is unnecessary?

In fact, the compiler does not see things this clearly; its role is to create "safe" code, avoiding (at least initially) any assumption and using standard sequences of instructions. Later on, if the proper optimization options are enabled, a second pass (or more) is performed to remove the redundant code. During the development and debugging phases of a project, though, it is always good practice to disable all optimizations because they might modify the structure of the code being analyzed and render single-stepping and breakpoint placement problematic. In the rest of this book we will consistently avoid using any compiler optimization option; we will verify that the required levels of performance are obtained regardless.

Testing

To test the code, we can choose to work with the simulator from the Disassembly Listing window itself, single-stepping on each assembly instruction. Or we can choose to work from the C source in the editor window, single-stepping through each C language statement (recommended). In both cases, we can:

1. Open the Local Variables window (**View | Locals**) to see immediately listed, in a small and convenient window, all the variables defined inside the current function (`main()`).

2. Open the Watch window (**View | Watch**) and **add** the `v0` and `v1` registers using the **Add SFR** combo box.

3. Single-step (**Debugger | Step Over** or **F8**) through the next few program lines, observing the effects on the variables in the Watch window. As we noted before, when the value of a variable in the Watch window or the Locals window changes, it is highlighted in red.

If you need to repeat the test, perform a Reset (**Debugger | Reset | Processor Reset**), but don't be surprised if the second time you run the code the contents of the local variables appear magically in place before you initialize them. Local variables (defined inside a function) are not cleared by the Startup code; therefore, if the RAM memory is not cleared between reruns, the RAM locations used to hold the variables `i`, `j`, and `k` will have preserved their contents.

Going `long long`

At this point, modifying only the first line of code, we can change the entire program to perform operations on 64-bit integer variables:

```
main ()
{
  long long i,j,k;

  i = 1234;     // assign an initial value to i
  j = 5678;     // assign an initial value to j
  k = i * j;    // multiply and store the result in k
}
```

Rebuilding the project, and switching again to the Disassembly Listing window (if you had the editor window maximized and you did not close the Disassembly Listing window, you could use the Ctrl+Tab command to quickly alternate between the editor and the Disassembly Listing), we can see how the newly generated code is a bit longer than the previous version. Though the initializations are still straightforward, the multiplication is now performed using several more instructions:

```
15:                           k = i*j;
9D00002C    8FC30000    lw          v1,0(s8)
9D000030    8FC20008    lw          v0,8(s8)
9D000034    00620019    multu       v1,v0
9D000038    00002012    mflo        a0
9D00003C    00002810    mfhi        a1
9D000040    8FC30000    lw          v1,0(s8)
9D000044    8FC2000C    lw          v0,12(s8)
9D000048    70621802    mul         v1,v1,v0
9D00004C    00A01021    addu        v0,a1,zero
9D000050    00431021    addu        v0,v0,v1
9D000054    8FC60008    lw          a2,8(s8)
9D000058    8FC30004    lw          v1,4(s8)
9D00005C    70C31802    mul         v1,a2,v1
9D000060    00431021    addu        v0,v0,v1
9D000064    00402821    addu        a1,v0,zero
9D000068    AFC40010    sw          a0,16(s8)
9D00006C    AFC50014    sw          a1,20(s8)
```

The PIC32 ALU can process only 32 bits at a time, so the 64-bit multiplication is actually performed as a sequence of 32-bit multiplications and additions. The sequence used by the compiler is generated with pretty much the same technique that we learned to use in elementary school, only performed on a 32-bit word at a time rather than one digit at a time. In practice, to perform a 64-bit multiplication using 32-bit instructions, there should be four multiplications and three additions, but you will note that the compiler has actually inserted only three multiplication instructions. What is going on here?

The fact is that multiplying two long long integers (64-bit each) will produce a 128-bit wide result. But in the previous example, we have specified that the result will be stored in yet another long long variable, therefore limiting the result to a maximum of 64 bits. Doing so, we have clearly left the door open for the possibility (not so remote) of an overflow, but we have also given the compiler the permission to safely ignore the most significant bits of the result. Knowing those bits are not going to be missed, the compiler has eliminated completely the fourth multiplication step, so in a way, this is already optimized code.

Note

Basic math tells us that the multiplication of two *n*-bit-wide integer values produces a 2*n*-bit-wide integer result. The C compiler knows this, but if we fail to provide a recipient with enough room to contain the result of the operation, or if there is simply no larger integer type available, as is the case of the multiplication of two `long long` integers, it has no choice but to discard (quietly) the most significant bits of the result. It is our responsibility not to let this happen by choosing the right integer types for the range of values used in our application. If necessary, you can predetermine the number of bits in the result of any product by finding the indexes of the first non-zero-bit (msb) for each operand and adding them together. If the sum is larger than the number of bits of the recipient type, you know there will be an overflow!

Integer Divisions

If we perform a similar analysis of the division operation on integer variables as in the previous examples, we will rapidly confirm how `char`, `short`, and `int` types are all treated the same as well:

```
main ()
{
  int i, j, k;

  i = 1234;
  j = 5678;
  k = i/j;
} // main
```

The code produced by the compiler is extremely compact and uses a single `div` assembly instruction.

```
15:                         k = i/j;
9D00001C    8FC30000    lw          v1,0(s8)
9D000020    8FC20004    lw          v0,4(s8)
9D000024    0062001A    div         v1,v0
9D000028    004001F4    teq         v0,zero
9D00002C    00001012    mflo        v0
9D000030    AFC20008    sw          v0,8(s8)
```

It is only when we analyze the case of a 64-bit division that we find that the compiler is using a different technique:

```
main ()
{
  long long i, j, k;
  i = 1234;
  j = 5678;
  k = i/j;
} // main
```

In fact, recompiling and inspecting the new code in the Disassembly Listing window we reveal a misleadingly short sequence of instructions leading to a subroutine call (jal).

```
15:                        k = i/j;
9D000030    8FC40010    lw       a0,16(s8)
9D000034    8FC50014    lw       a1,20(s8)
9D000038    8FC60018    lw       a2,24(s8)
9D00003C    8FC7001C    lw       a3,28(s8)
9D000040    0F40001A    jal      0x9d000068
9D000044    00000000    nop
9D000048    AFC20020    sw       v0,32(s8)
9D00004C    AFC30024    sw       v1,36(s8)
```

The subroutine itself will appear in the disassembly listing, after all the main function code. This subroutine is clearly separated and identified by a comment line that indicates it is part of a library, a module called libgcc2.c. The source for this routine is actually available as part of the complete documentation of the MPLAB C32 compiler and can be found in a subdirectory under the same directory tree where the MPLAB C32 compiler has been installed on your hard disk.

By selecting a subroutine in this case, the compiler has clearly made a compromise. Calling the subroutine means adding a few extra instructions and using extra space on the stack. On the other hand, fewer instructions will be added each time a new division (among long long integers) is required in the program; therefore, overall code space will be preserved.

Floating Point

Beyond integer data types, the MPLAB C32 compiler offers support for a few more data types that can capture fractional values—the floating-point data types. There are three types to choose from (see Table 4.2) corresponding to two levels of resolution: `float`, `double`, and `long double`.

Table 4.2: MPLAB C32 floating-point types comparison table.

Type	Bits
Float	32
Double	64
Long double	64

Notice how the MPLAB C32 compiler, by default, allocates for both the `double` and the `long double` types the same number of bits, using the double precision floating-point format defined in the IEEE754 standard.

Since the PIC32 doesn't have a hardware floating-point unit (FPU), all operations on floating-point types must be coded by the compiler using floating-point arithmetic libraries whose size and complexity are considerably larger/higher than any of the integer libraries. You should expect a major performance penalty if you choose to use these data types, but, again, if the problem calls for fractional quantities to be taken into account, the MPLAB C32 compiler certainly makes dealing with them easy.

Let's modify our previous example to use floating-point variables:

```
main ()
{
  float i,j,k;

  i = 12.34;        // assign an initial value to i
  j = 56.78;        // assign an initial value to j
  k = i * j;        // store the result in k
}
```

After recompiling and inspecting the Disassembly Listing window, you will immediately notice that the compiler has chosen to use a subroutine instead of inline code.

Changing the program again to use a double-precision floating-point type, long double, produces very similar results. Only the initial assignments seem to be affected, and all we can see is, once more, a subroutine call.

The C compiler makes using any data type so easy that we might be tempted to always use the largest integer or floating-point type available, just to stay on the safe side and avoid the risk of overflows and underflows. On the contrary, though, choosing the right data type for each application can be critical in embedded control to balance performance and optimize the use of resources. To make an informed decision, we need to know more about the level of performance we can expect when choosing the various precision data types.

Measuring Performance

Let's use what we have learned so far about simulation tools to measure the actual relative performance of the arithmetic libraries (integer and floating-point) used by the MPLAB C32 compiler. We can start by using the software simulator's (MPLAB SIM) built-in StopWatch tool, with the following code:

```
#include <p32xxxx.h>

main ()
{
  char         c1, c2, c3;
  short        s1, s2, s3;
  int          i1, i2, i3;
  long long    ll1, ll2, ll3;
  float        f1,f2, f3;
  long double  d1, d2, d3;

  c1 = 12;            // testing char integers (8-bit)
  c2 = 34;
  c3 = c1 * c2;

  s1 = 1234;          // testing short integers (16-bit)
  s2 = 5678;
  s3= s1 * s2;

  i1 = 1234567;       // testing (long) integers (32-bit)
  i2 = 3456789;
  i3= i1 * i2;
```

```
lll = 1234;          // testing long long integers (64-bit)
ll2 = 5678;
ll3= lll * ll2;

f1 = 12.34;          // testing single precision floating point
f2 = 56.78;
f3= f1 * f2;

d1 = 12.34;          // testing double precision floating point
d2 = 56.78;
d3= d1 * d2;
} // main
```

After compiling and linking the project, open the StopWatch window (**Debugger |
StopWatch**) and position the window according to your preferences (see Figure 4.1).
(Personally I like it docked to the bottom of the screen so that it does not overlap with the
editor window and it is always visible and accessible.)

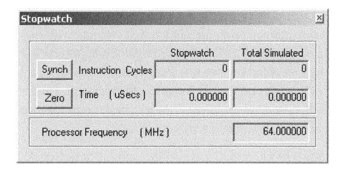

Figure 4.1: The MPLAB SIM StopWatch window.

Zero the StopWatch timer and execute a Step-Over command (**Debug | StepOver** or
press **F8**). As the simulator completes updating the StopWatch window, you can manually
record the execution time required to perform the integer operation. The time is provided
by the simulator in the form of a cycle count and an indication in microseconds derived
by the cycle count multiplied by the simulated clock frequency, a parameter specified in
the Debugger Settings (the **Debugger | Settings | Osc/Trace** tab).

Proceed by setting the cursor over the next multiplication, and execute a **Run To Cursor**
command or simply continue **StepOver** until you reach it. Again, **Zero** the StopWatch,

execute a **Step-Over**, and record the second time. Continue until all five types have been tested (see Table 4.3).

Table 4.3: Relative performance test results using MPLAB C32 rev. 0.20 (all optimizations disabled).

Multiplication Test	Width (Bits)	Cycle Count	Performance Relative to:	
			Int	Float
Char integer (`char`)	8	6	1	—
Short integer (`short`)	16	6	1	—
Integer (`int, long`)	32	6	1	—
Long integer (`long long`)	64	21	3.5	—
Single precision FP (`float`)	32	71	11.8	1
Double precision FP (`long double`)	64	159	26.5	2.23

Table 4.3 records the results (cycle counts) in the first column, with two more columns showing the relative performance ratios obtained by dividing the cycle count of each row by the cycle count recorded for two reference types. Don't be alarmed if you happen to record different values; several factors can affect the measure. Future versions of the compiler could possibly use more efficient libraries, and/or optimization features could be introduced or enabled at the time of testing.

Keep in mind that this type of test lacks any of the rigorousness required by a true performance benchmark. What we are looking for here is just a basic understanding of the impact on performance that we can expect from choosing to perform our calculations using one data type versus another. We are looking for the big picture—relative orders of magnitude. For that purpose, the table we just obtained can already give us some interesting indications.

As expected, 32-bit operations appear to be the fastest, whereas `long long` integer (64-bit) multiplications are about four times slower. Single precision floating-point operations require more effort than integer operations. Multiplying 32-bit floating-point numbers requires one order of magnitude more effort than multiplying 32-bit integers. From here, going to double precision floating-point (64-bit) about doubles the number of cycles required.

So, when should we use floating-point, and when should we use integer arithmetic?

Beyond the obvious, from the little we have learned so far we can perhaps extract the following rules:

1. Use integers every time you can, i.e. when fractions are not required or when the algorithm can be rewritten for integer arithmetic.

2. Use the smallest integer type that will not produce an overflow or underflow if you want to save on RAM memory space, but once you are not using 64-bit integers, you will not see any further performance improvement from going to any integer type smaller than 32-bit.

3. If you have to use a floating-point type (fractions are required), expect an order-of-magnitude reduction in the performance of the compiled program.

4. Double precision floating-point (long double) seems to only reduce the performance further, by a factor of two.

Keep in mind also that floating-point types offer the largest value ranges but also are always introducing approximations. As a consequence, floating-point types are not recommended for financial calculations. Use `long long` integers, if necessary, and perform all operations in cents (instead of dollars and fractions).

Debriefing

In this lesson, we have learned not only what data types are available and how much memory is allocated to them but also how they affect the resulting compiled program in terms of code size and execution speed. We used the MPLAB SIM simulator StopWatch tool to measure the number of instruction cycles required for the execution of a series of basic arithmetic operations. Some of the information we gathered will be useful to guide our actions in the future when we're balancing our needs for precision and performance in embedded-control applications.

Notes for the Assembly Experts

The brave few assembly experts that have attempted to deal with floating-point numbers in their applications tend to be extremely pleased and forever thankful for the great simplification achieved by the use of the C compiler. Single or double precision arithmetic becomes just as easy to code as integer arithmetic has always been.

When using integer numbers, though, there is sometimes a sense of loss of control, because the compiler hides the details of the implementation and some operations might become obscure or much less intuitive/readable. Here are some examples of conversion and byte manipulation operations that can induce some anxiety:

- Converting an integer type into a smaller or larger one

- Extracting or setting the most or least significant byte of a 16-bit or 32-bit data type

- Extracting or setting one bit out of an integer variable

The C language offers convenient mechanisms for covering all such cases via implicit type conversions, as in:

```
short   s;         // 16-bit
int     i;         // 32-bit
i = s;
```

The value of s is transferred into the two LSBs of i, and the two MSBs of i are cleared.

Explicit conversions (called *type casting*) might be required in some cases where the compiler would otherwise assume an error, as in:

```
short   s;         // 16-bit
int     i;         // 32-bit
s = (short) i;
```

(short) is a type cast that results in the two MSBs of i to be discarded as i is forced into a 16-bit value.

Bit fields are used to cover the conversion to and from integer types that are smaller than 1 byte. The PIC32 library files contain numerous examples of definitions of bit fields for the manipulation of all the control bits in the peripheral's SFRs.

Here is an example extracted from the include file used in our project, where the Timer1 module control register T1CON is defined and each individual control bit is exposed in a structure defined as T1CONbits:

```
extern unsigned int T1CON;
extern union {
  struct {
    unsigned :1;
    unsigned TCS:1;
    unsigned TSYNC:1;
    unsigned :1;
    unsigned TCKPS0:1;
    unsigned TCKPS1:1;
    unsigned TGATE:1;
    unsigned :6;
    unsigned TSIDL:1;
    unsigned :1;
    unsigned TON:1;
  };
  struct {
    unsigned :4;
    unsigned TCKPS:2;
  };
} T1CONbits;
```

You can access each bit field using the "dot" notation, as in the following example:
```
T1CONbits.ON = 1;
```

Notes for the 8-Bit PIC® Microcontroller Experts

The PIC microcontroller user who is familiar with the 8-bit PIC microcontrollers and their respective compilers will notice a considerable improvement in performance, both with integer arithmetic and with floating-point arithmetic. The 32-bit ALU available in the PIC32 architecture is clearly providing a great advantage by manipulating up to four times the number of bits per cycle, but the performance improvement is further accentuated by the availability of up to 32 working registers, which make the coding of critical arithmetic routines and numerical algorithms more efficient.

Notes for the 16-Bit PIC and dsPIC® Microcontroller Experts

Users of the MPLAB C30 compiler will have probably noticed by now how the new MPLAB C32 compiler assigns different widths to common integer types. For example, the int and short types used to be synonyms of 16-bit integers for the MPLAB C30 compiler. Although short is still a 16-bit integer, for the MPLAB C32 compiler int is now really a synonym of the long integer type. In other words, int has doubled its size. You might be wondering what happens to the portability of code when such a dramatic change is factored in.

The answer depends on which way you are looking at the problem. If you are porting the code "up," or, in other words, you are taking code written for a 16-bit PIC architecture to a 32-bit PIC architecture, most probably you are going to be fine. Global variables will use a bit more RAM space and the stack might grow as well, but it is also likely that the PIC32 microcontroller model you are going to use has much more RAM to offer. Since the new integer type is larger than that used in the original code, if the code was properly written, you don't have to worry about overflows and underflows.

On the contrary, if you are planning on porting some code "down," even if this is just being contemplated as a future option, you might want to be careful. If you are writing code for a PIC32 and rely on the int type to be 32-bit large, you might have a surprise later when the same code will be compiled into a 16-bit wide integer type by the MPLAB C30 compiler. The best way to avoid any ambiguity on the width of your integers is to use *exact-width* types.

A special set of exact-width integer types is offered by the inttypes.h library. They include the following types:

int8_t   Always an 8-bit signed type.

uint8_t   Always an 8-bit unsigned type.

int16_t   Always a 16-bit signed type.

uint16_t Always a 16-bit unsigned type.

int32_t   Always a 32-bit signed type.

uint32_t Always a 32-bit unsigned type.

int64_t   Always a 64-bit signed type.

uint64_t Always a 64-bit unsigned type.

If you use them when necessary, you can make your code more portable but also more readable because they will help highlight the portions of your code that are dependent on integer size.

Note

Another useful and sometimes misunderstood integer type is `size_t`, defined in the stddef.h library. It is meant to be used every time you need a variable to contain the size of an object in memory expressed in bytes. It is guaranteed by each ANSI compiler to have the right range so that it's always able to contain the size of the largest object possible for a given architecture. As expected, the function `sizeof()`, but also most of the functions in the string.h library, makes ample use of it.

Tips & Tricks

Math Libraries

The MPLAB C32 compiler supports several standard ANSI C libraries, including these:

- limits.h contains many useful macros defining implementation-dependent limits, such as, for example, the number of bits composing a `char` type (CHAR_BIT) or the largest integer value (INT_MAX).

- float.h contains similar implementation-dependent limits for floating-point data types, such as, for example, the largest exponent for a single precision floating-point variable (FLT_MAX_EXP).

- math.h contains trigonometric functions, rounding functions, logarithms, and exponentials but also many useful constants like pi (`M_PI`).

Complex Data Types

The MPLAB C32 compiler supports complex data types as an extension of both integer and floating-point types. Here is an example declaration for a single precision floating-point type:

```
__complex__ float z;
```

Note

Notice the use of a double underscore before and after the keyword complex.

The variable z so defined has now a *real* and an *imaginary* part that can be individually addressed using, respectively, the syntax:

```
__real__ z
```

and

```
__imag__ z
```

Similarly, the next declaration produces a complex variable of 32-bit integer type:

```
__complex__ int x;
```

Complex constants are easily created adding the suffix i or j, as in the following examples:

```
x = 2 + 3j;
z = 2.0f + 3.0fj;
```

All standard arithmetic operations (+, -, *, /) are performed correctly on complex data types. Additionally, the ~ operator produces the complex conjugate.

Complex types could be pretty handy in some types of applications, making the code more readable and helping avoid trivial errors. Unfortunately, as of this writing, the MPLAB IDE support of complex variables during debugging is only partial, giving access only to the "real" part through the Watch window and the mouse-over function.

Exercises

1. Write a program that uses Timer2 as a stopwatch for real-time performance measurements.

2. If the width of Timer2 is not sufficient, use Timer2 and Timer3 joined in the new 32-bit timer mode.

3. Test the relative performance of the division for the various data types.

4. Test the performance of the trigonometric functions relative to standard arithmetic operations.

5. Test the relative performance of the multiplication for complex data types

Books

Britton, Robert., *MIPS Assembly Language Programming* (Prentice Hall, 2003). It might seem strange to you that I am suggesting a book about assembly programming. Sure, we set off with the intention to learn programming in C, but if you're like me, you won't resist the curiosity and you will want to learn the assembly of the PIC32 MIPS core as well.

Links

http://en.wikipedia.org/wiki/Taylor_series. If you are curious, this site shows how the C compiler can approximate some of the functions in the math library.

Interrupts

The Plan

For reasons of efficiency, size, and ultimately cost, in the embedded-control world the smallest applications, which happen to be implemented in the highest volumes, most often cannot afford the "luxury" of a multitasking operating system and use the interrupt mechanisms instead to "divide their attention" among the many tasks at hand. Interrupts provide a very strong mechanism for *real-time* control, allowing our applications to deal with asynchronous external events. Unfortunately, the C programming language does not incorporate the concept of interrupts in its model, leaving the embedded-control programmer with the only choice of defining interrupts as a special kind of function.

Today we will see how the MPLAB© C32 compiler allows us to easily manage the interrupt mechanisms offered by the PIC32 microcontroller architecture.

Preparation

This entire lesson will be performed exclusively with software tools, including the MPLAB IDE, the MPLAB C32 compiler, and the MPLAB SIM simulator.

Use the New Project Setup checklist to create a new project called Interrupts and a new source file, similarly called interrupts.c.

The Exploration

An *interrupt* is an internal or external event that requires quick attention from the CPU. The PIC32 architecture provides a rich interrupt system that can manage as many as

64 distinct sources of interrupts. If necessary, each interrupt source can have a unique piece of code, called the *interrupt service routine (ISR)* or *interrupt handler*, directly associated with it, to provide the required response action. Interrupts can be completely asynchronous with the execution flow of the main program. They can be triggered at any point in time and in an unpredictable order.

Responding quickly to interrupts is essential to allow prompt reaction to the trigger event and a fast return to the main program execution flow. Therefore, the goal is to minimize the *interrupt latency*, defined as the time between the triggering event and the execution of the first instruction of the ISR. In the PIC32 architecture, the latency is extremely short. Although it is fixed for each given interrupt source—only three or four instruction cycles—other mechanisms common among all 32-bit architectures, such as the cache and the bus arbitration module that we will review in detail in future expeditions, may affect the overall response time, adding a small amount of nondeterminism. A deep understanding of the interrupt mechanism will help us minimize and possibly cancel its effect on our applications.

The MPLAB C32 compiler will help us manage the complexity of the interrupt system by providing a few language extensions and a rich set of functions included in the plib.h library.

Interrupts and Exceptions

To the MIPS core running inside the PIC32, all interrupts fall generally under the category of *exceptions*. This is a very broad category of events that gathers pretty much anything that can disrupt the normal flow of a program. A reset command produces an exception, an error in a division can produce an exception, but also access to a memory address that is not implemented (or restricted) will produce an exception, and the list goes on and on. Interrupts, after all, are the most benign kind of exception that can occur. The MIPS core relies on a few *vectors* (pointers to functions) located conveniently in separate RAM, program memory, or both regions to cover all possible types of exceptions (see Table 5.1). It is once more the role of the Startup code to place such vectors and offer default handlers for all the essential exceptions an embedded control application might need.

Don't worry if not all the entries in Table 5.1 make sense to you. Some of them refer to advanced features that we will encounter and discuss in a later chapter. Some are related to features, part of the MIPS architecture, that have no practical application in the PIC32MX implementation.

Table 5.1: Exception vectors table of the PIC32 architecture.

Exception Source	Memory Region	Description
Reset and NMI	Program	Normal reset and nonmaskable interrupt entry point.
On-chip debug	Program	Used by the ICD and EJTAG interfaces to enable in circuit debugging features.
Cache error	RAM or Program	Error condition specific to the cache mechanism.
TLB refill	RAM or Program	Not used on PIC32 because a fixed address translation scheme (FMT) is used in place of a full MMU.
General exception	RAM or Program	All other types of exceptions.
Interrupt	RAM or Program	The proper interrupt vector.

The basic MIPS interrupt mechanism provides for a single vector inside the exception table, and therefore a single interrupt service routine, to be dedicated to all possible interrupts events. Once the interrupt (exception) occurs, the content of a special register (known as cause) gives the service routine all the information necessary to identify the trigger event and the most appropriate action to take in response. To be able to resume execution after the interrupt has been dealt with, it is fundamental for an interrupt service routine to be able to save the processor context (*prologue*) before taking any action and to be able to restore it (*epilogue*) later exactly as it was before the interruption. The exact prologue and epilogue sequences can be somewhat convoluted and their analysis is beyond the scope of our exploration. For now, it will suffice to know that the MPLAB C32 compiler makes all this automatic and safe by allowing us to define "special" C functions for use as interrupt handlers, as long as a few limitations are kept in consideration, such as:

- Interrupt service functions are not supposed to return any value (use type void).

- No parameter can be passed to the function (use parameter void).

- They cannot be called directly by other functions.

- Ideally, they should not call any other function.

The first three limitations should be pretty obvious given the nature of the interrupt mechanism—since it is triggered by an asynchronous event, there cannot be parameters or a return value because there is no proper function call in the first place. The last is more of a recommendation to keep in mind for efficiency considerations.

Sources of Interrupt

The following events can be used to trigger an interrupt. Among the external sources available for the PIC32FJ512MX360L, there are:

- 5 × external pins with level trigger detection
- 22 × external pins connected to the Change Notification module
- 5 × Input Capture modules
- 5 × Output Compare modules
- 2 × serial port interfaces (UARTs)
- 4 × synchronous serial interfaces (SPI and I^{2}C)
- 1 × Parallel Master Port

Among the internal sources, we count:

- 1 × 32 internal (core) timer
- 5 × 16-bit timers
- 1 × analog-to-digital converter
- 1 × Analog Comparators module
- 1 × real-time clock and calendar
- 1 × Flash controller
- 1 × fail-safe clock monitor
- 2 × software interrupts
- 4 × DMA channels

Other models of PIC32 may have a different mix of internal and external interrupt sources. Many of these sources in their turn can generate several different interrupts. For example, a serial port interface peripheral (UART) can generate three types of interrupt:

- When new data has been received and is available in the receive buffer for processing

- When data in the transmit buffer has been sent and the buffer is empty, ready and available to transmit more

- When an error condition has been generated and action might be required to reestablish communication

By design, up to a total of 96 independent events could be managed by the PIC32 interrupt control module. That's a lot of interrupts!

Of course, when multiple sources of interrupts are enabled and used by an application, there is a need for the ISR to identify the specific one at hand and to be able to branch to an appropriate segment of code to deal with it. As we will see shortly, several flags and additional control mechanisms assist the programmer with this task.

Interrupt Priorities

Each interrupt source has seven associated control bits, grouped logically in various special-function registers:

- The *Interrupt Enable* bit (typically represented with the name of the interrupt source peripheral followed by the suffix −IE in the device datasheet), a single bit of data:

  1. When cleared, the specific trigger event is prevented from generating interrupts.

  2. When set, it allows the interrupt to be processed.

  At power-on, all interrupt sources are disabled by default.

- The *Interrupt Flag* (typically represented with a suffix -IF), a single bit of data, is set each time the specific trigger event is activated, independently of the status of the enable bit. Notice that, once set, it must be cleared (manually) by the user. In other words it must be cleared before exiting the ISR, or the same interrupt service routine will be immediately called again.

- The *Group Priority Level* (typically represented with a suffix -IP). Interrupts can have up to seven levels of priority (from ipl1 to ipl7). Should two interrupt events occur at the same time, the highest priority event will be served first. Three bits encode the priority level of each interrupt source. At any given point, the PIC32 execution priority-level value is kept in the MIPS core status register.

Interrupts with a priority level lower than the current value will be ignored. At power-on, all interrupt sources are assigned a default level of ip10, once more assuring that all interrupts are disabled.

- The *Subpriority Level*. Two more bits are allocated to define four more possible levels of priority within a priority group. If two events of the same priority level occur simultaneously, the one with the highest subpriority will be selected first. Once an interrupt of a given priority group is selected, though, any following interrupts of the same level (even if of higher subpriority) will be ignored until the current interrupt (flag) has been cleared.

Within an assigned priority level, a relative (default) priority among the various sources in a fixed order of appearance is defined for any given PIC32 model. When everything else fails (both group and subgroup priorities are identical), it is the natural order to decide between two simultaneous events (see Table 5.2).

Table 5.2: Interrupt sources of the PIC32FJ512MX360L.

Natural Order	Macro Abbreviation	IRQ Symbol	Description
0 (highest)	CT	_CORE_TIMER_IRQ	Core Timer Interrupt
1	CS0	_CORE_SOFTWARE_0_IRQ	Core Software Interrupt 0
2	CS1	_CORE_SOFTWARE_1_IRQ	Core Software Interrupt 1
3	INT0	_EXTERNAL_0_IRQ	External Interrupt 0
4	T1	_TIMER_1_IRQ	Timer 1 Interrupt
5	IC1	_INPUT_CAPTURE_1_IRQ	Input Capture 1 Interrupt
6	OC1	_OUTPUT_COMPARE_1_IRQ	Output Compare 1 Interrupt
7	INT1	_EXTERNAL_1_IRQ	External Interrupt 1
8	T2	_TIMER_2_IRQ	Timer 2 Interrupt
9	IC2	_INPUT_CAPTURE_2_IRQ	Input Capture 2 Interrupt
10	OC2	_OUTPUT_COMPARE_2_IRQ	Output Compare 2 Interrupt
11	INT2	_EXTERNAL_2_IRQ	External Interrupt 2

(*continued*)

Table 5.2: (Continued)

Natural Order	Macro Abbreviation	IRQ Symbol	Description
12	T3	_TIMER_3_IRQ	Timer 3 Interrupt
13	IC3	_INPUT_CAPTURE_3_IRQ	Input Capture 3 Interrupt
14	OC3	_OUTPUT_COMPARE_3_IRQ	Output Compare 3 Interrupt
15	INT3	_EXTERNAL_3_IRQ	External Interrupt 3
16	T4	_TIMER_4_IRQ	Timer 4 Interrupt
17	IC4	_INPUT_CAPTURE_4_IRQ	Input Capture 4 Interrupt
18	OC4	_OUTPUT_COMPARE_4_IRQ	Output Compare 4 Interrupt
19	INT4	_EXTERNAL_4_IRQ	External Interrupt 4
20	T5	_TIMER_5_IRQ	Timer 5 Interrupt
21	IC5	_INPUT_CAPTURE_5_IRQ	Input Capture 5 Interrupt
22	OC5	_OUTPUT_COMPARE_5_IRQ	Output Compare 5 Interrupt
23	SPI1E	_SPI1_ERR_IRQ	SPI 1 Fault
24	SPI1TX	_SPI1_TX_IRQ	SPI 1 Transfer Done
25	SPI1RX	_SPI1_RX_IRQ	SPI 1 Receiver Done
26	U1E	_UART1_ERR_IRQ	UART 1 Error
27	U1RX	_UART1_RX_IRQ	UART 1 Receiver
28	U1TX	_UART1_TX_IRQ	UART 1 Transmitter
29	I2C1B	_I2C1_BUS_IRQ	I2C 1 Bus Collision Event
30	I2C1S	_I2C1_SLAVE_IRQ	I2C 1 Slave Event
31	I2C1M	_I2C1_MASTER_IRQ	I2C 1 Master Event
32	CN	_CHANGE_NOTICE_IRQ	Input Change Interrupt
33	AD1	_ADC_IRQ	ADC Convert Done
34	PMP	_PMP_IRQ	Parallel Master Port Interrupt
35	CMP1	_COMPARATOR_1_IRQ	Comparator 1 Interrupt

(continued)

Table 5.2: (Continued)

Natural Order	Macro Abbreviation	IRQ Symbol	Description
36	CMP2	_COMPARATOR_2_IRQ	Comparator 2 Interrupt
37	SPI2E	_SPI2_ERR_IRQ	SPI 2 Fault
38	SPI2TX	_SPI2_TX_IRQ	SPI 2 Transfer Done
39	SPI2RX	_SPI2_RX_IRQ	SPI 2 Receiver Done
40	U2E	_UART2_ERR_IRQ	UART 2 Error
41	U2RX	_UART2_RX_IRQ	UART 2 Receiver
42	U2TX	_UART2_TX_IRQ	UART 2 Transmitter
43	I2C2B	_I2C2_BUS_IRQ	I2C 2 Bus Collision Event
44	I2C2S	_I2C2_SLAVE_IRQ	I2C 2 Slave Event
45	I2C2M	_I2C2_MASTER_IRQ	I2C 2 Master Event
46	FSCM	_FAIL_SAFE_MONITOR_IRQ	Fail-safe Clock Monitor Interrupt
47	RTCC	_RTCC_IRQ	Real Time Clock Interrupt
48	DMA0	_DMA0_IRQ	DMA Channel 0 Interrupt
49	DMA1	_DMA1_IRQ	DMA Channel 1 Interrupt
50	DMA2	_DMA2_IRQ	DMA Channel 2 Interrupt
51	DMA3	_DMA3_IRQ	DMA Channel 3 Interrupt
. . .			
56 (lowest)	FCE	_FLASH_CONTROL_IRQ	Flash Control Event

Interrupt Handlers Declaration

The MPLAB C32 compiler gives us two options to declare a function as "the" default interrupt handler (vector 0) at a given interrupt priority (ipl1, for example), using either the *attribute syntax* as follows:

```
void __attribute__ (( interrupt(ipl1),vector(0)))
InterruptHandler( void)
```

```
{
  // your interrupt service routine code here. . .
} // interrupt handler
```

or the *pragma syntax*, as follows:

```
#pragma interrupt InterruptHandler ipl1 vector 0
void InterruptHandler( void)
{
  // interrupt service routine code here. . .
} // interrupt handler
```

In both cases the result is that the compiler treats the function `InterruptHandler()` with the respect due to a proper ISR, including prologue and epilogue code sequences that provide safe context save and restore.

The MPLAB C32 compiler uses the `__attribute__ (( ))` mechanism in this and many other circumstances as a way to specify special attributes that modify the behavior of the compiler without violating the C language syntax. Personally, I find this syntax too cryptic; the double underscore, before and after, and the double parentheses in particular are hard on my eyes. My preferred way around the problem is to use a macro (defined in sys/attribs.h) that has the additional advantage of resembling the one found in previous PIC24 and dsPIC libraries:

```
__ISR( v, ipl)
```

In the following example, the `__ISR` macro is used to the same effect of the previous code snippet:

```
void __ISR( 0, ipl1) InterruptHandler (void)
{
  // interrupt service routine code here. . .
} // interrupt handler
```

The choice between the two syntax styles is yours and might well depend on your very personal preferences and previous experiences. Further, should you ever need to port code from a different compiler, chances are that one of the two methods will match your original source code more closely. So keep both in mind; you never know when they might come in handy.

The Interrupt Management Library

With up to 96 possible sources of interrupts, to manage the sophisticated priority mechanisms made available by the PIC32 interrupt controller module, we can definitely use a little help in the shape of a small library int.h provided as part of the standard PIC32 toolset.

We can invoke it directly, as in:

```
#include <int.h>
```

or indirectly as part of the entire peripherals support library:

```
#include <plib.h>
```

In both cases we gain access to a good number of precious little functions and macros (recognizable by the lower case m- prefix), including these:

- INTEnableSystemSingleVectoredInt(); is a function that follows a precise sequence of initialization of the interrupt control module (as prescribed in the device datasheet) to enable the basic interrupt management mode of the PIC32. The unusually long function name is worth typing because it relieves us from a considerable burden, making our code easy and safe.

- mXXSetIntPriority(x); is actually just a placeholder for a long list of similar macros (replace the *XX* with the interrupt source abbreviations from Table 5.2 to obtain each macro name). It assigns a given priority level (from 0 to 7) to the chosen interrupt source. The amount of work performed is not much in this case, but there is a considerable convenience factor because we are spared the painful search on the device datasheet for the correct IPCxx register where the -IP bits corresponding to the chosen interrupt source can be selected.

- mXXClearIntFlag(); is a macro that is, once more, representative of an entire class of macros that allow us to clear the interrupt flag (-IF bit) of the chosen interrupt source.

Single Vector Interrupt Management

Without any further hesitation, let's start laying out a first example that will use an ISR to service a timer. We will enable the Timer2 module, setting its period to a count of 15 and requesting that an interrupt be generated. The global variable count will be incremented at each period by the interrupt service routine:

```
/*
** Single Interrupt Vector test
*/
#include <p32xxxx.h>
#include <plib.h>

int count;

#pragma interrupt InterruptHandler ipl1 vector 0
void InterruptHandler( void)
{
  count++;
  mT2ClearIntFlag();
} // Interrupt Handler

main()
{
  // 1. init timers
  PR2 = 15;
  T2CON = 0x8030;

  // 2. init interrupts
  mT2SetIntPriority( 1);
  INTEnableSystemSingleVectoredInt();
  mT2IntEnable( 1);

  // 3. main loop
  while( 1);
} // main
```

There is one fundamental action that each interrupt handler (no matter how simple) is responsible for, and that is clearing the interrupt flag before returning. This is pretty much all our ISR is required to do beside incrementing count.

Notice also that in the main() function, after the the initialization (//1.) of the timer control register and period register, the interrupt configuration (//2.) is completed before enabling the interrupt source. Also, the Timer2 interrupt priority (1) must match the priority level declared by the #pragma syntax (ipl1).

> **Note**
>
> The compiler needs to know the priority level of the interrupt routine in order to use the correct prologue and epilogue. In fact, as we will learn shortly, interrupts of ipl7 should be given a special treatment, shorter prologue/epilogue, since they benefit from the availability of the alternate register set for a fast context switch.

The same code can obviously be written the "hard way," without using the int.h library but making direct access to the special function registers responsible for the configuration of the interrupt controller:

```
/*
** Single Interrupt vector test
*/
#include <p32xxxx.h>

#define _T2IE IEC0bits.T2IE
#define _T2IF IFS0bits.T2IF
#define _T2IP IPC2bits.T2IP

int count;
void __ISR( 0, ipl1) InterruptHandler( void)
{
    count++;
    _T2IF = 0;
} // interrupt handler

main()
{
    // 1. init timers
    PR2 = 15;
    T2CON = 0x8030;

    // 2. init interrupts
    _T2IP = 1;
    INTEnableSystemSingleVectoredInt();
    _T2IE = 1;

    // 3. main loop
    while( 1);
} // main
```

Note for the PIC24 AND dsPIC Experts

Unfortunately, the "shortcut" symbols _T2IF, _T2IE, and _T2IP that used to be so conveniently defined in the standard include files for the PIC24 and dsPIC architectures are no more part of the standard include files of the MPLAB C32 compiler. If you are porting some 16-bit code and need the compatibility, you will have to follow my example and redefine the shortcuts you need by hand on a case-by-case basis.

It is once more a matter of personal choice. Feel free to choose the style you like or that you find more intuitive and readable for your application.

Now it is time to get a new project ready for some hands-on interrupt testing:

1. Save the source file (of your choice) as **single.c** and, using the **New Project** checklist, create a new project **single.mcp** and add the source file to it.

2. Prepare the MPLAB SIM simulator for use as the debugging tool using the **MPLAB SIM Setup** checklist.

3. Now build the project using the **Project | Build** command (or the **Ctrl + F10** shortcut).

4. Open the Watch window (**View | Watch**) and add the global symbol count, selecting it in the combo box and clicking the **Add Symbol** button.

5. Select the TMR2 register in the SFR combo box and click the **Add SFR** button to add it to the Watch window.

6. Place a breakpoint, inside the interrupt handler routine, on the line where count is incremented and choose **Animate** (or **Run**) to execute the code.

If all went well, you should see that the program execution has stopped after a short while, reaching the breakpoint inside the interrupt handler. Although the code had been "stuck" for a while inside the (empty) main loop, upon reaching its period (set in the PR2 register), the Timer2 generated an interrupt request and the interrupt handler was transferred control.

Continuing with the animation (or running again) you will see that count keeps being incremented each time the execution of the main loop is briefly "interrupted."

Notice that each time you reach the breakpoint, in the Watch window the value of count is constantly updated and shown in red (since it keeps changing), but the value

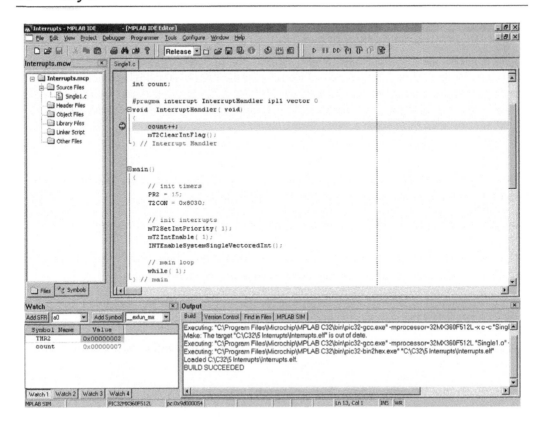

Figure 5.1: Screenshot of the single.c project.

of TMR2 is always the same and, perhaps surprisingly, not zero. In fact, when the Timer2 module reaches the value set in the period register (PR2), it does reset while it generates a new interrupt, but it also proceeds counting while the PIC32 starts the execution of the interrupt handler. By the time the interrupt handler prologue is completed and the program counter reaches the breakpoint, Timer2 is already showing a value of 2. What we have just done, perhaps involuntarily, is to obtain a rough measure of the interrupt handler overhead. Since we chose to use a prescaler of 1:8 for the Timer2 clock input, a count of 2 indicates that the prologue to the interrupt service routines occupies (at least) 16 clock cycles, equivalent to the execution of 16 instructions. You can verify it, if you are curious, by inspecting the code produced by the compiler in the Disassembly window.

But what would have happened if we had not selected a large prescale value (1:8) or if we had selected a shorter period? Of course you can test it by yourself with minor modifications to the example code. You will see how the interrupt routine gets called over and over and there is no more time spent inside the main loop. Not a big loss in our simple example, I agree, but in a practical application this would be a disaster. When the interrupts are too many, too frequent, or simply poorly managed, the main program can be stalled completely. It is our responsibility to make sure that the interrupt handler routine, including its prologue and epilogue, is not using up all the available processor cycles.

Managing Multiple Interrupts

If multiple sources of interrupts are used by an application, assigning different priority levels to each source solves only one part of the problem. The priority decides who gets served first if two or more interrupt events happen simultaneously. But when one of the (many) interrupts is being served, the others will have to wait for their turn to be served. However, in some cases the application requires not only multiple interrupts but the ability to *nest* the interrupt calls. When a lower-priority interrupt is being served and the ISR is being executed, a higher-priority interrupt might require immediate attention, in its turn interrupting the handler.

To enable nesting of interrupt calls, you will have to "manually" reenable interrupts immediately upon entry in the interrupt handler (using a MIPS assembly instruction) instead of waiting for the epilogue code to do it automatically upon exit.

Here is a simple example that extends our first project in an imaginary application where Timer3 is used to produce a second periodic interrupt of high(er) priority (level 3):

```
/*
** Single Vector Interrupt Nesting
*/
#include <p32xxxx.h>
#include <plib.h>

int count;

void __ISR( 0, ipl1) InterruptHandler( void)
{
  // 1. re-enable interrupts immediately (nesting)
  asm("ei");

  // 2. check and serve the highest priority first
  if ( mT3GetIntFlag())
```

```
  {
    count++;
    // clear the flag and exit
    mT3ClearIntFlag();
  } // _T3

  // 3. check and serve the lower priority
  else if ( mT2GetIntFlag())
  {
    // spend a LOT of time here!
    while( 1);

    // before clearing the flag and exiting
    mT2ClearIntFlag();
  } // _T2
} // Interrupt Handler

main()
{
  // 4. init timers
  PR3 = 20;
  PR2 = 15;
  T3CON = 0x8030;
  T2CON = 0x8030;

  // 5. init interrupts
  mT2SetIntPriority( 1);
  mT3SetIntPriority( 3);
  INTEnableSystemSingleVectoredInt();
  mT2IntEnable( 1);
  mT3IntEnable( 1);

  // main loop
  while( 1);
} // main
```

Notice how in // 1. the `ei` MIPS assembly instruction is used to reenable interruptions immediately upon entry in the handler. Omit this line of code and your interrupts will be queued automatically and served sequentially.

Also, in // 2. we use for the first time the new macro `mT3GetIntFlag()` from the int.h library that, intuitively enough, allows us to test the Timer3 interrupt flag. Since multiple interrupts are enabled, we need such a test to verify which one caused the interruption at hand. We test the highest-priority interrupt source first, and we proceed down the priority list in // 3. until all the sources enabled by the application are considered.

To build and test the new code, follow these simple steps:

1. Save this code as **nesting.c** and add it to the project using one of the many options (checklists) available.

2. Remove **single.c** from the project.

3. **Build** the project.

4. Place a **breakpoint** on the line where count is incremented.

5. Add **TMR3** to the Watch window to keep an eye on the new timer value.

6. Click **Animate** and observe what happens.

If all goes as planned, you will observe the following sequence of events unfold under your eyes:

1. The main initialization code in //4. and // 5. is executed straight through.

2. The application main loop is entered, and there we stay while the timers keep counting.

3. Timer2 reaches its period first, resets, and generates the first interrupt (level 1).

4. The interrupt handler is called and the selection process begins.

5. After the test in //3. succeeds, the culprit is found, and the handler portion relative to the Timer2 interrupt is executed.

6. This is a "long" loop, and the processor is stuck here for a while.

7. Timer3 reaches its period, resets, and generates a new interrupt of higher priority (level 3).

8. The first interrupt handler is . . . interrupted, and a new interrupt handler begins.

9. The selection process takes us immediately inside the handler portion that takes care of Timer3, where count is incremented and the breakpoint puts an end to the simulation.

So we did observe an interrupt . . . interrupting an interrupt handler. If you proceed with the animation from here, now you will see the whole process unroll back.

10. The Timer3 interrupt flag is cleared.

11. The (nested) handler terminates.

12. Control returns to the first handler.

13. From here, in a normal application, we would see the Timer2 handler terminate and return to the main loop where it all started.

But don't hold your breath; this is not going to happen this time, as you might have noticed. To make things more "interesting," I have designed the portion of the interrupt handler that takes care of the Timer2 interrupt (marked as // 3.) to be an infinite loop. This is clearly an exaggeration meant to give us ample opportunity to observe the higher-priority interrupt kicking in.

The nesting scheme can be repeated at multiple levels for as long as the stack has room and your mind can follow the nesting Russian dolls' game. In practice, I strongly discourage you to ever indulge in more than a two-level nesting scheme. It is just too easy to get into some pretty convoluted situations where it is going to be very hard for you to debug your way out. If you find yourself considering such a case, stop immediately, take a deep breath, and think again. This is probably a sign that you don't have your priority scheme well thought out, your handlers are too long, or both things at once. Most probably, there is a better and cleaner way to arrange things.

Multivectored Interrupt Management

The basic PIC32 interrupt service mechanism, we have seen so far, is not too dissimilar from the early 8-bit PIC® architectures, where all interrupt sources were funneled by a single interrupt vector into a single interrupt service routine. This arrangement allows for a great simplicity, but even considering the exceptional speed of the PIC32 (and its ability to execute one instruction per clock cycle), the need to save the processor context followed by the need to proceed through a sequential review of all enabled sources of interrupts can produce considerable overhead. As a consequence, a noticeable delay might be added in responding to a critical event.

To provide the smallest possible overhead and give lightning response to high-priority interrupts, the PIC32 offers an alternative mechanism that uses *vectored interrupts* and *multiple register sets*. In particular, the PIC32MX family offers a 64-vector table and two complete sets of 32 working registers that can be swapped automatically.

Notice that, although there can be as many as 96 interrupt sources in the PIC32 architecture, the maximum number of vectors is limited to 64 by the underlying MIPS core. As a consequence, the PIC32 designers have arranged for some interrupts that belong to the same peripheral to be grouped into the same vector (see Table 5.3).

Table 5.3: Vector table for the PIC32MX360F512L.

Vector Number	Vector Symbol	Notes
0	_CORE_TIMER_VECTOR	
1	_CORE_SOFTWARE_0_VECTOR	
2	_CORE_SOFTWARE_1_VECTOR	
3	_EXTERNAL_0_VECTOR	
4	_TIMER_1_VECTOR	
5	_INPUT_CAPTURE_1_VECTOR	
6	_OUTPUT_COMPARE_1_VECTOR	
7	_EXTERNAL_1_VECTOR	
8	_TIMER_2_VECTOR	
9	_INPUT_CAPTURE_2_VECTOR	
10	_OUTPUT_COMPARE_2_VECTOR	
11	_EXTERNAL_2_VECTOR	
12	_TIMER_3_VECTOR	
13	_INPUT_CAPTURE_3_VECTOR	
14	_OUTPUT_COMPARE_3_VECTOR	
15	_EXTERNAL_3_VECTOR	
16	_TIMER_4_VECTOR	
17	_INPUT_CAPTURE_4_VECTOR	
18	_OUTPUT_COMPARE_4_VECTOR	

(continued)

Table 5.3: (Continued)

Vector Number	Vector Symbol	Notes
19	_EXTERNAL_4_VECTOR	
20	_TIMER_5_VECTOR	
21	_INPUT_CAPTURE_5_VECTOR	
22	_OUTPUT_COMPARE_5_VECTOR	
23	_SPI1_VECTOR	Groups all three SPI1 interrupts.
24	_UART1_VECTOR	Groups all three UART1 interrupts.
25	_I2C1_VECTOR	Groups all I2C1 interrupts.
26	_CHANGE_NOTICE_VECTOR	
27	_ADC_VECTOR	
28	_PMP_VECTOR	
29	_COMPARATOR_1_VECTOR	
30	_COMPARATOR_2_VECTOR	
31	_SPI2_VECTOR	Groups all three SPI2 interrupts.
32	_UART2_VECTOR	Groups all three UART2 interrupts.
33	_I2C2_VECTOR	Groups all I2C2 interrupts.
34	_FAIL_SAFE_MONITOR_VECTOR	
35	_RTCC_VECTOR	
36	_DMA0_VECTOR	
37	_DMA1_VECTOR	
38	_DMA2_VECTOR	
39	_DMA3_VECTOR	
. . .		
44	_FCE_VECTOR	

Assigning a separate vector (pointing to a separate handler function) to each group of interrupt sources eliminates the need to test sequentially all possible sources of interrupt to find the one that needs to be served. But for a greater boost to the response time, the alternate register set can be a real bonus. Upon entry into the interrupt handler, the PIC32 can now simply swap the entire working registers set with a "fresh" new one instead of having to save the entire context on the stack with the long (standard) prologue sequence.

Further, nesting vectored interrupts is still a valid option to increase the responsiveness of the system when one or more lower-priority interrupts need to give way to higher-priority ones. But, since there is only one alternate set of registers, often referred to as the *shadow registers*, it would be dangerous to perform the swap twice. To prevent this kind of situation, the register set "swap" is performed automatically but only for interrupt sources of the highest level (ipl7).

With little effort, we should be able to transform the previous example to take advantage of the multivectored interrupt mode:

1. Split the single interrupt handler into two separate functions.

2. In the __ISR macro, replace the single default **vector** 0 with the appropriate vector number (found in Table 5.3) for each interrupt source/handler.

3. Remove the interrupt flag test; it is now implicit, and each handler is called only when the related interrupt source has raised the flag.

4. Set the Timer3 interrupt priority to level 7 to use the alternate register set feature. Remember to match the assigned level with the __ISR() declaration.

5. Replace the initialization function call with the new multivectored version: INTEnableSystemMultiVectoredInt();.

6. Send me an email if you managed to type the preceding function call without any typo on your first try. Courtesy of the PIC32 libraries' team, you could be the winner of a yet-to-be-determined grand prize for the "longest-functioncallspelledwithouterrorsatfirsttry" contest!

Here is the new code that you will save as multiple.c and replace as the main file in our project:

```
/*
** Multiple Vector Interrupt
*/
```

```
#include <p32xxxx.h>
#include <plib.h>

int count;

void __ISR( _TIMER_3_VECTOR, ipl7) T3InterruptHandler( void)
{
  // 1. T3 handler is responsible for incrementing count
  count++;

  // 2. clear the flag and exit
  mT3ClearIntFlag();
} // T3 Interrupt Handler

void __ISR( _TIMER_2_VECTOR, ipl1) T2InterruptHandler(void)
{
  // 3. re-enable interrupts immediately (nesting)
  asm("ei");

  // 4. T2 handler code here
  while( 1);

  // 5. clear the flag and exit
  mT2ClearIntFlag();
} // T2 Interrupt Handler

main()
{
  // 5. init timers
  PR3 = 20;
  PR2 = 15;
  T3CON = 0x8030;
  T2CON = 0x8030;

  // 6. init interrupts
  mT2SetIntPriority( 1);
  mT3SetIntPriority( 7);
  INTEnableSystemMultiVectoredInt();
  mT2IntEnable( 1);
  mT3IntEnable( 1);

  // 7. main loop
  while( 1);
} // main
```

If you build and animate the project, just as we did in the previous exercise, you should be able to verify that things are now working very much the same.

The Timer2 interrupt kicks in first and keeps the processor busy for . . . well, a very long time. But a Timer3 interrupt manages to interrupt the handler once more and update the count variable. In both cases, you will have noticed how the execution was transferred immediately and very efficiently to the right routine (if we have used the right vector numbers). What is not immediately obvious is how the response to the Timer3 interrupt has been faster than that to Timer2 (and any previous example) because of a much shorter handler prologue. If you want proof, you can switch to the Disassembly window and directly compare the two interrupt handler prologues. You will verify that the Timer3 interrupt handler requires half the instructions (and therefore time) than the low priority Timer2 handler prologue. The difference will only increase, in a practical application, as the main program grows in complexity and more registers need to be saved in the prologue.

Note

Even when we use the alternate register set feature, there is a need for a short prologue. In fact, when we enter a high-priority handler (ipl7) with a fresh register set, we have to initialize at least the stack pointer (one of the registers itself), copying it from the previous set. We also need to modify the interrupt priority mask (IM) of the PIC32, in the Status register, to disable lower-priority interrupts. The resulting (shortest possible) prologue still requires about seven assembly instructions.

A Simple Application

Adding a few more lines of code, we can transform our previous examples into a more practical application where Timer1 is used to maintain a real-time clock keeping track of tenths of a second, seconds, and minutes. As a simple visual feedback, we will use the lower 8 bits of PortA as a binary display showing the running seconds. Here is how to proceed:

- Declare a few new integer variables that will act as the seconds and minutes counters:

```
int dSec = 0;
int Sec = 0;
int Min = 0;
```

- Have the interrupt service routine increment the tenths of a second counter:

```
dSec++;
```

Note: For simplicity in this chapter we will assume the PIC32 is configured for operation with a single 16MHz system and pheripheral clock. In Chapter 7 we will review in more details the oscillator module and we will learn how to operate at much higher clock frequencies squeezing the maximum performance out of the device.

A few additional lines of code will be added to take care of the carryover into seconds and minutes.

- Set the Timer1 prescaler to 1:64 to help achieve the desired period:

```
T1CON=0x8020;
```

- Set the period register for Timer1 to a value that (assuming a 16 MHz peripheral clock with a 62.5ns period) will give us a 1/10th of a second period between interrupts:

```
PR1=25000-1; // 25,000 * 64 * 62.5ns=0.1 s
```

- Set PortA (LSB) as output and disable the JTAG port to gain full control of all LEDs:

```
DDPCONbits.JTAGEN = 0;

TRISA = 0xff00;
```

- Add code inside the main loop to continuously refresh the content of PortA (LSB) with the current value of the seconds counter:

```
PORTA = Sec;
```

Save the new code as clock.c and replace it as the new project source file. Here is what it should look like:

```
/*
** A real time clock
**
** example 5
*/

#include <p32xxxx.h>
#include <plib.h>

int dSec = 0;
int Sec = 0;
int Min = 0;

// 1. Timer1 interrupt service routine
void __ISR( 0, ipl1) T1Interrupt( void)
```

```
{
    // 1.1 increment the tens of a second counter

    dSec++;

    if ( dSec > 9)       // 10 tens in a second
    {
      dSec = 0;
      Sec++;                     // increment the seconds counter

      if ( Sec > 59)    // 60 seconds make a minute
      {
        Sec = 0;
        Min++;    // increment the minute counter

        if ( Min > 59)  // 59 minutes in an hour
            Min = 0;
      } // minutes
    } // seconds
  // 1.2 clear the interrupt flag
  mT1ClearIntFlag();
} //T1Interrupt

main()
{
  // 2.1 init I/Os
  DDPCONbits.JTAGEN = 0;       // disable JTAG port
  TRISA=0xff00;                // set PORTA LSB as output

  // 2.2 configure Timer1 module
  PR1 = 25000-1;  // set the period register
  T1CON = 0x8020; // enabled, prescaler 1:64, internal clock

  // 2.3 init interrupts
  mT1SetIntPriority( 1);
  mT1ClearIntFlag();
  INTEnableSystemSingleVectoredInt();
  mT1IntEnable( 1);

  // 2.4. main loop
  while( 1)
  {
    // your main code here
```

```
    PORTA=Sec;
  } // main loop
} // main
```

To test the new project using the MPLAB SIM simulator, follow these simple steps:

1. Open the Watch window (dock it to your favorite spot).

2. Add the following variables:

    *   dSec, select from the Symbol pull-down box, then click **Add Symbol**.

    *   TMR1, select from the SFR pull-down box, then click **Add SFR**.

    *   Status, select from the SFR pull-down box, then click **Add SFR**.

3. Open the Simulator StopWatch window (**Debugger | StopWatch**).

4. Set a breakpoint on the first instruction of the interrupt response routine after 1.1. Set the cursor on the line and from the right-click menu, select **Set Breakpoint**, or simply double-click. By setting the breakpoint here, we will be able to observe whether the interrupt is actually being triggered.

5. Execute a Run (**Debugger | Run** or press **F9**). The simulation should stop relatively quickly, with the program counter cursor (the green arrow) pointing right at the breakpoint inside the ISR.

So we did stop inside the interrupt service routine! This means that the trigger event was activated; that is, the Timer1 reached a count of 24,999 (remember, though, that the Timer1 count starts with 0; therefore, 25,000 counts have been performed), which, multiplied by the prescaler value, means that $25,000 \times 64$, or exactly 1.6 million, cycles have elapsed.

The StopWatch window will confirm that the total number of cycles executed so far is, in fact, slightly higher than 1.6 million. The StopWatch count includes the time required by the initialization part of our program, too. At the PIC32's execution rate (16 million instructions per second), this all happened in a tenth of a second!

From the Watch window, we can now observe the current value of the processor interrupt priority mask (IM), a bit field inside the Status register. Since we are inside an ISR that was configured to operate at level ipl1, we should be able to verify that bits 10 thru 15 of the status register (Status) contain the value 1.

In Figure 5.2, I have circled the portion of the `Status` register containing the interrupt mask (`IM`) bit field, as shown in the Watch window. Also, the StopWatch shows the time lapsed (in milliseconds) from start to the first breakpoint. Single-stepping from the current position (using either the `StepOver` or the `StepIn` commands), we can monitor the execution of the next few instructions inside the ISR. Upon its completion, we can observe how the interrupt mask returns back to zero:

1. After executing another **Run** command, we should find ourselves again with the program counter (represented graphically by the green arrow) pointing inside the ISR. This time, you will notice that exactly 1.6 million cycles have been added to the previous count.

2. Add the **Sec** and **Min** variables to the Watch window.

3. Execute the **Run** command a few more times to verify that, after 10 iterations, the seconds counter `Sec` is incremented.

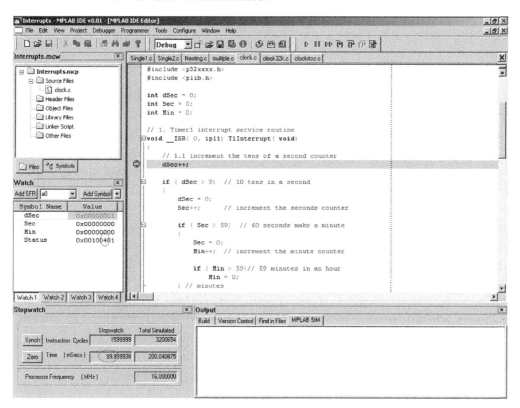

Figure 5.2: Screenshot Clock.c simulation.

To test the minutes increment, you might want to remove the current breakpoint and place a new one a few lines below; otherwise you will have to execute the Run command exactly 600 times!

1. Place the new breakpoint on the **Min++** statement in 1.2.

2. Execute **Run** once and observe that the seconds counter has already been cleared.

3. Execute the **Step Over** command once and the minute counter will be incremented.

The interrupt routine has been executed 600 times, in total, at precise intervals of one tenth of a second. Meanwhile, the code present in the main loop has been executed continuously to use the vast majority of the grand total of 960 billion cycles. In all honesty, our demo program did not make much use of all those cycles, wasting them all in a continuous update of the PortA content. In a real application, we could have performed a lot of work, all the while maintaining a precise real-time clock count.

The Secondary Oscillator

There is another feature of the PIC32 Timer1 module (common to all previous generations of 8-bit and 16-bit PIC microcontrollers) that we could have used to obtain a real-time clock. In fact, there is a low-frequency oscillator (known as the *secondary oscillator*) that can be used to feed the Timer1 module instead of the high-frequency clock. Since it is designed for low-frequency operation (typically it is used in conjunction with an inexpensive 32,768 Hz crystal), it requires very little power to operate. And since it is independent from the main clock circuit, it can be maintained in operation when the main clock is disabled and the processor enters one of the many possible low-power modes. In fact, the secondary oscillator is an essential part for many of those low-power modes. In some cases it is used to replace the main clock, in others it remains active only to feed the Timer1 or a selected group of peripherals.

To convert our previous example for use with the secondary oscillator, we will need to perform only a few minor modifications, such as:

- Change the interrupt routine to count only seconds and minutes; the much slower clock rate does not require the extra step for the tenth of a second:

```
// 1. Timer1 interrupt service routine
void __ISR( 0, ipl1) T1Interrupt( void)
```

```
    {
      // 1.1
      Sec++;      // increment the seconds counter
      if ( Sec > 59)   // 60 seconds make a minute
      {
      Sec = 0;
      Min++;      // increment the minute counter
      if ( Min > 59)   // 59 minutes in an hour
         Min = 0;
      } // minutes

      // 1.2 clear the interrupt flag
      mT1ClearIntFlag();
    } //T1Interrupt
```

- Change the period register to generate one interrupt every 32,768 cycles:

```
PR1 = 32768-1;     // set the period register
```

- Change the Timer1 configuration word (the prescaler is not required anymore):

```
T1CON = 0x8002;    // enabled, prescaler 1:1, use secondary
oscillator
```

Unfortunately, you will not be able to immediately test this new configuration with the simulator, since the secondary oscillator input is not fully supported by MPLAB SIM.

In a later lesson we will learn how a new set of tools will help us generate a *stimulus file* that could also be used to provide a convenient emulation of a 32 kHz crystal connected to the T1CK and SOSCI pins of the PIC32.

The Real-Time Clock Calendar (RTCC)

Building on the previous two examples, we could evolve the real-time clock implementations to include the complete functionality of a calendar, adding the count of days, days of the week, months, and years.

These few new lines of code would be executed only once a day, once a month, or once a year and therefore would produce no decrease whatsoever in the performance of the overall application. Although it would be somewhat entertaining to develop such code once, considering leap years and working out all the details, the PIC32MX family already has a complete Real-Time Clock and Calendar (RTCC) module built in and ready for use.

How convenient! Not only does it work from the same low-power secondary oscillator, but it comes with all the bells and whistles, including a built in Alarm function that can generate interrupts. In other words, once the module is initialized, it is possible to activate the RTCC module and wait for an interrupt to be generated. For example, the interrupt can be set for the exact month, day, hour, minute, and second you desire once a year (or, if set on February 29, even once every four years!).

This is what the interrupt service routine would look like:

```
// 1. RTCC interrupt service routine
void __ISR( 0, ipl1) RTCCInterrupt( void)
{
  // 1.1 your code here, will be executed only once a year
  // or once every 365 x 24 x 60 x 60 x 16,000,000 MCU cycles
  // that is once every 504,576,000,000,000 MCU cycles

  // 1.2 clear the interrupt flag
  mRTCCClearIntFlag();
} // RTCCInterrupt
```

To initialize the RTCC module, though, we will need to substantially modify the main program. The proper configuration of the RTCC module requires a number of registers to be accessed in the right order and filled with the correct data. Fortunately, as part of the standard PIC32 peripheral libraries including plib.h, we gain access to a powerful set of functions that make the entire process quite painless. Here is all the code required:

```
main()
{
  // 2.1 init I/Os
  DDPCONbits.JTAGEN = 0;        // disable JTAG port
  TRISA = 0xff00;               // set PORTA LSB as output

  // 2.2 configure RTCC module
  RtccInit();                   // inits the RTCC
  // set present time
  rtccTime tm; tm.sec=0x15; tm.min=0x30; tm.hour=01;
  // set present date
  rtccDate dt;
  dt.wday=0; dt.mday=0x15; dt.mon=0x10; dt.year=0x07;
  RtccSetTimeDate(tm.l, dt.l);
```

```
// set desired alarm to Feb 29th
dt.wday=0; dt.mday=0x29; dt.mon=0x2;
RtccSetAlarmTimeDate(tm.l, dt.l);

// 2.2 init interrupts,
mRTCCSetIntPriority( 1);
mRTCCClearIntFlag();
INTEnableSystemSingleVectoredInt();
mRTCCIntEnable( 1);

// 2.3. main loop
while( 1)
{
  // your main code here
  // . . .
} // main loop
} // main
```

Debriefing

In this lesson, we have seen how an interrupt service routine can be simple to code, thanks to the language extensions built into the MPLAB C32 compiler and the powerful interrupt control mechanisms offered by the PIC32 architecture. Interrupts can be an extremely efficient tool in the hands of the embedded-control programmer to help manage multiple tasks while maintaining precious timing and resources constraints. At the same time, they can be an extremely powerful source of trouble. In the PIC32 reference manual and the MPLAB C32 User Guide, you will find much more useful information than we could possibly cram into one single day of exploration. Today we took the opportunity to learn more about the uses of Timer1 and the secondary low-power oscillator, and we got a glimpse of the features of the powerful Real-Time Clock and Calendar (RTCC) module.

Notes for the PIC Microcontroller Experts

Notice that on the PIC32 architecture, a pair of convenient instructions allow enabling and disabling of all interrupts at once. If there are portions of code that require all interrupts to be temporarily disabled, you can use the following inline assembly commands:

```
asm("di");
. . .                        // protected code here
asm("ei");
```

But if the portion of code you want to protect from interrupts could be used at times when you don't know whether interrupts are already enabled/disabled, you might want to use a bit more caution and call one of the following two functions from the plib.h library:

- `INTDisableInterrupts()`; not only disables interrupts but also returns a value corresponding to the original interrupts status.

- When you're finished, use `INTRestoreInterrupts( status);` to restore the original system status.

Tips & Tricks

According to the PIC32 datasheet, to activate the secondary low-power oscillator, you need to set the `SOSCEN` bit in the `OSCCON` register. But before you rush to type the code in the last example and try to execute it on a real target board, notice that the `OSCCON` register, containing vital controls for the MCU affecting the choice of the main active oscillator and its speed, is protected by a locking mechanism. As a safety measure, you will have to perform a special unlock sequence first or your command will be ignored. The PIC32MX peripheral libraries come to our rescue in this case with a number of useful functions that manipulate the oscillator module configuration and perform all the necessary lock and unlock sequences, including:

- `mOSCEnableSOSC()`, lets us enable or disable (`mOSCDisableSOSC()`) the external secondary oscillator (SOSC) at run time.

- `OSCConfig()`, can change dynamically (during program execution) the desired clock source, the PLL multiplier, PLL postscaler, and/or the FRC divisor.

- `mOSCSetPBDIV()`, lets us change the Peripheral Bus clock divider dynamically. Use this function with great caution because it will simultaneously affect the operation of all your peripherals.

Note

Changing the clock source will succeed only if the Clock Switching configuration bit is enabled. Check your settings in the **Configure | Configuration bits** menu or your configuration bit #pragmas.

Two additional functions take care of reconfiguring the PIC32MX for IDLE and SLEEP mode operation:

- `mPowerSaveSleep()`, stops both the system clock and the peripheral bus clock of the PIC32 and the device goes into an ultra low-power mode. Any reset and active asynchronous (remember the peripheral clock is stopped) peripheral's event will wake up the device, even if the corresponding interrupt is not enabled. Examples of valid wakeup sources are Change Notification module inputs, External Interrupt pins, Reset, and Brown Out signals.

- `mPowerSaveIdle()`, stops the system clock but leaves the peripheral clock running. Any active peripheral interrupt source can wake up the device. Examples of valid wakeup sources are UART, SPI, Timers, Input Capture, Output Compare, and most other peripherals.

Exercises

Write interrupt-based routines for the following peripherals:

1. Edge selectable interrupts

2. Change notification interrupts

3. Output compare

Books

Curtis, Keith E., *Embedded Multitasking* (Newnes, Burlington, MA, 2006). Keith knows multitasking and what it takes to create small and efficient embedded-control applications.

Links

http://en.wikipedia.org/wiki/Interrupts. This is a great starting point to learn about interrupts.

http://en.wikipedia.org/wiki/Computer_multitasking. To continue with multitasking, especially keeping an eye on real-time multitasking and asynchronous events handling.

Memory

The Plan

The beauty of using a completely integrated, single-chip microcontroller device lies in its reduced size, its increased robustness, and the convenience of having a complete set of peripherals harmoniously preassembled for us, ready for use. Unfortunately, as most embedded-control designers quickly realize, it is the amount of available memory (Flash and RAM) that most often seems to dictate the cost and availability of a product. Learning how to make the most use of both is imperative.

Today we will review the basics of string declaration and manipulation in C language as an excuse to investigate the memory allocation techniques used by the MPLAB® C32 compiler. The PIC32 core offers some pretty advanced features never before seen on 8- or 16-bit PIC® architectures. These include the ability to remap memory spaces, to cache memory contents, and to share the memory bus with a direct memory access (DMA) mechanism. We will use several tools, including the Disassembly Listing window, the Memory window, and the Map file, to investigate how the MPLAB C32 compiler and linker operate in combination to generate the most compact and efficient code.

Preparation

This lesson will be performed exclusively with software tools, including the MPLAB IDE, the MPLAB C32 compiler, and the MPLAB SIM simulator.

Use the New Project Setup checklist to create a new project called **Strings** and a new source file, similarly called **strings.c**.

The Exploration

Strings are treated in C language as simple ASCII character arrays. Every character composing a string is assumed to be stored sequentially in memory in consecutive 8-bit integer elements of the array. After the last character of the string, an additional byte containing a value of 0 (represented in a character notation with '\0') is added as a termination flag.

Note

This is just a convention that applies to the standard C string manipulation library string.h. It would be entirely possible, for example, to define a different library that, for example, stores strings in arrays where the first element is used to record the length of the string. In fact, Pascal programmers will be very familiar with this method.

Let's get started by reviewing the declaration of a variable containing a single character:

```
char c;
```

As we have seen from the previous lessons, this is how we declare an 8-bit integer (character) that is treated as a signed value ($-128 .. +127$) by default.

We can declare and initialize it with a numerical value:

```
char c = 0x41;
```

Or we can declare and initialize it with an ASCII value:

```
char c = 'a';
```

Note the use of the single quotes for ASCII character constants. The result is the same, and to the C compiler there is absolutely no distinction between the two declarations; characters *are* numbers.

We can now declare and initialize a string as an array of 8-bit integers (characters):

```
char s[5] = { 'H', 'E', 'L', 'L', 'O'};
```

In this example, we initialized the array using the standard notation for numerical arrays. But we could have also used a far more convenient notation (a shortcut) specifically created for string initializations:

```
char s[5] = "HELLO";
```

To further simplify things and save you from having to count the number of characters composing the string (thus preventing human errors), you can use the following notation:

```
char s[] = "HELLO";
```

The MPLAB C32 compiler will automatically determine the number of characters required to store the string while automatically adding a termination character (zero) that will be useful to the string manipulation routines later to correctly identify the string length. So, the preceding example is, in truth, equivalent to the following declaration:

```
char s[6] = { 'H', 'E', 'L', 'L', 'O', '\0' };
```

Assigning a value to a char (8-bit integer) variable and performing arithmetic on it is no different than performing the same operation on any integer type:

```
char c;          // declare c as an 8-bit signed integer
c = 'a';         // assign the value 'a' from the ASCII table
c ++;            // increment it. . .
                 // it will represent the ASCII character 'b'
```

The same operations can be performed on any element of an array of characters (string), but there is no simple shortcut, similar to the one used above, for the initialization that can assign a new value to an entire string:

```
char s[15];      // declare s as a string of 15 characters
s = "Hello!";    // Error! This does not work!
```

Including the string.h file at the top of your source file, you'll gain access to numerous useful functions that will allow you to:

- Copy the content of a string onto another:

  ```
  strcpy( s, "HELLO");      // s : "HELLO"
  ```

- Append (or concatenate) two strings:

  ```
  strcat( s, "WORLD");      // s : "HELLO WORLD"
  ```

- Determine the length of a string:

  ```
  i = strlen( s);           // i : 11
  ```

and many more.

Memory Space Allocation

Though a compiler's job is that of generating the code that manipulates variables, it is the linker that is responsible for deciding *where* variables are to be placed in memory, finding a physical address for every object in the memory space(s) available. Just as with numerical initializations, every time a string variable is declared and initialized, as in:

```
char s[] = "Exploring the PIC32";
```

three things happen:

- The MPLAB C32 linker reserves a contiguous set of memory locations (in RAM space) to contain the variable—20 bytes in the preceding example. This space is part of the so-called data section.

- The MPLAB C32 linker stores the initialization value in a 20-byte-long table (in Flash program space). This space is part of the rodata code section or read-only section.

- The MPLAB C32 compiler creates a small routine that will be called before the main() function (part of the Startup code we mentioned in previous chapters) to copy the values into RAM, therefore initializing the variable.

In other words, the string "Exploring the PIC32" ends up using twice the space you would expect, because a copy of it is stored in Flash program memory and space is reserved for it in RAM memory, too. Additionally, you must consider the initialization code and the time spent in the actual copying process. If the string is not supposed to be manipulated during the program execution but is only used "as is," transmitted to a serial port or sent to a display, there is no need to waste precious resources. Declaring the string as a *constant* will save RAM space and initialization code and time:

```
const char s[] = "Exploring the PIC32";
```

Now the MPLAB C32 linker will only allocate space in program memory, in the rodata code section, where the string will be directly accessible. The string will be treated by the compiler as a direct pointer into program memory and, as a consequence, there will be no need to waste RAM space.

In the previous examples of this lesson, we saw other strings implicitly defined as constants—for example, when we wrote:

```
strcpy( s, "HELLO");
```

The string "HELLO" was *implicitly* defined as of `const char` type and similarly assigned to the `rodata` section in program memory.

Note

If the same constant string is used multiple times throughout the program, the MPLAB C32 compiler will automatically store only one copy in the `rodata` section to optimize memory use, even if all optimization features of the compiler have been turned off.

We will start investigating these issues with the MPLAB SIM simulator and the following short snippet of code:

```
/*
** Strings
*/
#include <p32xxxx.h>
#include <string.h>

// 1. variable declarations
const char a[] = "Exploring the PIC32";
char b[100] = "Initialized";

// 2. main program
main()
{
  strcpy( b, "MPLAB C32");    // assign new content to b
} // main
```

1. Build the project using the **Project Build** checklist.

2. Add the Watch window (and dock it to the preferred position).

3. Select the two variables *a* and *b* from the symbol selection box and click the **Add Symbol** button to add them to the Watch window (see Figure 6.1).

A little + symbol enclosed in a box will identify these variables as arrays and will allow you to expand the view to identify each individual element (see Figure 6.2).

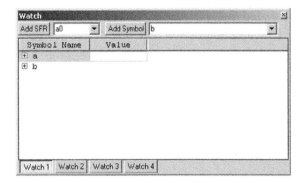

Figure 6.1: Watch window containing two strings.

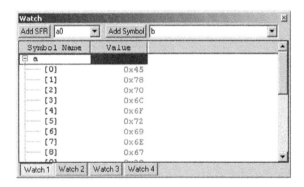

Figure 6.2: String Expanded view.

By default, MPLAB shows each element of the array as hex values, but you can change the display to ASCII characters or to reflect your personal preferences:

1. Select one element of the array with the **left button** of your mouse.

2. **Right-click** to show the Watch window menu.

3. Select **Properties** (the last item in the menu).

You will be presented with the Watch window Properties dialog box (see Figure 6.3).

From this dialog box you can change the format used to display the content of the selected array element, but you can also observe the Memory field (grayed) that tells you where the selected variable is allocated: data or code space.

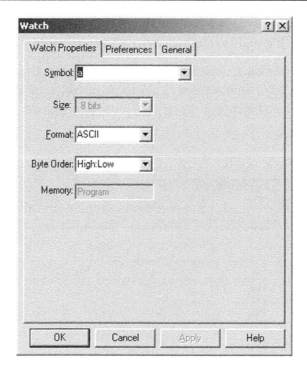

Figure 6.3: The Watch window Properties dialog box.

If you select the Properties dialog box for the constant string *a*, you will notice that the memory space is indicated as Program, confirming that the constant string is using only the minimum amount of space required in the Flash program memory of the PIC32 and no RAM needs to be assigned to it.

On the contrary, the Properties dialog box will reveal how the string **b** is allocated in a File Register, or in other words RAM memory.

Each variable value can be simultaneously presented in multiple formats by adding new columns to the table inside the Watch window:

1. Select the **top row** of the table inside the Watch window (in the column to the right of the default Value column).

2. Select any of the additional formats (check **Char**, for example).

3. Repeat for as many formats as you want, or have space for, inside the window.

Continuing our investigation, notice how the string **a** appears to be already initialized; the Watch window shows it's ready to use, right after the project build.

The string *b*, on the other hand, appears to be still empty, uninitialized. Only when we enable the MPLAB SIM simulator and we click the reset button for the first time to reach the beginning of the main function is the string *b* initialized with the proper value (see Figure 6.4).

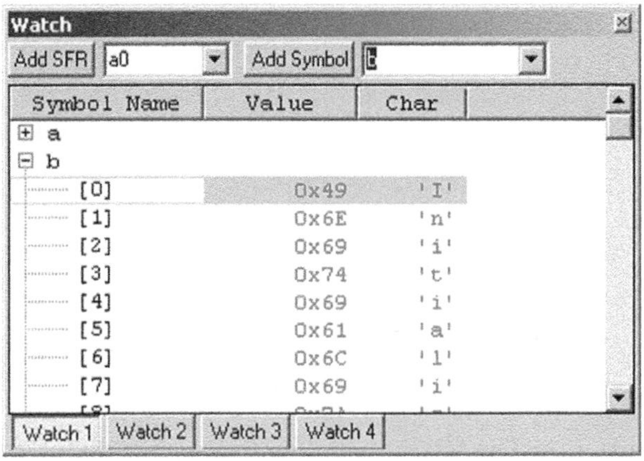

Figure 6.4: The string b after the Startup code execution.

As we have seen, *b* is allocated in RAM space, and the Startup code must be executed first for the variable to be initialized and "ready for use."

Once more we can use the Disassembly Listing window to observe the code produced by the compiler:

```
14:                  // 2. main program
15: main()
16: {
9D000018    27BDFFE8    addiu       sp,sp,-24
9D00001C    AFBF0014    sw          ra,20(sp)
9D000020    AFBE0010    sw          s8,16(sp)
9D000024    03A0F021    addu        s8,sp,zero
17:         strcpy( b, "MPLAB C32");     // assign new content to b
9D000028    3C02A000    lui         v0,0xa000
```

```
9D00002C    24440000    addiu       a0,v0,0
9D000030    3C029D00    lui         v0,0x9d00
9D000034    2445074C    addiu       a1,v0,1868
q9D000038   0F400016    jal         0x9d000058
9D00003C    00000000    nop
18:                     } // main
9D000040    03C0E821    addu        sp,s8,zero
9D000044    8FBF0014    lw          ra,20(sp)
9D000048    8FBE0010    lw          s8,16(sp)
9D00004C    27BD0018    addiu       sp,sp,24
9D000050    03E00008    jr          ra
9D000054    00000000    nop
```

We can see that the main() function is short and followed by the strcpy() library function full disassembly appended at the bottom of the listing. Don't let the length and apparent complexity of the function distract you; it is a pretty optimized piece of code that is designed to take maximum advantage of the 32-bit bus and cache system used by the PIC32. Its analysis is beyond the scope of our explorations today.

You should instead appreciate that this is the only routine attached. Although the string.h library contains dozens of functions, and the include file string.h contains the declarations for all of them, the linker is wisely appending only the functions that are actually being used.

Looking at the Map

Another tool we have at our disposal to help us understand how strings (and in general any array variable) are initialized and allocated in memory is the *.map file*. This text file, produced by the MPLAB C32 linker, can be easily inspected with the MPLAB editor and is designed specifically to help you understand and resolve memory allocation issues.

To find this file, look for it in the main project directory where all the project source files are. Select **File | Open** and then browse until you reach the project directory. By default the MPLAB editor will list all the .c files, but you can change the File Type field to .map (see Figure 6.5).

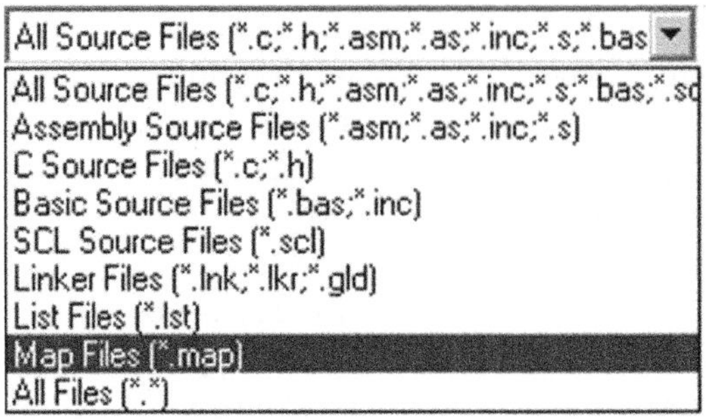

Figure 6.5: Selecting the .map file type.

Map files tend to be pretty long and verbose, but by learning to inspect only a few critical sections, you will be able to find a lot of useful data. Essentially this file is composed of three parts:

- *The List of Included Archive Members.* This is a list of filenames of all the library modules and object files the linker considered to build the project, followed by the file that caused it to be included and the specific symbol that was required. Most of these files are included automatically by the linker script, but you will promptly recognize a line containing our main object file strings.o, where we called the function strcpy() that in turn caused strcpy.o to be linked in. Here is the line that documents it:

```
C:/Program Files/Microchip/../pic32mx/lib\libc.a(strcpy.o)
                                  Strings.o (strcpy)
```

- *The Memory Configuration Table.* This contains the position and size of each memory area, both data and program, used by the project. This is supposed to fit the configuration of the specific PIC32 device chosen. Here is the table:

```
Memory Configuration

Name                       Origin              Length
Attributes
kseg0_program_mem    0x9d000000    0x00080000    xr
kseg0_boot_mem       0x9fc00490    0x00000970
exception_mem0       x9fc01000     0x00001000
kseg1_boot_mem       0xbfc00000    0x00000490
debug_exec_mem       0xbfc02000    0x00000ff0
config3              0xbfc02ff0    0x00000004
```

```
config2                0xbfc02ff4          0x00000004
config1                0xbfc02ff8          0x00000004
config0                0xbfc02ffc          0x00000004
kseg1_data_mem         0xa0000000          0x00008000       w !x
sfrs                   0xbf800000          0x00100000
*default*              0x00000000          0xffffffff
```

You will find some of the area names to be intuitively understandable, whereas others (that follow a long MIPS tradition) will look rather arcane.

- *The Linker Script and Memory Map.* This is the longest part containing a seemingly interminable list of *memory section* names. Each one of the memory sections is eventually placed by the linker in one of the memory areas listed previously, according to strict rules defined in the linker script. The sections we are most interested in are the following:

1. .reset section, containing the code that will be placed by the linker at the reset vector. This is normally filled with a default handler (_reset()):
```
.reset     0xbfc00000     0x10 C:/. . ./pic32mx/lib/crt0.o
           0xbfc00000            _reset
```

2. .vector_x sections—there are 64 of them, each associated to the corresponding interrupt handler. They will be empty unless your program is using the specific interrupt handler.
```
.vector_0     0x9fc01200     0x0
```

3. .startup section, where the C0 initialization code is placed.
```
.startup     0x9fc00490     0x1e0 C:/. . ./lib/crt0.o
```

4. .text sections—you will find many of them, where all the code generated by the MPLAB C32 compiler from your source files is placed. Here is the specific part produced by our main() function:
```
.text        0x9d000018     0x40 Strings.o
             0x9d000018          main
```

Note

The name of this section (.text), although somewhat misleading, follows a long tradition among C compilers. It has been used since the original implementation of the very first C compiler.

5. `.rodata` section, where read-only (constant) data is placed in program memory space. Here we can find space for our constant string *a*, for example:

    ```
    .rodata         0x9d000738         0x20 Strings.o
                    0x9d000738                a
    ```

6. `.data` section, where RAM memory is allocated for global variables.

    ```
    .data           0xa0000000         0x64 Strings.o
                    0xa0000000                b
    ```

7. And finally a pointer to the `.data1` section, where the initialization value, ready for the C0 code to load into the *b* variable, is placed, once more, in program memory space:

    ```
    *(.data1)
                    0x9d00076c         _data_image_begin=LOADADDR(data)
    ```

To verify what can be really found at such addresses, we will need to use the Memory window (select **View | Memory**). Here select the **Data View** tab to visualize the memory contents in classic *hex dump* format. Then **right-click** with the mouse pointer inside the Memory window and choose **Go To** from the context menu (or press **Ctrl + G**) to activate the Go To dialog box (Figure 6.6).

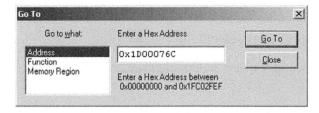

Figure 6.6: The Memory window Go To dialog box.

In the Hex Address field, type the address found above (`0x9d0076c`) and press the **Go To** button. The Memory window will center around the selected address where you will be able to recognize the initialization value we have been looking for.

```
Address       00        04        08        0C        ASCII
1D00_0760  9D0003AC  9D0004F4  9D000578  74696E49  ........ x...Init
1D00_0770  696C6169  0064657A  00000000  00000000  ialized. ........
```

Pointers

Pointers are variables used to refer indirectly (point to) other variables or part of their contents. Pointers and strings go hand in hand in C programming; in general they are a powerful mechanism to work on any array data type. They're so powerful, in fact, that they are also one of the most dangerous tools in a programmer's hands and a source of a disproportionately large share of programming bugs. Some programming languages, such as Java, have gone to the extreme of completely banning the use of pointers in an effort to make the language more robust and verifiable.

The MPLAB C32 compiler takes advantage of the PIC32 architecture to manage with ease large amounts of data memory and program memory (up to 4GB). The MPLAB C32 compiler makes no distinction between pointers to data memory objects and const objects allocated in program memory space. This allows a single set of standard functions to manipulate variables and/or generic memory blocks as needed from both spaces.

The following classic program example compares the use of pointers versus indexing to perform sequential access to an array of integers:

```
int    *pi;        // define a pointer to an integer
int     i;         // index/counter
int a[10];         // the array of integers
// 1. sequential access using array indexing
for( i=0; i<10; i++)
    a[ i] = i;
// 2. sequential access using a pointer
pi=a;
for( i=0; i<10; i++)
{
    *pi = i;
    pi++;
}
```

In 1. we performed a simple `for` loop, and at each round in the loop we used *i* as an index in the array. To perform the assignment, the compiler will have to multiply the value of *i* by the size of the array element in bytes (4) and add the resulting offset to the initial address of the array *a*.

In 2. we initialized a pointer to point to the initial address of the array *a*. At each round in the loop we simply used the pointer indirection operator (*) to perform the assignment; then we simply incremented the pointer.

Comparing the two cases, we see how, by using the pointer, we can save at least one multiplication step for each round in the loop. If inside the loop the array element is used more times, the performance improvement will be proportionally greater.

Pointer syntax can become very "concise" in C, allowing for some pretty effective code to be written but also opening the door to more bugs.

At a minimum, you should become familiar with the most common contractions. The previous snippet of code is more often reduced to the following:

```
// 2. sequential access to array using pointers
for( i=0, p=a; i<10; i++)
  *pi++ = i;
```

Also note that an empty pointer—that is, a pointer without a target—is assigned a special value NULL, which is implementation specific and defined in stddef.h.

The Heap

One of the advantages offered by the use of pointers is the ability to manipulate objects that are defined dynamically (at run time) in memory. The *heap* is the area of data memory reserved for such use, and a set of functions, part of the standard C library stdlib.h, provides the tools to allocate and free the memory blocks. They include at a minimum the two fundamental functions: malloc( ) and free( ).

```
void *malloc(size_t size);
```

The first function takes a block of memory of requested size from the heap and returns a pointer to it.

```
void free(void *ptr);
```

The second function returns the block of memory pointed to by ptr to the heap.

The MPLAB C32 linker places the heap in the RAM memory space left unused above all project global variables and the reserved stack space. Although the amount of memory left unused is known to the linker, you will have to explicitly instruct the linker to reserve an exact amount for use by the heap, the default size being zero.

Use the **Project | BuildOptions | Project** menu command to open the Build Options dialog box, select the **MPLAB PIC32 Linker** tab, and define the heap size in bytes.

As a general rule, allocate the largest amount of memory possible. This will allow the `malloc()` function to make the most efficient use of available memory. After all, if it is not assigned to the heap, it will remain unused.

The PIC32MX Bus

If the previous section, exploring techniques employed by the MPLAB C32 compiler and linker for the allocation of variables, had your head spinning and you feel a little dizzy, you might want to take a break now!

If on the contrary it only served to increase your curiosity, follow me for a little longer as we continue the exploration to investigate the reasons for the architectural foundations of the PIC32 memory bus.

The PIC32 architecture is different from all previous PIC microcontroller architectures (both 8- and 16-bit) with which you might be familiar. The PIC32 follows the more traditional Von Neumann model instead of the classic (PIC) Harvard model. The big difference is that two completely separate and independent buses are no longer available. A single large (32-bit) bus gives access to both the Program Memory (Flash) and Data Memory (RAM) now.

The Von Neumann approach allows for a more economical implementation (two separate 32-bit buses would have been very expensive) and at the same time provides a simpler unified programming model, eliminating the need for the many "tricks" used by 8- and 16-bit Harvard architectures to allow access to data tables in program memory and finally removing important barriers, allowing for the first time a PIC processor to execute code from RAM memory!

It would seem that all these advantages would be immediately offset by a reduction in performance, but this is not the case. In fact a *five-stage pipeline* mechanism and a *pre-fetch cache* mechanism are used to allow efficient access to the bus while maintaining an unprecedented sustained execution rate of *one instruction per clock cycle*.

Note

Later, in the next chapter, we will have the opportunity to look in detail at the operation of the memory cache module and analyze its impact on device performance. Without anticipating too much here, I would like to point out an important detail. The PIC32 core and the cache module are actually connected by two separate 32-bit buses called *I* and *D*. They allow the processor to simultaneously request instructions and data from the cache. So the PIC32 is really a Harvard or a Von Neumann machine? I'll leave you to decide. What matters to me is that it is just so fast and efficient!

Given the same clock frequency—say, 20 MHz—a PIC32 can execute up to *four* times more instructions per second than a PIC16 or PIC18. That is 20 million instructions per second where a PIC16 or PIC18 would only execute 5 million instructions per second. It also means that it can execute *twice* the number of instructions per second that a PIC24, dsPIC30 or dsPIC33 would, given the same clock. If you consider that each one of the PIC32 instructions can now directly manipulate an integer quantity that is 32 bits wide (rather than 8 bits or 16 bits), you can start to get a sense of the effective increase in computational power provided by the PIC32.

In the next chapter we will look further into the operation of the PIC32 oscillator and clock management circuits. We will also review in more detail the operation of the instruction pre-fetch and data cache to help us understand where the new performance limits of the PIC32 architecture are and how we can configure the device for optimal performance and power consumption levels.

PIC32MX Memory Mapping

The MIPS core at the heart of the PIC32 has a number of advanced features designed to allow the separation of the memory space dedicated to an application or applications from that of an operating system via the use of a *memory management unit* (MMU) and two distinct modes of operation: *user* and *kernel*. Since the PIC32MX family of devices is clearly targeting embedded-control applications that most likely would not require much of that complexity, the PIC32 designers replaced the MMU with a simpler *fixed mapping translation* (FMT) unit and a *bus matrix* (BMX) control mechanism.

The FMT allows the PIC32 to conform to the programming model used by all other MIPS-based designs so that standardized address spaces are used. This fixed but compatible scheme simplifies the design of tools and application and the porting of code to the PIC32 while considerably reducing the size and therefore cost of the device.

The BMX allows a level of flexibility in partitioning the main memory areas. It also helps control the arbitration of access to memory between the CPU data and instruction fetch requests, the DMA peripheral requests, and the In-Circuit Debugger (ICD) logic.

Table 6.1 illustrates the relatively complex translation table and the resulting memory map of the PIC32MX family of devices. It could be intimidating at first look, but if you follow me through the next few paragraphs you will find it . . . well, understandable.

Table 6.1: PIC32MX translation table and memory map.

	Memory Type	Virtual Addresses		Physical Addresses		Size in Bytes
		Begin Address	End Address	Begin Address	End Address	Calculation
Kernal Address Space	Boot Flash	0xBFC00000	0xBFC02FFF	0x1FC00000	0x1FC02FFF	12 KB
	Program Flash[1]	0xBD000000	0xBD000000 + BMXPUPBA − 1	0x1D000000	0x1D00000 + BMXPUPBA − 1	BMXPUPBA
	Program Flash[2]	0x9D000000	0x9D000000 + BMXPUPBA − 1	0x1D000000	0x1D000000 + BMXPUPBA − 1	BMXPUPBA
	RAM (Data)	0x80000000	0x80000000 + BMXDKPBA − 1	0x00000000	BMXDKPBA − 1	BMXDKPBA
	RAM (Prog)	0x80000000 + BMXDKPBA	0x80000000 + BMXDUDBA − 1	BMXDKPBA	BMXDUDBA − 1	BMXDUDBA− BMXDKPBA
	Peripheral	0xBF800000	0xBF8FFFFF	0x1F800000	0x1F8FFFFF	1 MB
User Address Space	Program Flash	0x7D000000 + BMXPUPBA	0x7D000000 + PFM Size − 1	0xBD000000+ BMXPUBPA	0xBD000000 + PFM Size − 1	PFM Size − BMXPUBPA
	RAM (Data)	0x7F000000 + BMXDUDBA	0x7F000000 + BMXDUPBA − 1	0xBF000000+ BMXDUDBA	0xBF000000 + BMXDUPBA − 1	BMXDUPBA − BMXDUDBA
	RAM (Prog)	0x7F000000 + BMXDUPBA	0x7F000000 + RAM Size[3] − 1	0xBF000000 + BMXDUPBA	0xBF000000 + RAM Size[3] − 1	DRM Size − BMXDUPBA

Notes:

[1] Program Flash virtual addresses in the non-cacheable range (KSEG1).

[2] Program Flash virtual addresses in the cacheable and prefetchable range (KSEG0).

[3] The RAM size varies between PIC32MX device variants.

First, let's find out where the main memory blocks (RAM and Flash memory) of the PIC32 are physically located inside the 32-bit addressing space (see Figure 6.7). Check the physical address column and you will find that RAM begins at address 0x00000000, and Flash memory begins at 0x1D000000. Finally, all peripherals (SFRs) are found in the block that begins at address 0x1F800000, and a 12 K portion of Flash memory is found at address 0x1FC00000 for use by a bootloader.

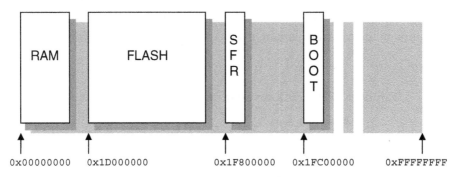

Figure 6.7: PIC32 physical addressing space.

Access to those memory areas can be required for different purposes. The PIC32 designers wanted to make sure that we would be able to impose special "rules" to protect the applications from common (programming) errors isolating regions of memory. For example, when running an operating system (OS), we might desire to prevent application code to touch data (RAM) areas that are part of the OS. In other words, *user code* must not be allowed to access the *kernel data*. The BMX control unit is the one that performs the first layer of manipulation (see Figure 6.8). Through some of its control registers, we can split the main physical memory areas in slices of variable size. For example, using the BMXPUPBA register, we can split a portion of the Flash memory to be remapped for use only in user mode at physical address 0xBD000000 and higher. Similarly, RAM data memory can be split into four slices using the registers BMXDKPBA and BMXDUDBA, separating kernel data from user data memory and then splitting further each piece of memory for programs that want to execute from RAM to achieve higher performance; RAM maximum access speed is typically much higher than Flash memory, even when a cache mechanism is taken into account.

This is where the FMT (or more generically, an MMU) adds a new layer of complexity to the entire system, translating all *physical addresses* into *virtual addresses* and shuffling

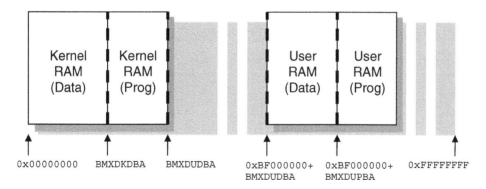

Figure 6.8: Bus matrix RAM partitioning.

things around a bit. This is meant to create two widely separate address spaces where your programs can run: one for user applications in the lower half of the 32-bit addressing space (below 0x80000000) and one for kernel (above 0x80000000) in accordance with the standard practice of all MIPS-based processors. These correspond to the two halves of Table 6.1, where the first two columns show you the new virtual addresses assigned to each memory area in the corresponding mode.

> **Note**
>
> The only addresses the MPLAB C32 compiler and linker are concerned with, as seen in the early part of this chapter, are virtual addresses!

For clarity, Figure 6.9 illustrates the resulting virtual memory map as seen by an application program running in user mode.

Notice how the Boot Flash memory is *not* mapped at all in user mode. There is no virtual address that will allow a user program to touch the protected area. No matter how bad, the code is running in user mode; it cannot harm the underlying operating system (or bootloader).

Similarly, notice how the peripherals (SFRs) don't have a corresponding mapping in the user virtual address space. Again, no matter how bad the user code is, it cannot reach the hardware and modify or cripple the device configuration.

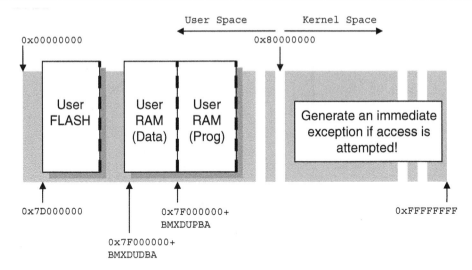

Figure 6.9: User mode virtual memory map.

The Embedded-Control Memory Map

All this is nice and dandy if you are planning to run a heavyweight OS with all the bells and whistles, but in most embedded-control applications you will *not* use all these features. All your code will most likely always be running in kernel mode only, at the same level as an OS would. And even when you're using an OS, you will find that most *real-time operating systems* (RTOSs) don't use these features either, favoring speed of execution and efficiency over "protection." This is a reasonable choice for embedded control. The application code is "well known"; it is supposed to be robust and well tested and should therefore be trusted!

This is great news because it means that from now on, we can completely ignore the bottom half of Table 6.1 and concentrate all our attention on only the kernel mode virtual map (see Figure 6.10)!

A final note is required to clarify the reason for two virtual address spaces being dedicated to the kernel program Flash memory. They are traditionally referred to as *kseg0* and *kseg1* in the MIPS literature. If you look at the Physical Addresses columns in Table 6.1, you will notice that eventually both point to the same physical memory space. The difference is only in the way the memory cache mechanism will manage the two virtual addresses. If a program is executing from the first virtual address space (*kseg1*), the memory cache is automatically disabled. Vice versa, portions of code that are placed

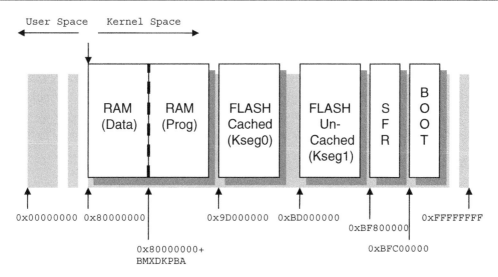

Figure 6.10: PIC32 Embedded-control (kernel mode) virtual memory map.

in the *kseg0* segment will be accessible by the cache mechanism. We will learn more in the next few chapters about the reason for this choice and the consequences for your code performance.

Debriefing

Today we have quickly reviewed the basics of string declaration and manipulation. We have also touched briefly on the use of pointers and dynamic memory allocation. We have seen how the .map file can help us identify where and how the memory of the PIC32 will be used by our applications. But today we have also explored the bus matrix module of the PIC32 and learned how it provides us with a very flexible mechanism to control the segmentation and access to blocks of Flash and RAM memory. Although many embedded-control applications will only use the most basic (default) configuration, the PIC32MX architecture offers a standard address space layout that makes it compatible with a wide range of tools and operating systems already available for the MIPS architecture.

Notes for the C Experts

In the C language, strings are defined as simple arrays of characters. The C language model has no concept of different memory regions (RAM vs. Flash). The `const` attribute is normally used in C language, together with most other variable types, only to assist the compiler in catching common parameter usage errors. When a parameter is passed to a function as a const or a variable is declared as a const, the compiler can in fact help flag any following attempt to modify it. The MPLAB C32 compiler extends this semantic in a very natural way, allowing us to provide hints to the compiler and linker to make more efficient use of the memory resources.

Notes for the Assembly Experts

The string.h library contains many block manipulation functions that can be useful, via the use of pointers, to perform operations on *any* type of data arrays, not just strings. They are:

- memcpy(), to copy the content of any block of memory to a new address

- memmove(), to move the contents of a block of memory to a new location

- memcmp(), to compare the contents of two blocks of memory

- memset(), to initialize the contents of a block of memory

The ctype.h library instead contains functions that help discriminate individual characters according to their positions in the ASCII table, to discriminate lowercase from uppercase, and/or to convert between the two.

Notes for the PIC Microcontroller Experts

Since the PIC32MX program memory is implemented using (single-voltage) Flash technology, programmable at run time during code execution, it is possible to design *bootloader*-based applications—that is, applications that automatically "update" part or all of their own code.

It is also possible to utilize a section of the Flash program memory as a nonvolatile memory (NVM) storage area. Some pretty basic limitations apply, though. For example, Flash memory can only be deleted in large blocks, called *pages*, composed of 1,024 words before data can be written one word at a time or in smaller blocks called *rows* composed of 128 words.

The PIC32 peripheral library comes to our assistance, offering a small set of functions (NVM.H) dedicated to the manipulation of on-chip Flash memory. Perhaps the most powerful function of the lot is NVMProgram(), capable of writing a block of arbitrary length to a given virtual address, automatically performing the necessary partitioning when page boundaries are crossed.

Tips & Tricks

String manipulation can be fun in C once you realize how to make the zero termination character work for you efficiently. Take, for example, the mycpy() function:

```
void mycpy( char *dest, char * src)
{
  while( *dest++ = *src++);
}
```

This is quite a dangerous piece of code, since there is no limit to how many characters could be copied, there is no check as to whether the dest pointer is pointing to a buffer that is large enough, and you can imagine what would happen should the src string be missing the termination character. It would be very easy for this code to continue beyond the allocated variable spaces and to corrupt the entire contents of the data memory. Ah, the power of pointers!

Soon we will explore the DMA module and we'll discover its ability to share the PIC32 memory bus to perform fast data transfers between memory and peripherals. We'll also explore using the DMA module to move large blocks of data between different memory buffers very efficiently. In fact, a few of the DMA functions in the PIC32 peripheral library are dedicated to the use of DMA channels to perform string and block manipulations, including DmaChnMemcpy(), DmaChnStrcpy(), and DmaChnStrncpy(). In the same set of functions can be found DmaChnMemCrc(), which does not transfer any data but feeds the CRC module with the contents of a given (no matter how large) block of data. Alternatively, a CRC calculation can automatically be performed during any block transfer performed by the DMA module by calling the CrcAttachChannel() function.

Exercises

You can develop new string manipulation functions to perform the following operations:

1. Search sequentially for a string in an array of strings.

2. Implement a binary string search.

3. Develop a simple hash table management library.

Books

Wirth, N., *Algorithms+Data Structures=Programs* (Prentice-Hall, Englewood Cliffs, NJ, 1976). With unparalleled simplicity, Wirth, the father of the Pascal programming language, takes you from the basics of programming all the way up to writing your own compiler. They tell me this book is no longer easy to find; however hard it might be to locate a copy, I promise you it will be worth the effort!

Links

http://en.wikipedia.org/wiki/Pointers#Support_in_various_programming_languages.
Learn more about pointers and see how they are managed in various programming languages.

Experimenting

Congratulations! You have endured the first six days of exploration and gained the necessary confidence to complete simple projects using the MPLAB PIC32 software tool suite. As a consequence, in the next group of lessons, more is going to be expected of you!

In the second part of this book, we continue exploring one by one the fundamental peripherals that allow a PIC32 to interface with the outside world. Since the complexity of the examples will grow a little bit, having a PIC32 chip at hand is highly recommended so that you will be able to test the many practical example projects. A PIC32 Starter Kit with a PIM adapter and/or an actual PIC32MX processor module (PIM) and any of the compatible in-circuit debuggers will do. I will also refer often to the Explorer 16 demonstration board, but any compatible third-party tool that offers similar features or allows for a small prototyping area can be used just as effectively.

Running

The Plan

In the six previous days of exploration, we have gradually begun reviewing the most basic concepts of C programming as they apply to embedded control and in particular as they apply to the PIC32MX architecture. We have also started to familiarize ourselves with the basic features of the PIC32 that affect its performance, such as the 32-bit multiplier, the interrupt system, the register set(s), and the memory management module. But so far, we have only been counting the number of assembly instructions looking inside the disassembly window, or counting the instruction cycles, using the MPLAB® SIM simulator StopWatch. In all cases we avoided any direct reference to time when considering the execution of code, using peripherals (timers) when necessary to provide delays of any length. Even when discussing interrupts or comparing the efficiency of various numeric types, we have not yet established any hard relationship with the actual *speed of execution* of our code. This was done on purpose, to isolate different subjects and keep the level of complexity growing gradually. Before we can understand how fast we can make a PIC32 truly "run," we need to study two new critical systems: the clock system and the memory cache system. Both are new to the PIC® architecture and are essential if you want to fine-tune the PIC32 engine for maximum performance.

Preparation

Today, in addition to the usual software tools, including the MPLAB IDE and the MPLAB C32 compiler, you will need real hardware to be able to perform our experiments. It does not matter if you have a PIC32 Starter Kit or any of the other in-circuit debuggers connected to an Explorer 16 demonstration board. You will need the real thing—a PIC32MX chip "running" on the hardware platform of your choice.

Use the New Project Setup checklist to create a new project called **Running** and a new source file, similarly called **running.c**.

The Exploration

Let's start by taking a look at the main clock circuit of the PIC32MX family. As you can see from the block diagram in Figure 7.1, this is a complex piece of hardware with which it will require some time to become familiar.

For those of you already knowledgeable about the previous generations of 8-bit PIC microcontrollers, most of this diagram will look somewhat familiar. For those of you familiar with the dsPIC33 and PIC24H families in particular, it will look exceptionally similar! This is of course no coincidence. All PIC microcontrollers, since the very first PIC16C54, have sported a flexible oscillator circuit, and this flexibility has been extended generation after generation, evolving gradually into the present form as offered on the PIC32MX. Let's see what can be done, and most importantly, why!

Looking at the left side of the block diagram, you will notice that there are five oscillators or clock sources. Two of them use internal oscillators and three of them require external crystals or oscillator circuits:

- Internal oscillator (FRC) is designed for high-speed operation with low power consumption. It requires no external components and provides a relatively accurate nominal 8 MHz clock ($\pm 2\%$) after calibration.

- Internal low-frequency and low-power oscillator (LPRC) is designed for low-speed operation with low power consumption. Requires no external components and provides a basic (low accuracy) 32 kHz clock.

- External primary oscillator (POSC) is designed for high-speed operation with accurate (quartz-based) operation. Up to 20 MHz crystals can be connected directly (to the OSCI, OSCO pins) while two gain settings are available: XT for typical quartzes below 10 MHz and HS for quartzes at or above the 10 MHz frequency.

- External low-frequency and low-power oscillator (also known as the secondary oscillator, SOSC) is designed for low-speed and low-power operation with external crystals of 32,768 Hz. It can be used as the main oscillator for the entire chip or just as the source for the Timer1 and RTCC modules. Its high accuracy makes it the ideal clock source for applications that need exact timekeeping.

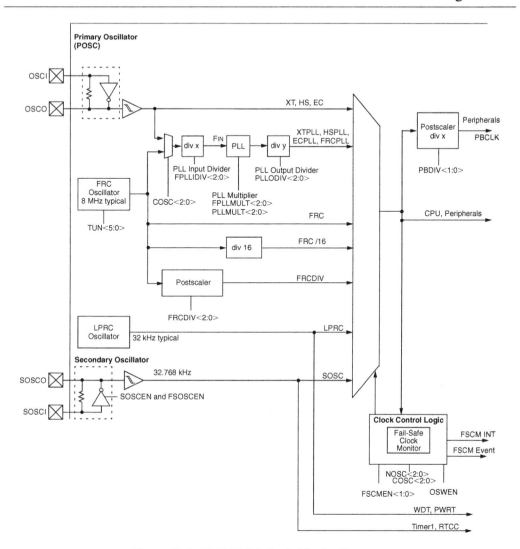

Figure 7.1: PIC32MX clock block diagram.

- External clock source (EC) mode allows an external circuit to completely replace the oscillator and provide the microcontroller a square wave input of any desired frequency.

These five sources offer a basic range of choices to generate an input clock signal of desired frequency, power consumption, and accuracy, but much more can be done with the following stages, illustrated on the right side of the block diagram. In fact, the clock produced by each source can be further multiplied and/or divided to offer an even wider selection of frequencies.

Performance vs. Power Consumption

It is beyond the scope of this book to illustrate all possible options for each clock source, but it is important that you understand the reason why the designers of the PIC32 went through all this effort to offer you so many different ways to produce what is, after all, a simple square wave.

In embedded control, but also in consumer applications, whether your application is portable—battery powered—or has a dedicated power supply of sorts, two important constraints apply:

- Power consumption will dictate the size and cost of the power supply circuit you will have to design. If battery operated, this parameter will dictate the size and cost of the battery, or vice versa, the life (hours of operation) of your application.

- Performance, however measured, will dictate how much work your application will be able to perform in a given amount of time. For some real-time applications, this parameter can be a total deal breaker.

As is often the case, in embedded-control application design, the two constraints are in direct conflict. To obtain a greater amount of work from a given circuit, we want to maximize the clock speed. But because of the laws of physics that govern the operation of any CMOS integrated circuit, the higher the clock speed provided, the higher is the power consumption of the device. The two entities are in fact linked inexorably in a linear relationship: Double the clock and you will double the amount of work produced, but you will also see a corresponding increase in the power consumption of the device.

Note

The power consumption will not double as you double the frequency. There is a *static* component and a *dynamic* component to the power consumption of each CMOS device. The first one remains constant independent from the clock frequency; it is only the dynamic part that will grow.

Much can and has been done inside the PIC32 to make sure that the greatest amount of work is produced for any given ounce of power. For example, the PIC32MX datasheet (only the advanced datasheet is available at the time of this writing) reports on the electrical characteristics of the device that, when operating at the frequency of 4 MHz, a typical current consumption of 11 mA will be observed (at 3.3 V and 25°C). But at 72 MHz and in the same conditions, the same device will consume just 64 mA.

As good as these numbers are, it is still our responsibility to find the correct balance between performance and power consumption for each application so to minimize cost, reduce size, or simply maximize the battery life (and, let me add, "fight global warming as well"!).

Not only does it make no sense to run an applications at 72 MHz when the same job can be done at 4 MHz, but also consider the fact that most applications operate in different modes at different times. Although it might seem overkill, I will make a parallel with a cell phone application. Most of the time, the cell phone is in standby just waiting for a button to be pressed to awake it. At other times it could be performing simple functions such as searching through a contact book and updating information on the internal memory. Then only a small fraction of the time will be spent performing some hard number crunching, digital signal processing, and running an algorithm to compress and decompress the audio input and output streams.

Similar conditions can be found in many embedded-control (and consumer) applications, and the higher the flexibility of the clock circuit, the better you will be able to manage the power consumption of the application. To help you obtain the most complete set of power management options, the PIC32 clock module offers the following features:

- Run-time switching between internal and external oscillator sources

- Run-time control over the clock dividers

- Run-time control over the PLL circuit (clock multiplier)

- IDLE modes, where the CPU is halted and individual peripherals continue to operate

- SLEEP mode, where the CPU and peripherals are halted and awaiting a specific event (set of) to awaken

- Separate control (divider) over the peripheral clock (PBCLK), so that when the CPU is required to operate with a high-frequency clock, the power consumption of the peripheral modules can be optimized

The Primary Oscillator Clock Chain

We will begin our exploration at the primary oscillator clock signal chain, since it is the most common and, in many of the following chapters, we will need to develop demonstration projects that will require either a high level of performance or high clock accuracy. As you can verify visually, on the Explorer 16 demonstration board and PIC32 Starter Kit, an 8 MHz crystal is connected across the OSCI and OSCO pins. At this frequency (below 10 MHz) it is recommended we set the primary oscillator for operation in XT mode.

Depending on the application, we are immediately confronted with two possibilities. We could use the 8 MHz input signal as is or feed it to a multiplier (PLL) circuit. The appeal of the second option is obvious, but with it comes the need to learn more about PLL circuits.

Phase locked loops (PLLs) are complex little circuits, but the designers have managed to hide all the complexity of the PIC32 PLL from us with the condition that we respect a few simple rules. First, we need to feed it with a specific input frequency range (<4 MHz). Second, we need to allow it time to stabilize or "lock" before we attempt to execute code and synchronize with it. A simple control mechanism is provided (via the OSCCON register illustrated in Figure 7.2) to select the frequency multiplication factor (PLLMULT) and to verify the proper locking (SLOCK).

U-0	U-0	R/W-x	R/W-x	R/W-x	R/W-x	R/W-x	R/W-x
—	—	PLLODIV<2:0>			FRCDIV<2:0>		
bit 31							bit 24

R/W-0	R-0	U-0	R/W-x	R/W-x	R/W-x	R/W-x	R/W-x
DRMEN	SOSCRDY	—	PBDIV<1:0>		PLLMULT<2:0>		
bit 23							bit 16

U-0	R-0	R-0	R-0	U-0	R/W-x	R/W-x	R/W-x
—	COSC<2:0>			—	NOSC<2:0>		
bit 15							bit 8

R/W-0	r-0	R-0	R/W-0	R/W-0	r-0	R/W-0	R/W-0
CLKLOCK	—	SLOCK	SLPEN	CF	—	SOSCEN	OSWEN
bit 7							bit 0

Figure 7.2: The OSCCON register.

So when using the Explorer 16 board or the PIC32 Starter Kit, to respect the first rule we will need to reduce the input frequency from 8 MHz to 4 MHz. Looking at the block diagram in Figure 7.1 or the simplified diagram in Figure 7.3, you will notice how the input divider is conveniently available to us to perform the first frequency reduction.

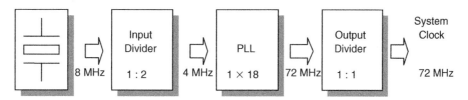

Figure 7.3: Primary oscillator clock chain.

The multiplication factor of the PLL can be selected among a number of values ranging from 15× all the way up to 24× and it is controlled by the PLLMULT bits. Since the maximum operating frequency of the PIC32MX is (at the time of this writing) restricted to 75 MHz, selecting a factor of 18× will give 72 MHz, the closest match compatible with the device operating specifications. The output divider block provides us with a final opportunity to manage the clock frequency. When we will need the maximum performance, we will leave the output divider set to a 1:1 ratio. Should our application require it, we will be able to reduce the power consumption by dividing the output frequency all the way down to 1:256th or approximately 280 kHz. Below this frequency we would be much better served by using the secondary oscillator (SOSC), its operating range is in fact between 32 kHz and 100 kHz, or by the low power internal oscillator (LPRC) operating at approximately 32 kHz. For our reference, from the advanced datasheet we learn that the typical power consumption of the PIC32 when operating off the LPRC would be just 200 µA!

The Peripheral Bus Clock

As another way to optimize performance and power consumption in an application, the PIC32 feeds a separate clock signal to all the peripherals. This is obtained by sending the System clock through yet another divide circuit (extending further the chain of modules illustrated in Figure 7.3), producing the PB clock signal. Very often a high processor speed means that a large prescaler is required in front of a timer to obtain the required timing, or a large baud rate divider is required for a serial port (more on this later). Thanks to the peripheral bus divider, the share of power consumed by the peripheral bus can be reduced while the processor is free to operate at maximum speed.

This feature is controlled by the PBDIV bits found, once more, inside the OSCCON register. A reasonable value that we have been using so far and we will continue to use consistently for the peripheral bus across all future example projects will be 36 MHz corresponding to 1:2 ratio between the system clock and the PB clock.

Initial Device Configuration

The ability to control the clock at run time gives us a great tool to manage power, but what happens when the device is first activated, at power-up?

As you might know, there is a group of bits known as the *configuration bits* stored in the nonvolatile (Flash) memory of the PIC32. These bits provide the initial configuration of the device. The oscillator module uses a few of those bits to get the initial setting for the OSCCON register. These are the configuration bits you can set using the MPLAB **Configure | Configuration Bits. . .** menu.

It is about time that we review the settings I have been recommending that you use since the beginning using the Device Configuration checklist.

My recommended configuration for all the exercises in this book is represented in Figure 7.4. It includes the following options, in order of importance for the oscillator configuration:

1. Use the primary oscillator with PLL circuit.

2. Select the XT mode for the primary oscillator.

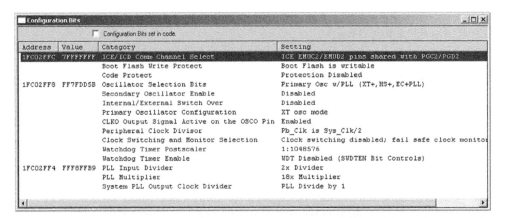

Figure 7.4: Device Configuration dialog box.

3. Set the PLL input divider to 1:2 ratio (to produce a 4 MHz input as we have seen).

4. Set the PLL multiplier to 18×.

5. Set the PLL output divider to 1:1 ratio (to produce a 72 MHz system clock output).

6. Set the peripheral clock divider to 1:2 ratio (to produce a 36 MHz PB clock output).

The following additional options complete the configuration:

7. Enable the clock output. This can be disabled when using any of the internal oscillators to gain control of an additional I/O pin.

8. Disable the secondary oscillator. (You will be able to enable it later, at run time.)

9. Disable the internal/external oscillator switchover. (We will use only the external crystal in all exercises, but you might experiment with other settings.)

Finally, the following options are commonly used during debugging and development:

10. Share DBG2 and PGM2 if you are using the ICD/ICSP interface. (This depends on your in circuit debugger of choice.)

11. Allow the Boot Flash to be modified (Bootloader write protection off).

12. Disable code protection (at least during development).

13. Disable the Watchdog timer.

14. Disable clock switching and FailSafe Clock Monitor.

Once set, these configuration bits are stored in the workspace file (.mcw) and will be programmed into the device configuration bits by your programming tool of choice each time new code is programmed into the device.

By comparing Figures 7.2 and 7.4, you will notice that the value of the PLL input divider is present only as a configuration bit option, but it cannot be modified via the OSCCON register. If you reflect on this, you will find it is logical. Since the external crystal value cannot change (unless the part is unsoldered from the PCB and a new one of different frequency is put in its place), there is no possible reason to modify the input divider value at run time. If the value set by the configuration bits was incorrect in the first place, the PLL multiplier would not be working and the PIC32 could not execute any code anyway.

Setting Configuration Bits in Code

As a way to make the project code self-documenting and to avoid any possible future mishap (should the project file be lost and the source code of an application used with the wrong settings), the MPLAB C32 compiler offers one additional mechanism to assign values to the device configuration bits. It is based on the use of the `#pragma config` directive.

Since the number of configuration bits and their values can change from device to device, MPLAB offers a list of the available options for each PIC32 device model as part of the Help system. Select **Help | Topic** to open the help system selection dialog box, and click **PIC32MX Config Settings** (see Figure 7.5).

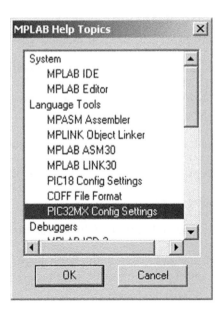

Figure 7.5: MPLAB Help Topics dialog box.

Select the device model that you are using, **PIC32MX360F512L**, and then identify the correct syntax to be used for each configuration bit. Table 7.1 shows the PLL output divider example.

Table 7.1: PLL output divider values

`FPLLODIV = DIV_1`	Divide by 1
`FPLLODIV = DIV_2`	Divide by 2
`FPLLODIV = DIV_4`	Divide by 4
`FPLLODIV = DIV_8`	Divide by 8
`FPLLODIV = DIV_16`	Divide by 16
`FPLLODIV = DIV_32`	Divide by 32
`FPLLODIV = DIV_64`	Divide by 64
`FPLLODIV = DIV_256`	Divide by 256

Multiple configuration bits can be set inside a `#pragma config` statement by separating them with a comma, as in the following example, where I have reproduced the standard oscillator settings as described previously:

```
#pragma config POSCMOD=XT, FNOSC=PRIPLL
#pragma config FPLLIDIV=DIV_2, FPLLMUL=MUL_18, FPLLODIV=DIV_1
```

Notice that if a parameter is not specified in the `#pragma`, it will assume the default value as specified in the device datasheet.

Let's complete the configuration with one more `#pragma` statement to set the peripheral bus clock divider, disable the watchdog and the code protection, and to enable programming of the boot memory as required for all our future projects (at least during the development phase):

```
#pragma config FPBDIV=DIV_2, FWDTEN=OFF, CP=OFF, BWP=OFF
```

My recommendation is that you place this code at the top of the source file containing the main function in each new project.

To avoid conflicts with the configuration bits set by MPLAB in the Configuration Bits dialog box (refer back to Figure 7.4), make sure to check the **Configuration Bits Set in Code** check box.

Note

When the Configuration Bits Set in Code check box is checked, the entire contents of the dialog box are grayed out. This is the default for every new project. Be careful, though—if you forget to set the #pragma config statement in your code, you'll end up with the default device configuration, as described in the device datasheet. This default configuration is designed for "safe" operation and most of its settings are conflicting or incorrect for use during development. I chose not to set the configuration bit in code in the first few chapters of the book to avoid the "distraction" in your code and to avoid having to anticipate too much too soon. From now on, the choice is yours!

Heavy Stuff

It is time to write some tough code, program it on a PIC32 Starter Kit or an Explorer 16 demonstration board, and start measuring the actual performance of the PIC32MX.

See what I found in my code archives! Buried in a remote subdirectory of my hard drive, back from the old days at university when I studied the basics of digital signal processing, I wrote this code:

```
// input vector
unsigned char inB[N_FFT];

// input complex vector
float  xr[N_FFT];
float  xi[N_FFT];

// Fast Fourier Transformation
void FFT(void)
{
  int    m, k, i, j;
  float  a, b, c, d, wwr, wwi, pr, pi;

    // FFT loop
    m = N_FFT/2;
    j = 0;
    while(m > 0)
    { /* log(N) cycle */
      k = 0;
      while(k < N_FFT)
      { // batterflies loop
```

```
    for(i = 0; i < m; i++)
    { // batterfly
      a = xr[i+k];        b = xi[i+k];
      c = xr[i+k+m];        d = xi[i+k+m];
      wwr=wr[i<<j];   wwi = wi[i<<j];
      pr=a-c;  pi = b-d;

        xr[i+k]      = a + c;
        xi[i+k]      = b + d;
        xr[i+k+m]    = pr * wwr - pi * wwi;
        xi[i+k+m]    = pr * wwi + pi * wwr;
      } // for i
    k += m<<1;
    } // while k
    m >>= 1;
    j++;
  } // while m
} // FFT
```

This is the Fast Fourier Transform (FFT) function, one of the most common digital signal processing tools, albeit in a simplified form designed to operate on a set of samples whose size is purposely chosen as a power of two. The FFT is an efficient algorithm to compute the discrete Fourier transform (DFT) and its inverse, that is, what takes us from a signal *time domain* representation to the same signal in the *frequency domain* representation and back. In other words, if you supply as input to an FFT function an array of values (inB[]) that represent samples of an input signal, the function will return a new array containing values corresponding to the amplitudes of each harmonic (sinusoidal component) of the input signal—i.e., the signal *frequency spectrum*. FFTs are of great importance to a wide variety of applications beyond digital signal processing, including solving partial differential equations and algorithms for quick multiplication of very large integers. Many studies have been done on how to optimize FFTs and determine the minimum possible number of arithmetic operations required to perform them on a given data set. But we are not interested in optimizing the algorithm here; on the contrary, we will use the "scholastic" implementation as an example of an algorithm requiring heavy floating-point arithmetic for our performance-testing purposes.

Actually, the algorithm illustrated previously represents only a part of the work that a complete discrete Fourier transform implementation requires. To obtain the necessary accuracy, the input data set must first be *windowed* before use. Think of it as though a segment of the input signal was cut abruptly and its sharp edges at the extremities need to be filed to smooth out the algorithm response:

```
// apply Hann window to input vector
void windowFFT(unsigned char *s)
{
  int i;
  float *xrp, *xip, *wwp;

  // apply window to input signal
  xrp = xr; xip = xi; wwp = ww;
  for(i=0; i<N_FFT; i++)
  {
    *xrp++ = (*s++ - 128) * (*wwp++);
    *xip++ = 0;
  } // for i
} // windowFFT
```

After the FFT, the modulus of the (complex) output must be taken and scaled back in place, in this case overwriting the input array:

```
void powerScale(unsigned char *r)
{
  int i, j;
  float t, max;
  float xrp, xip;

  // compute signal power (in place) and find maximum
  max = 0;
  for(i=0; i<N_FFT/2; i++)
  {
    j = rev[i];
    xrp = xr[j];
    xip = xi[j];
    t = xrp*xrp + xip*xip;
    xr[j] = t;
    if (t>max)
      max = t;
  }

  // bit reversal, scaling of output vector as unsigned char
```

```
   max = 255.0/max;
   for(i=0; i<N_FFT/2; i++)
   {
     t = xr[rev[i]] * max;
     *r++ = t;
   }
} // powerScale
```

To streamline operation and avoid obvious inefficiencies, a minimum of housekeeping is typically performed ahead of time by initializing a few arrays containing frequently used values such as the so-called *rotations array*, the *window array* itself, and the *bit reversal array*. Here is how we define them and the initialization function we can use:

```
// input vector
unsigned char inB[N_FFT];
volatile int inCount;

// rotation vectors
float  wr[N_FFT/2];
float  wi[N_FFT/2];

// bit reversal vector
short  rev[N_FFT/2];

// window
float  ww[N_FFT];
void initFFT(void)
{
  int i, m, t, k;
  float *wwp;

  for(i=0; i<N_FFT/2; i++)
  {
    // rotations
    wr[i] = cos(PI2N * i);
    wi[i] = sin(PI2N * i);

    // bit reversal
    t = i;
    m = 0;
    k = N_FFT-1;
    while (k>0)
```

```
   {
     m = (m << 1)+(t & 1);
     t = t >> 1;
     k = k >> 1;
   }
   rev[i]=m;
  } // for i

  // initialize Hanning window vector
  for(wwp=ww, i=0; i<N_FFT; i++)
    *wwp++ = 0.5 - 0.5 * cos(PI2N * i);

} // initFFT
```

Scared? Confused? Don't be! Take this code as is; it's heavy stuff. The larger N_FFT, the number of samples in your input array, the harder it gets for our PIC32 to work on it.

All we need to do, for now, is to package it nicely in a source file, save it as fft.c, and then add it to the source files of a new project that we will call Running.

To keep things clean and tidy, let's also prepare a small include file fft.h where we will define all the symbols required to use the fft.c module.

```
/*
**   FFT.h
**
**   power of two optimized algorithm
*/

#include <math.h>

#define N_FFT  256                // samples must be power of 2
#define PI2N   2 * M_PI/N_FFT

extern unsigned char inB[];
extern volatile int inCount;

// preparation of the rotation vectors
void initFFT(void);

// input window
void windowFFT(unsigned char *source);

// fast fourier transform
void FFT(void);
```

```
// compute power and scale output
void powerScale(unsigned char *dest);
```

Add fft.h to the include files of the Running project as well.

Next let's create our project main source file. How about run.c for a name (see Figure 7.6)?

Figure 7.6: The Running project's Project window.

Let's add the configuration bit settings at the very top of the source code for maximum visibility, and let's include the fft.h file as well since we will soon use all its functions:

```
/*
** Run.c
**
*/
// configuration bit settings
#pragma config POSCMOD=XT, FNOSC=PRIPLL
#pragma config FPLLIDIV=DIV_2, FPLLMUL=MUL_18, FPLLODIV=DIV_1
#pragma config FWDTEN=OFF, CP=OFF, BWP=OFF
```

```
#include <p32xxxx.h>
#include <plib.h>
#include "fft.h"
```

Now let's create a main function that, in order, will perform the following:

1. Initializations:

   1.1. The `initFFT()` function needs to be called first:

   1.1. Filling the input buffer (`inB[]`) with a test signal, a sinusoid for simplicity:

```
main()
{
    int i, t;
    double f;

 // 1. initializations
    initFFT();

 // test sinusoid
    for (i=0; i<N_FFT; i++)
    {
       f = sin(2 * PI2N * i);
       inB[i] = 128+(unsigned char) (120.0 * f);
    } // for
```

2. The actual FFT algorithm, composed of the sequence of three function calls:

```
// 2. perform FFT algorithm
   windowFFT(inB);
   FFT();
   powerScale(inB);
```

3. A main (infinite) loop where it can rest after the exhausting performance:

```
   // 3. infinite loop
   while( 1);
} // main
```

Ready, Set, Go!

At this point we could already build the project, program a device, and, using a couple of breakpoints and a manual stopwatch, we could try to capture the actual time required. But the effort would be extremely tedious and imprecise. I have a better idea: Why don't we make the PIC32 time itself?

We can use, once more, one of the five 16-bit timers available or, for the occasion, we could experiment using for the first time a "pair" of timer modules combined to form a 32-bit timer. This option is available for the pairs formed by Timer2 and Timer3 together as well as Timer4 and Timer5. The latter pair is used in the following example, to bracket the FFT sequence:

```
// init 32-bit timer4/5
    OpenTimer45(T4_ON | T4_SOURCE_INT, 0);
  // clear the 32-bit timer count
    WriteTimer45(0);

// insert FFT function calls here

  // read the timer count
  t = ReadTimer45();
```

Notice how I used the functions from the timer.h library, and including plib.h at the top of the program, we automatically included all the peripheral libraries at once.

The `OpenTimerXX()` function allows us to configure the timer, selecting the clock source and the prescaler value. It is equivalent to writing directly to the `TxCON` register as we did in the previous explorations, if only slightly more readable. The main drawback, as often is the case for these peripheral libraries, is that you won't find the list of valid parameters to use (such as `T4_SOURCE_INT`) inside the device datasheet where the timer module is described; you will have to rely instead on a separate document (the library manual) and often resort to inspecting personally the include file—timer.h in this case. It is in fact by inspecting this file (you can open it with the MPLAB Editor) that you will learn how, when used as a pair, the correct parameters to pass to the initialization function are taken from those of the first module of the pair (T4 in our case).

The function `WriteTimerXX()`, as you would expect, allows us to set the initial counter value and effectively start our stopwatch, while the function `ReadTimerXX()` will read a 32-bit count value. It won't stop our stopwatch, but it will take a reading at that precise moment; that is what we need.

Let's open the Watch window by selecting the **View** | **Watch** menu and **Add** the symbol **t** to it. Unless you have already configured the Watch window to use decimal as the default format, click with the **right mouse button** on top of the symbol **t** to activate the Watch window context menu, and choose **Properties**. Select **Decimal** as the default representation for this variable.

Now you are ready to build our project and program it onto the device with your development tool of choice. Set a **breakpoint** on the line containing the infinite loop, press **Run**, and sit back and relax while the PIC32 works hard to solve the problem for you. After a short while, MPLAB will come back alive as the PIC32 reaches the breakpoint, and we will be able to read the timed value from the 32-bit integer variable **t**. In my case it turned out to be 6,140,495!

Well, at least now you understand why I suggested we use a 32-bit timer. As fast as a fast Fourier transform can be, it is hard work, and a 16-bit timer would not suffice to keep track of such a large number of cycles.

Converting the timer count in actual seconds, milliseconds, and microseconds is not hard if we remember how we configured the oscillator and the primary clock path. The PIC32 system bus clock frequency was set to 72 MHz, while all the peripherals were provided a 36 MHz peripheral bus clock. Dividing the timer value by the peripheral bus frequency, we obtain:

$$T = t/Fpb = 6,142,543/36,000,000 = 0.17062\,s$$

We can automate the conversion by asking the PIC32 to do it for us from now on—just add the following line of code after the stopwatch capture:

$$f = t/36E6;$$

This will reuse the variable **f** to perform the division using floating-point arithmetic. Add **f** to the Watch window so that, from now on, we will get to see the result of our experiments expressed correctly in seconds and fractions (see Figure 7.7).

Fine-Tuning the PIC32: Configuring Flash Wait States

Whether you think that 170 ms is a good time in which to perform a 256-point FFT or not, of one thing I am sure: The PIC32 can do better. In fact, beyond selecting the fastest clock speed and properly configuring the oscillator module, a number of advanced mechanisms on the PIC32 still require our attention to achieve the fine tuning that will

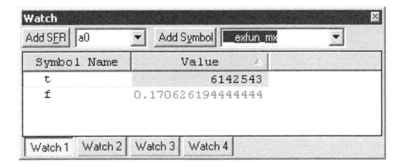

Figure 7.7: Testing the PIC32 performance using a 32-bit timer.

provide us with the highest possible level of performance. The number-one limitation to the performance of an embedded control processor is the speed of its Flash memory. Unfortunately, once more, there is a conflict of interest; the fastest available Flash memory banks are also the ones requiring the highest power consumption.

The designers of the PIC32 found that a perfect balance could be obtained by using a low-power Flash memory and decoupling the PIC32 core system bus from the memory bus by providing the ability to add a number of wait states (corresponding to up to seven clock cycles), during which the processor is stalled waiting for data to be fetched from the Flash memory. Depending on the difference in speed between memory and core, an increasing number of wait states might be required. By default, at power-up this mechanism is set for the safest possible condition that is reached by setting the maximum number of wait states. Hence there is an opportunity for us to reduce the number to the minimal possible value, given the actual operating specifications of the device. The number of wait states is controlled by the CHECON special function register (see Figure 7.8) and in particular by the PFMWS bits.

We could directly assign values, between 0 and 7, to the CHECON register's bits, as in the following example:

```
CHECONbits.PFMWS = 7;    // set max number of waitstates
```

But we would have to assume the responsibility for identifying the minimum safe number of wait states for the worst-case operating conditions of our application (relying on the electrical characteristics from the device datasheet). In fact, should we use the wrong number of wait states, the execution of code from Flash memory could become erratic,

U-0	U-0	U-0	U-0	U-0	U-0	U-0	U-0
—	—	—	—	—	—	—	—
31	30	29	28	27	26	25	24
U-0	U-0	U-0	U-0	U-0	U-0	U-0	R/W-0
—	—	—	—	—	—	—	CHECOH
23	22	21	20	19	18	17	16
U-0	U-0	r-0	r-0	U-0	U-0	R/W-0	R/W-0
—	—	—	—	—	—	DCSZ[1:0]	
15	14	13	12	11	10	9	8
U-0	U-0	R/W-0	R/W-0	U-0	R/W-1	R/W-1	R/W-1
—	—	PREFEN[1:0]		—		PFMWS[2:0]	
7	6	5	4	3	2	1	0

Figure 7.8: The CHECON control register.

and to make things worse, this would become detectable only under specific extreme conditions of power supply voltage and temperature.

As a better alternative, we can use an ad hoc library function provided with the PIC32MX peripheral libraries: SYSTEMConfigWaitStatesAndPB(freq). The function requires the system clock frequency to be passed as an integer parameter and was designed by the PIC32 application support team to set the "recommended" minimum wait states for the given system clock frequency, taking all the guesswork away.

Note

The ... AndPB part of the function name is supposed to remind us that the same function will also automatically modify the peripheral clock frequency setting of the PB divider as required to keep the peripheral bus always below 50 MHz. As it happens, this is exactly what we had the system configured for (at power-up) anyway.

So it is time to give a second try at our project, with the added "tuning" of the wait states performed by the following line of code (placed inside the initialization section of our main() function):

```
SYSTEMConfigWaitStatesAndPB(72000000L);
```

Rebuild the Running project and reprogram your development board. Let the application run once more until it reaches the breakpoint (see Figure 7.9).

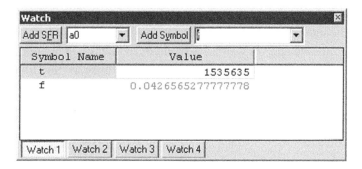

Figure 7.9: The PIC32 performance after wait states tuning.

Now, this is an improvement! We just reduced the FFT execution time from 170ms to 42 ms. This is better than a 4× speed improvement.

Fine-Tuning the PIC32: Enabling the Instruction and Data Cache

But there is so much more we can do. As we understand more of the PIC32 architecture, we notice that between the MIPS core bus and the memory bus there is actually an entirely new module: *the cache*. Think of it as a small but very fast block of RAM memory sitting between the processor and the Flash memory. Every time the processor fetches an instruction or a word of data from the Flash memory, the cache module will keep a copy but will also remember the address. When and if the processor needs the same data again (from the same address) in the (near) future, the cache will quickly be able to retrieve it, avoiding any new access to the Flash memory block (and avoiding all wait states eventually associated).

The larger a cache memory module, the higher the probability that a copy of a specific piece of data or instruction will be found in it. The reverse is also true: The shorter the inner loop of a given algorithm, the higher the impact that the availability of the cache module will have on its performance. This is because once all the cache is filled and a new instruction is fetched, the content of the cache must be "rotated," and the oldest or least recently used instruction/data needs to be overwritten by the new information.

Unfortunately, cache memory is, by its very nature, very expensive, and the PIC32MX designers had to balance costs and benefits by setting the maximum capacity of 16 lines of 16 bytes each, for a total of 64 complete 32-bit instructions, equivalent to 256 bytes.

There is much more flexibility (and therefore complexity) involved in the inner workings of the PIC32 cache module, but we don't need to know much more for now to decide that we like the cache module and we want to activate it. In fact, by default at power-on, it is disabled, and as in the previous case, there is a convenient library function (defined in the pcache.h module) awaiting our call:

```
CheKseg0CacheOn();
```

> **Note**
>
> The Kseg0 is the virtual memory space where MPLAB C32 allocates all the code segments produced by compiling our project codes by default. You will remember that code placed in this address space "can" be cached, whereas code place in Kseg1 will *not* be cached, regardless of the cache module settings and status.

Rebuild the Running project and reprogram your development board. Let the application run once more until it reaches the breakpoint (see Figure 7.10).

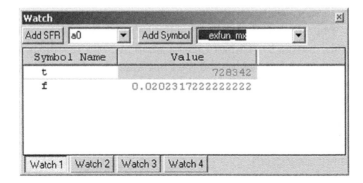

Figure 7.10: The PIC32 performance after enabling the cache.

Now, this is another important improvement! We just reduced the FFT execution time from 42 ms to 20 ms. This is a further 2× speed improvement.

Fine-Tuning the PIC32: Enabling the Instruction Pre-Fetch

But we are far from finished. The cache module of the PIC32 has another important feature to offer that promises similarly large rewards once enabled. It is the ability to perform instructions pre-fetching. That is, the cache module not only records the instructions being fetched by the PIC32 core; it also "runs ahead" and reads a whole block of four instructions (four words of 32 bits) at a time. If the code is executed sequentially, the next three memory fetches will be performed with the equivalent of zero wait states. Every time a branch is executed, breaking the sequential flow of the program, the pre-fetched cached data is discarded and the correct next instruction is loaded but without any additional penalty beyond the required wait states.

The cache pre-fetch is disabled by default at power-up, and the PREFEN bits in the CHECON register control the behavior of the module. They can be set by directly accessing the SFR or by using the macro mCheConfigure() defined in the pcache.h library:

```
mCheConfigure(CHECON | 0x30);
```

After appending this line of code to the list of initialization calls inside the main() function, let's rebuild the Running project and reprogram the development board. Let the application run once more until it reaches the breakpoint (see Figure 7.11).

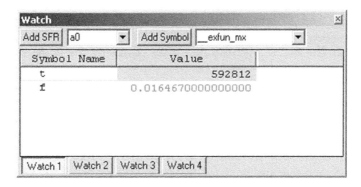

Figure 7.11: The PIC32 performance after enabling the cache.

We once more reduced the FFT execution time from 20 ms to 16.4 ms. This is a further 20-percent performance improvement.

Fine-Tuning the PIC32: Final Notes

As anticipated, the complexity of the cache module is considerable, and the number of additional possible "tricks" is practically unlimited if you dare dig deeper. I will mention only one last option related to accessing the RAM memory. As it happens, even regular RAM memory access is by default slowed by the presence of a single wait state. Its presence is already greatly mitigated by the cache, and the impact on the overall processor performance can be further reduced by the efficiency of the compiler and its use of the processor registers. Nonetheless, it is worth trying to disable it using the mBMXDisableDRMWaitState() function.

In my experiments, this produced an almost unnoticeable performance improvement, but the mileage can vary greatly with the nature of the application (see Figure 7.12).

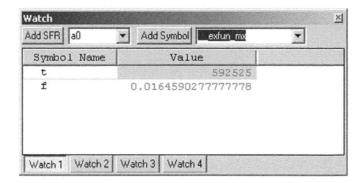

Figure 7.12: The PIC32 performance after removing the RAM wait states.

After rebuilding the project with the added last fine-tuning step, we obtained an additional 1-percent performance improvement.

In summary, in only four lines of code we have been able to produce an almost unimaginable performance improvement compared to our initial measurements using the default configuration at start-up. We went from 170.62 ms down to 16.45 ms, equivalent to a 10× speed performance boost to our FFT algorithm!

```
// configure PB frequency and the number of wait states
SYSTEMConfigWaitStatesAndPB(72000000L);

// enable the cache for max performance
CheKseg0CacheOn();

// enable instruction prefetch
mCheConfigure(CHECON | 0x30);

// disable RAM wait states
mBMXDisableDRMWaitState();
```

Fortunately, the PIC32 support team has been preparing a shortcut for us, a single simple library function that, from now on, will allow us to perform *all* of the above optimizations in a single function call:

```
SYSTEMConfigPerformance(72000000L);
```

A precious little function that fine-tunes the Flash memory and RAM access while unleashing the power of the cache and pre-fetch module of the PIC32. How about renaming it `SportTuning()` or `RacingMode()`?

Debriefing

Step by step, today we learned to tune up the engine of the PIC32, first in coarse steps, then gradually in finer steps, until we have been able to squeeze the most performance out of the machine. Keep in mind that the tuning process is very much dependent on the task at hand. Different applications will respond differently to each turn of the various "control knobs" we have touched today. Also, the result obtained is by no means representative of the fastest FFT implementation possible on a PIC32. In fact, we have deliberately chosen not to modify the original algorithm in any way, to highlight instead the relative performance gains obtained by our use of various hardware features available on the PIC32MX architecture. In the process we have also learned something new about the peripherals set and, in particular, the PIC32 timer modules that allow us to combine them in pairs to produce 32-bit timers.

Notes for the Assembly Experts

Once more we have resisted the temptation to use any hand optimization, avoiding any use of the assembly language. In reality, those of you who want to learn more about the PIC32 assembly will soon discover that there are powerful instructions in the PIC32 instruction set that we could have used to further boost the performance of the microcontroller in many signal processing applications. In particular, I am referring to the multiply and accumulate instructions, or multiply and add (MADD), as they are known in MIPS lingo.

Notes for the PIC® Microcontroller Experts

Thanks to the cache and the pre-fetch mechanism, the PIC32 can execute "almost" one instruction per clock cycle, even when operating at the maximum clock frequency while using a low-power Flash memory. The operative word here is "almost," since we cannot be sure that this happens all the time. The cache is inevitably going to generate *misses* here and there; for example, the MCU will have to wait from time to time while a group of words is loaded by the pre-fetch mechanism or a new word of data is loaded into the cache. The more your code revolves around a short loop that fits entirely in the PIC32 cache memory (256 bytes), the smaller the percentage of misses you will experience. By the way, although we don't have the time and space to cover the subject in the necessary depth in this book, most of the control registers inside the cache module are actually there to allow us some insight into the workings of the cache and to help us "profile" a specific piece of code.

So, can we claim that the PIC32 is a 72 MIPS machine, meaning that is it really executing 72 million instructions per second? I think the wise answer is "mostly" yes, but . . . it depends on your code and how well you can get the cache to work for you.

Tips & Tricks

One powerful tool, available as part of the MPLAB IDE, is the Data Monitor and Control Interface, or DCMI for friends and fans. You can activate it by selecting **Tools | DCMI** on the MPLAB IDE main menu. When used in combination with any of the in circuit debuggers and even the MPLAB SIM simulator, it can provide us with a window into the device data space by producing graphics but also letting us "interactively" modify the data with a sort of configurable graphical user interface (GUI). In particular, when playing with the FFT you might be interested in checking the shape of the input signal we synthesized (sinusoid) and in visualizing the output of the FFT routine. Once in the DCMI window, follow the next few steps in exact order:

1. Click the **Dynamic Data View** tab.
2. Check the **Graph1** check box.
3. Right-click with your mouse on the **first graph** to expose the context menu.
4. Select **Configure Data Source** (see Figure 7.13).
5. Select the **inB** buffer among the list of Global Symbols.
6. Click the **OK** button.

Now set a breakpoint on the line containing the `OpenTimer45()` call, just following the `inB[]` buffer initialization, and run the program.

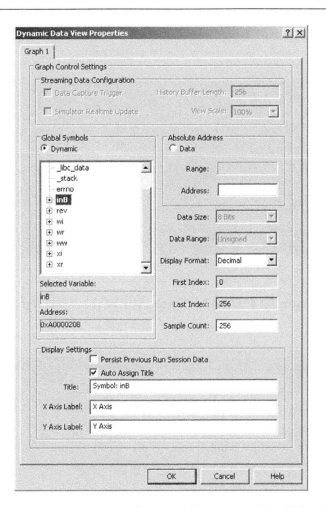

Figure 7.13: DCMI Dynamic Data View Properties dialog box.

As the program halts you should see the content of the inB[] buffer nicely visualized inside the Dynamic Data View window (see Figure 7.14).

It's a 2 Hz sinusoid, or I should say a sinusoid whose period is half the input sample count.

Now we can set a second breakpoint on the line where the ReadTimer45() function is called, after the FFT is performed and the scaling is performed to visualize the output. Remember that the output of an FFT contains only half the size of the input number of samples, so you can change the **Sample Count** field of the visualization to **128** instead of the default value (256) automatically offered by the DCMI. Also maximize the window to obtain a better detail (see Figure 7.15).

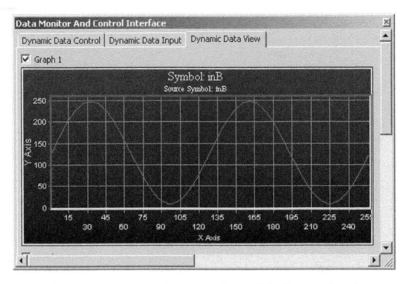

Figure 7.14: Dynamic Data View of the input signal.

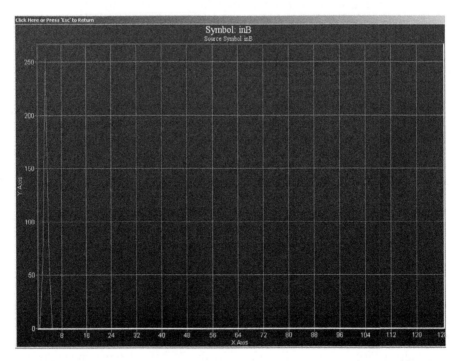

Figure 7.15: Dynamic Data View of the FFT output: The signal spectrum.

As you can see, the one and only peak in the signal power spectrum is easily found on the X-axis (considering the sample count starts from 1) at the position that would correspond to a frequency of 2 Hz (or two periods within the input sample count). Verify that this is exactly what we have designed the input test signal to be!

Exercises

1. Verify the shape and size of the output of the FFT (real and imaginary components) before the power scaling.

2. Remove the windowing and observe if and how the spectrum of the signal appears to change.

3. Use multiple input sinusoids to create a composite signal and observe the FFT output.

4. Experiment with allocating (more) cache space (lines) to the data space and observe the resulting performance changes.

Books

Sweetman, Dominic, *See MIPS Run*, second edition (2006). This is a must-read if you want to truly understand the most advanced features of the PIC32 MIPS core. The second edition is recommended because it focuses on the more modern implementations of the MIPS cores and adds notes on Linux kernel implementation details. (Don't try this at home on the PIC32MX . . . not just yet.)

Links

http://en.wikipedia.org/wiki/FFT. Helpful in learning more about uses of and methods to perform a Fast Fourier Transform.

http://en.wikipedia.org/wiki/Spectral_music. FFt can be fun! Think graphics, but also think music composition.

http://en.wikipedia.org/wiki/Window_function. No, we're not talking about *those* windows; these windows can dramatically change your views!

http://wn.wikipedia.org/wiki/CPU_cache. The PIC32MX is the first PIC microcontroller to use a cache mechanism. It is worth looking deeper in the subject to understand which decisions and compromises the designers of the PIC32 had to make to maximize performance while delivering an inexpensive product.

Communication

The Plan

Except for the most basic embedded-control applications, it is very likely that you will soon find that your application needs to communicate with other more or less intelligent devices. They could be personal computers, sensors, displays, or other microcontrollers on the same board or remote. To reduce cost, you will be looking for a solution that uses a small number of pins and wires and that will steer your search in the direction of a serial communication interface.

In embedded control, communication is equally a matter of understanding the protocols as well as the characteristics of the physical media available. Learning to choose the right communication interface for the application can be as important as knowing how to use it.

Today we will compare the basic communication peripherals available in all the general-purpose devices of the new PIC32MX family. In particular we will explore asynchronous serial communication interfaces (UART) and synchronous serial communication interfaces (SPI and I^2C), comparing their relative strengths and limitations for use in embedded control applications.

Preparation

In addition to the usual software tools, including the MPLAB® IDE, MPLAB C32 compiler student version, and the MPLAB SIM simulator, this lesson will require the use of the Explorer 16 demonstration board and one of the In-Circuit Debugging tools such as the MPLAB ICD2, MPLAB ICD3, MPLAB REAL ICE, or PIC32 Starter Kit. If you intend to use the latter, though, you will need the special PIC32 Starter Kit adapter (PIM).

The Exploration

The PIC32MX family offers seven communication peripherals that are designed to assist in all common embedded-control applications. As many as six of them are *serial* communication peripherals; they transmit and receive a single bit of information at a time. They are:

- 2 × the Universal Asynchronous Receiver and Transmitters (UARTs)

- 2 × the SPI synchronous serial interfaces

- 2 × the I²C synchronous serial interfaces

The main difference between a *synchronous* interface (like the SPI or I²C) and an *asynchronous* one (like the UART) is in the way the timing information is passed from transmitter to receiver. Synchronous communication peripherals need a physical line (a wire) to be dedicated to the *clock* signal, providing synchronization between the two devices. The device(s) that originates the clock signal is typically referred to as the *master*, as opposed to the device(s) that synchronizes with it, called the *slave(s)*.

Synchronous Serial Interfaces

The I²C interface (see Figure 8.1), for example, uses two wires and therefore two pins of the microcontroller: one for the clock (SCL) and one bidirectional for the data (SDA).

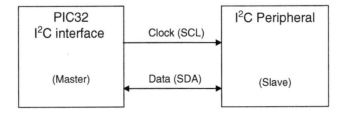

Figure 8.1: I²C interface block diagram.

The SPI interface (see Figure 8.2) instead separates the data line in two, one for the input (SDI) and one for the output (SDO), requiring one extra wire but allowing simultaneous (faster) data transfer in both directions.

To connect multiple devices to the same serial communication interfaces (a bus configuration), the I²C interface requires a 10-bit address to be sent over the data line

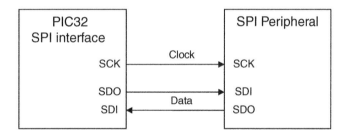

Figure 8.2: SPI interface block diagram.

before any actual data is transferred. This slows the communication but allows the same two wires (SCL and SDA) to be used for as many as (theoretically) 1,000 devices. Also, the I²C interface allows multiple devices to act as masters and share the bus using a simple arbitration protocol.

The SPI interface (see Figure 8.3), on the other side, requires an additional physical line, the slave select (SS), to be connected to each device. In practice this means that in using an SPI bus, as the number of connected devices grows, the number of I/O pins required on the PIC32 grows proportionally with them.

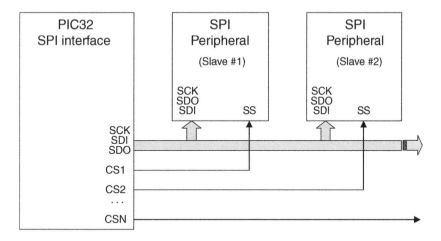

Figure 8.3: SPI bus block diagram.

Sharing an SPI bus among multiple masters is theoretically possible but practically very rare. The main advantages of the SPI interface are truly its simplicity and the speed that can be one order of magnitude higher than that of the fastest I²C bus (even without taking into consideration the details of the protocol-specific overhead).

Asynchronous Serial Interfaces

In asynchronous communication interfaces (see Figure 8.4), there is no clock line, whereas typically two data lines—TX and RX, respectively—are used for input and output, and optionally two more lines can be used to provide a hardware handshake. The synchronization between transmitter and receiver is obtained by extracting timing information from the data stream itself. *Start* and *stop bits* are added to the data, and precise formatting (with a fixed baud rate) must be set to allow reliable data transfers.

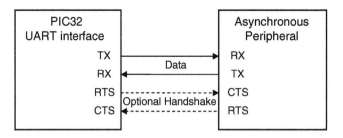

Figure 8.4: Asynchronous serial interface block diagram.

Several asynchronous serial interface standards dictate the use of special transceivers to improve the noise immunity, extending the physical connection distance up to several thousand feet.

Each serial communication interface has its advantages and disadvantages. Table 8.1 summarizes the most important ones as well as the most common applications.

Table 8.1: Serial interfaces comparison table.

Peripheral	Synchronous		Asynchronous
	SPI	I²C	UART
Max bit rate	20 Mbit/s	1 Mbit/s	500 kbit/s
Max bus size	Limited by number of pins	128 devices	Point to point (RS232), 256 devices (RS485)
Number of pins	3 + n × CS	2	2(+2)

(continued)

Table 8.1: (Continued)

Peripheral	Synchronous		Asynchronous
	SPI	I²C	UART
Pros	Simple, low cost, high speed	Small pin count, allows multiple masters	Longer distance (use transceivers for improved noise immunity)
Cons	Single master, short distance	Slowest, short distance	Requires accurate clock frequency
Typical application	Direct connection to many common peripherals on same PCB	Bus connection with peripherals on same PCB	Interface with terminals, personal computers, and other data acquisition systems
Examples	Serial EEPROMs (25CXXX series), MCP320X A/D converter, ENC28J60 Ethernet controller, MCP251X CAN controller . . .	Serial EEPROMs (24CXXX series), MCP98XX temperature sensors, MCP322x A/D converters . . .	RS232, RS422, RS485, LIN bus, MCP2550 IrDA interface . . .

Parallel Interfaces

The Parallel Master Port (PMP) completes the list of basic communication interfaces of the PIC32. The PMP has the ability to transfer up to 16 bits of information at a time while providing several address lines to interface directly to most commercially available LCD display modules (alphanumeric and graphic modules with integrated controller) as well as Compact Flash memory cards (or CF-I/O cards), printer ports, and an almost infinite number of other basic 8- and 16-bit parallel devices available on the market and featuring the standard control signals: -CS, -RD, and -WR.

Today we begin focusing specifically on the use of a synchronous serial interface, the SPI. In the next few days we will also cover the asynchronous serial interface and the PMP.

Synchronous Communication Using the SPI Modules

The SPI interface is perhaps the simplest of all the available interfaces, although the PIC32 implementation is particularly rich in options and interesting features.

The SPI interface (see Figure 8.5) is essentially composed of a shift register. Bits are simultaneously shifted in, most significant bit (MSb) first, from the SDI line and shifted out from the SDO line in synch with the clock on the SCK pin. The size of the shift register can vary from 8, 16, or 32 bits.

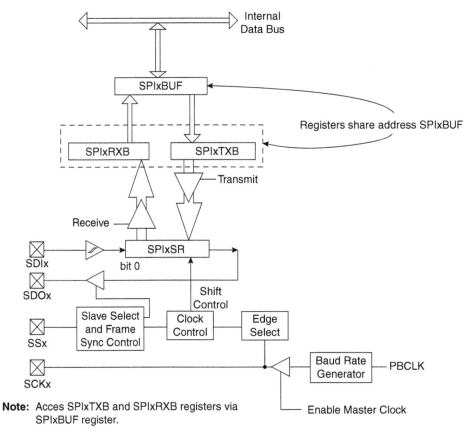

Note: Acces SPIxTXB and SPIxRXB registers via SPIxBUF register.

Figure 8.5: The SPI module block diagram.

If the device is configured as a bus master, the clock is generated internally, derived from the peripheral bus clock (Fpb) by a baud rate generator, and output on the SCK pin. Otherwise, the device is a bus slave and the clock is received from the SCK pin.

As for all other peripherals we will encounter, the essential configuration options are controlled by the SFR `SPIxCON` and the baud rate generator control register `SPIxBRG` (see Figure 8.6).

R/W-0	R/W-0	R/W-0	U-0	U-0	U-0	U-0	U-0
FRMEN	FRMSYNC	FRMPOL	—	—	—	—	—
bit 31							bit 24

U-0	U-0	U-0	U-0	U-0	U-0	R/W-0	U-0
—	—	—	—	—	—	SPIFE	—
bit 23							bit 16

R/W-0	R/W-0	R/W-0	R/W-0	R/W-0	R/W-0	R/W-0	R/W-0
ON	FRZ	SIDL	DISSDO	MODE32	MODE16	SMP	CKE
bit 15							bit 8

R/W-0	R/W-0	R/W-0	U-0	U-0	U-0	U-0	U-0
SSEN	CKP	MSTEN	—	—	—	—	—
bit 7							bit 0

Figure 8.6: The SPIxCON control register.

Notice in Figure 8.6 that the lower (least significant) 16 bits of the `SPIxCON` register contain all the essential configuration bits, whereas the top 16 bits contain control bits that refer only to advanced features of the SPI port (framed modes). This makes the `SPIxCON` control register compatible with the previous generations of 16-bit PIC® microcontrollers, since the top bits default to zero.

To demonstrate the basic functionality of the SPI peripheral we will use the Explorer 16 demo board, on which the PIC32 SPI2 module is connected to a 25LC256 EEPROM device, often referred to as a Serial EEPROM (or SEE, or sometimes just E^2, pronounced *e-squared*). This is a small and inexpensive device that contains 256 Kbits, or 32 Kbytes, of nonvolatile high-endurance memory.

Use the New Project Setup checklist to create a new project called **SPI** and a new source file, similarly called **spi2.c**.

The most direct way to configure the SPI2 module for communication with the serial memory device is by manually assigning the correct value to each bit of the `SPI2CON` register. According to the 25LC256 device datasheet (DS21822), downloadable from the Microchip Web site, the SEE responds to a short list of 8-bit commands that must be supplied via the SPI interface with the following setting (notice in parentheses the corresponding values of the control bits in the SPI2CON register):

- 8-bit mode (MODE16 = 0, MODE32 = 0)

- Clock IDLE level is low, clock ACTIVE is high (CKP = 0)

- Serial output changes on transition from ACTIVE to IDLE (CKE = 1)

We will also need to configure the PIC32 to act as the SPI bus master (MSTEN = 1), since the memory is a slave-only device—in other words, it expects to receive a clock signal on the SCK pin.

The resulting configuration value can be defined as a constant that will be later assigned to the SPI2CON register:

```
// peripheral configurations

#define SPI_CONF 0x8120  // SPI on, 8-bit master, CKE=1,CKP=0
```

To determine the baud rate, we will use Equation 8.1 (from the PIC32 datasheet):

Equation 8.1: Formula to determine SPI clock frequency.

$$F_{SCK} = \frac{F_{PB}}{2 * (SPIxBRG + 1)}$$

We can either use the SPI2BRG default value (0 at power-up, giving a baud rate divider of 1:2) or assign an appropriate value to slow the communication and correspondingly help reduce the EEPROM power consumption—for example:

```
#define SPI_BAUD  15  // clock divider Fpb/(2 * (15+1))
```

With such settings, the baud rate divider is set to 1:32 of Fpb, corresponding to about 280 kHz when the PIC32 is configured for a 9 MHz peripheral bus as set and documented by the following few lines that we will place at the top of our source code:

```
// configuration bit settings,  Fcy=72MHz,  Fpb=9MHz
#pragma config POSCMOD=XT, FNOSC=PRIPLL
#pragma config FPLLIDIV=DIV_2,  FPLLMUL=MUL_18,  FPLLODIV=DIV_1
#pragma config FPBDIV=DIV_8,  FWDTEN=OFF,  CP=OFF,  BWP=OFF
```

From the Explorer 16 User Guide (DS51589 Appendix A board schematic), we learn that pin 12 of PortD (RD12) is connected to the memory chip-select (CS) pin. Notice that this is an active low input. A couple of definitions will help make our code more readable:

```
// I/O definitions
#define CSEE        _RD12        // select line for EEPROM
#define TCSEE       _TRISD12     // tris control for CSEE pin
```

We can now write the peripheral initialization part of our demonstration program:

```
// 1.init the SPI peripheral
TCSEE = 0;                  // make SSEE pin output
CSEE = 1;                   // de-select the EEPROM
SPI2CON = SPI_CONF;         // select mode and enable
SPI2SPI2BRG = SPI_BAUD;     // select clock speed
```

We can now write a small function that will be used to transfer data to and from the serial EEPROM device:

```
// send one byte of data and receive one back at the same time
int writeSPI2( int i)
{
  SPI2BUF = i;                      // write to buffer for TX
  while( !SPI2STATbits.SPIRBF);     // wait for transfer complete
  return SPI2BUF;                   // read the received value
}//writeSPI2
```

The `writeSPI2()` is a truly bidirectional transfer function. It immediately writes a character to the transmit buffer and then enters a loop to wait for the *receive* flag to be set to indicate that the transmission was completed as well as data was received back from the device. The data received is then returned as the value of the function.

When we're communicating with the memory device, though, there are situations when a command is sent to the memory, but there is no immediate response. There are also cases when data is read from the memory device, but no further commands need to be sent by the PIC32. In the first case (for example, the write command), the return value of the function can simply be ignored. In the second case (for example, the read command), a dummy value can be sent to the memory while shifting in data from the device.

The 25LC256 datasheet contains accurate depictions of all seven possible command sequences that can be used to read or write data to and from the memory device. A small table of constants can help encode and document all such commands in our code:

```
// 25LC256 Serial EEPROM commands
#define SEE_WRSR    1           // write status register
#define SEE_WRITE   2           // write command
#define SEE_READ    3           // read command
#define SEE_WDI     4           // write disable
#define SEE_STAT    5           // read status register
#define SEE_WEN     6           // write enable
```

Before we attempt any more complex task, let's test the little code we have assembled so far to verify that communication with the device can be properly established. For example, we can use the Read Status Register (SEE_STAT) command to interrogate the EEPROM and obtain the value of its internal status register.

Testing the Read Status Register Command

After sending the appropriate command byte (SEE_STAT) with a first call to the writeSPI2() function, we will need to send a second (dummy) byte to capture the response from the memory device (see Figure 8.7).

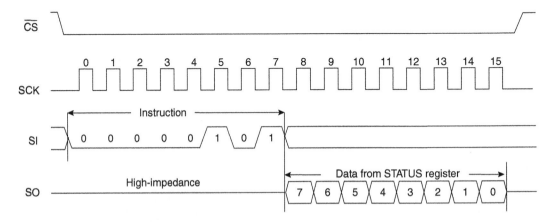

Figure 8.7: The complete Read Status Register command timing sequence.

Sending any command to the SEE requires, at a minimum, the following four-step sequence:

1. Activate the memory, taking the **CS pin** low.

2. Shift out the **8-bit** command.

3. Depending on the specific command, send or receive multiple bytes of data.

4. Deactivate the memory (taking high the **CS pin**) to complete the command. After this step the memory will go back to a low-power consumption standby mode.

In practice, the following code is required to perform the complete Read Status Register operation:

```
// Check the Serial EEPROM status
CSEE = 0;                       // select the Serial EEPROM
writeSPI2( SEE_STAT);           // send a READ STATUS COMMAND
i = writeSPI2( 0);              // send dummy, read data
CSEE = 1;                       // deselect to complete command
```

The complete project listing should look like:

```
/*
** SPI2
**
*/
#include <p32xxxx.h>

// configuration bit settings, Fcy=72 MHz, Fpb=9 MHz
#pragma config POSCMOD=XT, FNOSC=PRIPLL
#pragma config FPLLIDIV=DIV_2, FPLLMUL=MUL_18, FPLLODIV=DIV_1
#pragma config FPBDIV=DIV_8, FWDTEN=OFF, CP=OFF, BWP=OFF

// I/O definitions
#define CSEE  _RD12           // select line for Serial EEPROM
#define TCSEE _TRISD12        // tris control for CSEE pin

// peripheral configurations
#define SPI_CONF   0x8120     // SPI on, 8-bit master,CKE=1,CKP=0
#define SPI_BAUD   15         // clock divider Fpb/(2 * (15+1))
```

```
// 25LC256 Serial EEPROM commands
#define SEE_WRSR    1       // write status register
#define SEE_WRITE   2       // write command
#define SEE_READ    3       // read command
#define SEE_WDI     4       // write disable
#define SEE_STAT    5       // read status register
#define SEE_WEN     6       // write enable

// send one byte of data and receive one back at the same time
int writeSPI2( int i)
{
    SPI2BUF = i;                        // write to buffer for TX
    while( !SPI2STATbits.SPIRBF);  // wait for transfer complete
    return SPI2BUF;                     // read the received value
}//writeSPI2

main ()
{
  int i;
  // 1. init the SPI peripheral
  TCSEE = 0;                        // make SSEE pin output
  CSEE = 1;                         // de-select the Serial EEPROM
  SPI2CON = SPI_CONF;               // select mode and enable SPI2
  SPI2BRG = SPI_BAUD;               // select clock speed
  // main loop
  while( 1)
  {
    // 2. Check the Serial EEPROM status
    CSEE = 0;                       // select the Serial EEPROM
    writeSPI2( SEE_STAT);           // send a READ STATUS COMMAND
    i=writeSPI2( 0);                // send/receive
    CSEE=1;                         // deselect terminate command
  } // main loop
} // main
```

Follow the Debugger Setup checklist appropriate for your tool of choice to enable the In-Circuit Debugger and prepare the project configuration. Then follow the Project Build checklist to compile and link the demo code. Then:

1. After connecting the Explorer 16 demo board, program the PIC32 selecting the **Debugger | Program** option. By default MPLAB will choose the smallest

range of memory required to transfer the project code into the device so that programming time will be minimized. After a few seconds, the PIC32 should be programmed, verified, and ready to execute.

2. Add the **Watch** window to the project.

3. Select **i** in the symbol selection box, then click the **Add Symbol** button.

4. Set the cursor on the last line of code in the main loop (containing the CSEE deselect) and set a **breakpoint** (double-click).

5. Start the execution by selecting the **Debugger | Run** command.

6. When the execution terminates, the contents of the 25LC256 memory Status Register should have been transferred to the variable *i*, visible in the Watch window.

Unfortunately, you will be disappointed to learn that the default status of the 25LC256 memory (at power-on) is represented by the value 0×00 (see Table 8.2).

Table 8.2: The 25LC256 Serial EEPROM status register.

7	6	5	4	3	2	1	0
W/R	–	–	–	W/R	W/R	R	R
WPEN	x	x	x	BP1	BP0	WEL	WIP
W/R = writable/readable; R = read-only.							

In fact, from Table 8.2, which illustrates the contents of the EEPROM status register, and from the device datasheet we learn that, at power-on, the Block Protection bits (BP1 and BP0) are supposed to be cleared unless a block code protection had been set, the Write Enable Latch (WEL) is disabled, and no Write In Progress (WIP) flag should be active.

Not a very telling result for our little test program. So, to spice up things a little we could start by setting the Write Enable Latch before interrogating the Status Register; it would be great to see bit 1 set.

Let's insert the following code before Section 2 that we will promptly renumber to 2.2:

```
// 2.1 send a Write Enable command
CSEE = 0;                  // select the Serial EEPROM
writeSPI2( SEE_WEN);       // send command, ignore data
CSEE=1;
```

1. Rebuild the project.

2. Reprogram the device.

3. Run (or Run to Cursor).

If everything went well, you will see now the variable i in the Watch window turn red and show a value of 2. Now, these are the great satisfactions that you can get only by developing code for a powerful 32-bit embedded controller!

More seriously, now that the Write Enable latch has been set, we can add a write command and start "modifying" the contents of the EEPROM device. We can write a single byte at a time, or we can write a long string, up to a maximum of 64 bytes, all in a single sequence/command called Page Write. Read more on the datasheet about address restrictions that apply to this mode of operation, though.

Writing Data to the EEPROM

After sending the write command, 2 bytes of address must be supplied before the actual data is shifted out. The following code exemplifies the correct write sequence:

```
// send a Write command
CSEE = 0;                       // select the Serial EEPROM
writeSPI2( SEE_WRITE);          // send command, ignore data
writeSPI2( ADDR_MSB);           // send MSB of memory address
writeSPI2( ADDR_LSB);           // send LSB of memory address
writeSPI2( data);               // send the actual data
// send more data here to perform a page write
CSEE = 1;                       // start actual EEPROM write cycle
```

Notice how the actual EEPROM write cycle initiates only after the CS line is brought high again. Also, it will be necessary to wait for a time (Twc) specified in the memory device datasheet for the cycle to complete before a new command can be issued.

There are two methods to make sure that the memory is allowed the right amount of time to complete the write command. The simplest one consists of inserting a fixed delay after the write sequence. The length of such a delay should be longer than the maximum cycle time specified in the memory device datasheet (Twc max = 5 ms).

A better method consists of checking the Status Register contents before issuing any further read/write command, then waiting for the Write In Progress (WIP) flag to be cleared; this will also coincide with the Write Enable bit (WEN) being reset. By doing so, we will be waiting only the exact minimum amount of time required by the memory device in the current operating conditions.

Reading the Memory Contents

Reading back the memory contents is even simpler. Here is a snippet of code that will perform the necessary sequence:

```
// send a Write command
CSEE = 0;                 // select the Serial EEPROM
writeSPI2( SEE_READ);     // send command, ignore data
writeSPI2( ADDR_MSB);     // send MSB of memory address
writeSPI2( ADDR_LSB);     // send LSB of memory address
data=writeSPI2( 0);       // send dummy, read data
// read more data here sequentially incrementing the address
CSEE = 1;                 // terminate the read sequence
                          // and return to low power
```

The read sequence can be indefinitely extended by sequentially reading the entire memory contents if necessary and, upon reaching the last memory address (0x7FFF), rolling over and starting from 0x0000 again.

A 32-Bit Serial EEPROM Library

We can now assemble a small library of functions dedicated to accessing the 25LC256 serial EEPROM. The library will hide all the details of the implementation, such as the SPI port used, specific sequences, and timing details. It will expose instead only two basic commands to read and write integer data types (32-bit) to a generic (black box) nonvolatile storage device.

Let's create a new project using the Project Wizard and the usual checklist. An appropriate name could be **SEE**. After creating a new source file **see.c**, we can copy most of the definitions we prepared in the SPI project:

```
/*
** SEE Access Library
*/

#include "p32xxxx.h"
#include "see.h"

// I/O definitions
#define CSEE   _RD12          // select line for Serial EEPROM
#define TCSEE  _TRISD12       // tris control for CSEE pin

// peripheral configurations
#define SPI_CONF  0x8120      // SPI on, 8-bit master,CKE=1,CKP=0
#define SPI_BAUD  15          // clock divider Fpb/(2 * (15+1))

// 25LC256 Serial EEPROM commands
#define SEE_WRSR    1         // write status register
#define SEE_WRITE   2         // write command
#define SEE_READ    3         // read command
#define SEE_WDI     4         // write disable
#define SEE_STAT    5         // read status register
#define SEE_WEN     6         // write enable
```

Let's also copy the initialization code, the write function, and the status register read commands. Each one will become a separate function:

```
// send one byte of data and receive one back at the same time
int writeSPI2( int i)
{
  SPI2BUF = i;                     // write to buffer for TX
  while( !SPI2STATbits.SPIRBF);    // wait for transfer complete
  return SPI2BUF;                  // read the received value
}// writeSPI2

void initSEE( void)
{
  // init the SPI2 peripheral
  CSEE = 1;                        // de-select the Serial EEPROM
  TCSEE = 0;                       // make SSEE pin output
```

```
  SPI2CON = SPI_CONF;              // enable the peripheral
  SPI2BRG = SPI_BAUD;              // select clock speed
}// initSEE

int readStatus( void)
{
  // Check the Serial EEPROM status register
  int i;
  CSEE = 0;                        // select the Serial EEPROM
  writeSPI2( SEE_STAT);            // send a READ STATUS COMMAND
  i = writeSPI2( 0);               // send/receive
  CSEE = 1;                        // deselect terminate command
  return i;
} // readStatus
```

To create a function that reads an integer value from nonvolatile memory, first
we verify that any previous command (write) has been correctly terminated by
reading the status register. A sequential read of 2 bytes is used to assemble an integer
value:

```
int readSEE( int address)
{ // read a 32-bit value starting at an even address

  int i;

  // wait until any work in progress is completed
  while ( readStatus() & 0x1); // check WIP

  // perform a 16-bit read sequence (two byte sequential read)
  CSEE = 0;                        // select the Serial EEPROM
  writeSPI2( SEE_READ);            // read command
  writeSPI2( address >>8);         // address MSB first
  writeSPI2( address & 0xfc);      // address LSB (word aligned)
  i = writeSPI2( 0);               // send dummy, read msb
  i = (i<<8)+ writeSPI2( 0);       // send dummy, read lsb
  i = (i<<8)+ writeSPI2( 0);       // send dummy, read lsb
  i = (i<<8)+ writeSPI2( 0);       // send dummy, read lsb
  CSEE = 1;
  return ( i);
}// readSEE
```

Finally, the write enable function can be created by extracting the short segment of code used to access the Write Enable latch from our previous project and adding a page write sequence:

```c
void writeEnable( void)
  {
  // send a Write Enable command
  CSEE = 0;                    // select the Serial EEPROM
  writeSPI2( SEE_WEN);         // write enable command
  CSEE = 1;                    // deselect to complete the command
}// writeEnable

void writeSEE( int address, int data)
{ // write a 32-bit value starting at an even address

  // wait until any work in progress is completed

  while ( readStatus() & 0x1) // check the WIP flag

  // Set the Write Enable Latch
  writeEnable ();

  // perform a 32-bit write sequence (4 byte page write)
  CSEE = 0;                    // select the Serial EEPROM
  writeSPI2( SEE_WRITE);       // write command
  writeSPI2( address>>8);      // address MSB first
  writeSPI2( address & 0xfc);  // address LSB (word aligned)
  writeSPI2( data >>24);       // send msb
  writeSPI2( data >>16);       // send msb
  writeSPI2( data >>8);        // send msb
  writeSPI2( data);            // send lsb
  CSEE = 1;
}// writeSEE
```

More functions could be added at this point to access short (16-bit) or long long (64-bit) data types, for example, but for our proof of concept this will suffice.

Note that the page write operation (see the 25LC256 memory datasheet for details) requires the address to be aligned on a power of two boundaries (in our case, just an address divisible by 4 will do). The requirement must be extended to the read function for consistency.

Save the code in the **see.c** file and add it to the project using one of the three methods shown in the checklists. You can either use the editor right-click menu and select

Add to Project or by right-clicking on the project window on the Source Files branch and choosing **Add Files**, then selecting the **see.c** file from the current project directory.

To make a few selected functions from this module accessible to other applications, create a new file, see.h, and insert the following declarations:

```
/*
** SEE Access library
**
** encapsulates 25LC256 Serial EEPROM
** as a NVM storage device for PIC32 + Explorer16 applications
*/
// initialize access to memory device
void initSEE(void);

// 32-bit integer read and write functions
// NOTE: address must be an even value between 0x0000 and 0x3ffc
// (see page write restrictions on the device datasheet)
int  readSEE ( int address);
void writeSEE( int address, int data);
```

This will expose only the initialization function and the integer read/write functions, hiding all other details of the implementation.

Add the see.h file to the project by right-clicking in the project windows on the Header Files icon and selecting it from the current project directory.

Testing the New SEE Library

To test the functionality of the library, we can create a test application containing a few lines of code that repeatedly read the contents of a memory location (at address 16), increment its value, and write it back to the memory.

```
/*
** SEE Library test
*/
// configuration bit settings, Fcy=72MHz, Fpb=9MHz
#pragma config POSCMOD=XT, FNOSC=PRIPLL
#pragma config FPLLIDIV=DIV_2, FPLLMUL=MUL_18, FPLLODIV=DIV_1
#pragma config FPBDIV=DIV_8, FWDTEN=OFF, CP=OFF, BWP=OFF
#include <p32xxxx.h>
#include "see.h"
```

```
main ()
{
  int data;

  // initialize the SPI2 port and CS to access the 25LC256
  initSEE();
  // main loop
  while ( 1)
  {
    // read current content of memory location
    data = readSEE( 16);
    // increment current value
    data++;      // <-set brkpt here

    // write back the new value
    writeSEE( 16, data);
    //address++;

  } // main loop
} //main
```

   1. Save this file as **SEEtest.c** and add it to the current project, too.

Invoking the **Build All** command, you will observe the MPLAB C32 compiler to work sequentially on the two source files (.c) and later the linker to combine the object codes to produce an output executable (.hex).

   2. Add **data** to the Watch window.

   3. Set a breakpoint on the line immediately following the read command to allow us to test the proper operation of the SEE library.

   4. Click the **Run** command and watch the program stop after the first read.

   5. Note the value of data and then **Run** again. It should increment continuously, and even when resetting the program or completely disconnecting the board from the power supply to reconnect it later, we will observe that the contents of location 16 will be preserved and successively incremented.

Careful—if the main program loop is left running indefinitely without any breakpoint, the library test program will quickly turn into a test of the Serial EEPROM endurance. In fact, the loop will continue to reprogram location 16 at a rate that will be mostly dependent on the actual Twc of the device. In a best-case scenario (maximum Twc = 5 ms), this will mean 200 updates every second. Or, in other terms, the theoretical endurance limit of the EEPROM (1,000,000 cycles) will be reached in 5,000 seconds, or slightly less than one hour and a half of continuous operation.

Debriefing

Today we have just started our exploration of the serial interfaces, comparing the basic differences among them and reviewing some of their most common uses in embedded control. In particular, we have experimented briefly with the SPI module in its simplest configuration to gain access to a 25LC256 Serial EEPROM memory, one of the most common types of nonvolatile memory peripherals used in embedded-control applications. The small library module we developed will hopefully be useful to you in future applications, to provide additional nonvolatile storage (32 K bytes) to your applications on the Explorer 16.

Notes for the C Experts

The C programmer used to developing code for large workstations and personal computers will be tempted to further develop the library to include the most flexible and comprehensive set of functions. My word of advice is to resist, hold your breath, and count to 10, especially before you start adding any new parameter to the library functions. In the embedded-control world, passing more parameters means using up more stack space, spending more time copying data to and from the stack, and in general producing a larger output code. Keep the libraries simple and therefore easy to test and maintain. This does not mean that proper object-oriented programming practices should not be followed. On the contrary, our example can be considered an example of *object encapsulation*, since most of the details of the SPI interface and Serial EEPROM internal workings have been hidden from the user, who is provided instead with a simple interface to a generic storage device organized in 32-bit words.

Notes for the Explorer 16 Experts

One of the least-known features of the Explorer 16 board is related to the use of two digital multiplexer devices (74HCT4053) present on the board and marked U6 and U7. The first one in particular was added to the board to allow the swap of the SDI and SDO lines of the SPI1 port reaching the PICTail™ connectors so that two

Explorer 16 boards could be cross-connected and the two microcontrollers could exchange data. The swap is controlled by the RB12 pin when configured as a digital output and pulling low (otherwise a pull-up resistor takes care of things). Proper connection requires, of course, that one of the two microcontrollers be configured as master, therefore producing the SCK signal, and the other as slave. Also keep in mind that only one of the two boards can be connected to the power supply; the other will be powered via the PIC Tail connector. Similarly, RB13 and RB14, in conjunction with the U7 multiplexer, are designed to allow cross-connection via the UART1 serial interface.

Notes for the PIC24 Experts

The SPI module of the PIC32 is mostly identical to the PIC24 peripheral, yet some important enhancements have been included in its design. Here are the major differences that will affect your code while you're porting an application to the PIC32:

1. The SPIxCON register control bits layout has been updated to resemble more closely the layout of most other peripherals so that the module ON, FRZ, and IDL bits are now located in the standard position (bit 15, bit 14, bit 13). They used to be found in the SPIxSTAT register.

2. The upper half of the SPIxCON register (being now expanded to 32 bits) provides room for the framing control bits (FRMEN, SPIFSD...) previously located in a second control register SPIxCON2.

3. The new MODE32 bit now selects the 32-bit mode operation.

4. The clock prescaler/divider of the SPI module (which used to be a two-tier 3 + 2 bit prescaler) is expanded to a full 9-bit baud rate generator module cleanly controlled by a separate register SPIxBRG.

Tips & Tricks

If you store important data in an external nonvolatile memory (SEE), you might want to put some additional safety measures in place (both hardware and software). From a hardware perspective, make sure that:

*   Adequate power supply decoupling (capacitor) is provided close to the device.

*   A pull-up resistor (10 k Ohm) is provided on the Chip Select line, to avoid floating during the microcontroller power-up and reset.

- An additional pull-down resistor (10 k Ohm) can be provided on the SCK clock line to avoid clocking of the peripheral during power-up, when the PIC32 I/Os might be floating (tri-state).

- Verify clean and fast power-up and down slopes are provided to the microcontroller to guarantee reliable Power-On Reset (POR) operation. If necessary, add an external voltage supervisor (see MCP809 devices for example).

A number of software methods can then be employed to prevent even the most remote possibility that a program bug or the proverbial cosmic ray might trigger the write routine. Here are some suggestions:

- Avoid reading and especially updating the SEE content right after power-up. Allow a few milliseconds for the power supply to stabilize (this will be heavily application dependent).

- Add a software write-enable flag, and demand that the calling application set the flag before calling the write routine, possibly after verifying some application-specific entry condition.

- Add a stack-level counter; each function in the stack of calls implemented by the library should increment the counter upon entry and decrement it on exit. The write routine should refuse to perform if the counter is not at the expected level.

- Some users refuse to use the SEE memory locations corresponding to the first address (0x0000) and/or the last address (0xffff), believing they could be statistically more likely to be subject to corruption.

- More seriously, store two copies of each essential piece of data, performing two separate calls to the write routine. If each copy contains a checksum or, simply by comparison, when reading it back, it will be easy to identify a memory corruption problem and possibly recover.

Exercises

Although the PIC32 SPI peripheral module operates off the peripheral clock system that could be ticking as fast as 50 MHz, few peripherals can operate at such speeds at 3 V. Specifically, the 25LC256 series Serial EEPROMs, operate with a maximum clock rate of 5 MHz when the power supply is in the 2.5 V to 4.5 V range. This means that the fastest

SPI port configuration compatible with the memory device can be obtained with a baud rate generator configured for 1:8 operation (36 MHz/8 = 4.5 MHz). A sequential read command could therefore provide a maximum throughput close to 4 Mbit per second or 512 Kbytes per second. Even at such a rate, the PIC32 would be able to execute 140 instructions before each new byte of data is received. This means that in our simple SEE application example, a lot of processing power is wasted sitting in loops and waiting for each byte to be transferred.

1. Develop a more advanced library based on an interrupt-driven state machine and/ or using the DMA to make a more efficient use of the PIC32 processing power. We explore the use of the DMA in conjunction with the SPI port in Chapter 13, although it won't be to interface to a serial EEPROM but for more mundane and fun applications.

2. Try enabling the new 32-bit mode of the SPI module to accelerate basic read and write word operation. But watch out: The SEE commands are byte wide, so you will probably need to switch back and forth between 8- and 32-bit modes. Are you really going to save any time/code?

Books

Eady, F., *Networking and Internetworking with Microcontrollers* (Newnes, Burlington, MA, 2004). An entertaining introduction to serial communication in embedded control. The author explores the basic synchronous and asynchronous communication interfaces to help 8-bit microcontrollers communicate.

Links

www.microchip.com/stellent/idcplg?IdcService=SS_GET_PAGE&nodeId=1406&dDoc Name=en010003. Use this link or search the Microchip Website for a free software tool called Total Endurance Software. It will help you estimate the endurance you can expect from a given NVM device in your actual application conditions. It will give you an indication of the total number of e/w cycles or the number of expected years of your application life before a certain target failure rate is reached.

Asynchronous Communication

The Plan

Once you remove the clock line from the serial interface between two devices, what you obtain is an asynchronous communication interface. Whether you want full bidirectional (duplex) communication or just half-duplex (one direction at a time), multipoint, or point-to-point communication, there are many asynchronous protocols that can make communication possible and allow for efficient use of the media. In this lesson we will review the PIC32 asynchronous serial communication interface modules, UART1 and UART2, to implement a basic RS232 interface. We will develop a console library that will be handy in future projects for interface and debugging purposes.

Preparation

In addition to the usual software tools, including the MPLAB® IDE, the MPLAB C32 compiler, and the MPLAB SIM simulator, this lesson will require the use of the Explorer 16 demonstration board, your In-Circuit Debugger of choice, and a PC with an RS232 serial port (or a serial to USB adapter). You will also need a terminal emulation program. If you are using the Microsoft Windows operating system, the HyperTerminal application will suffice (**Start|Programs | Accessories | Communication | HyperTerminal**).

The Exploration

The UART interface is perhaps the oldest interface used in the embedded-control world. Some of its features were dictated by the need for compatibility with the first mechanical teletypewriters. This means that at least some of its technology has centuries'-old roots.

On the other hand, nowadays finding an asynchronous serial port on a new computer (and especially on a laptop) is becoming a challenge. The serial port has been declared a "legacy interface," and for several years now, strong pressure has been placed on computer manufacturers to replace it with the USB interface. Despite the decline in their popularity and the clearly superior performance and characteristics of the USB interface, asynchronous serial interfaces are strenuously resisting in the world of embedded applications because of their great simplicity and extremely low cost of implementation.

Four main classes of asynchronous serial application are still being used:

1. *RS232 point-to-point connection.* Often simply referred to as "the serial port"; used by terminals, modems, and personal computers using +12V/−12V transceivers.

2. *RS485 (EIA-485) multi-point serial connection.* Used in industrial applications; uses a 9-bit word and special half-duplex transceivers.

3. *LIN bus.* A low-cost, low-voltage bus designed for noncritical automotive applications. It requires a UART capable of baud rate autodetection.

4. *Infrared wireless communication.* Requires a 38–40 kHz signal modulation and optical transceivers.

The PIC32's UART modules can support all four major application classes and packs a few more interesting features, too.

To demonstrate the basic functionality of a UART peripheral, we will use the Explorer16 demo board where the UART2 module is connected to an RS232 transceiver device and to a standard 9 poles D female connector. This can be connected to any PC serial port or, in absence of the "legacy interface" as mentioned above, to an RS232 to USB converter device. In both cases the Windows HyperTerminal program will be able to exchange data with the Explorer16 board with a basic configuration setting.

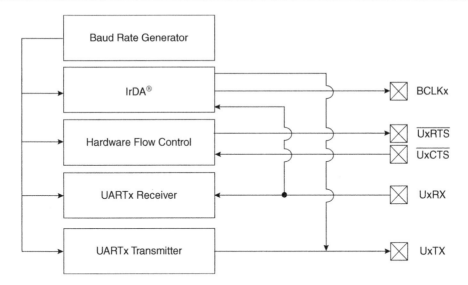

Figure 9.1: Simplified UART modules block diagram.

The first step is the definition of the transmission parameters. The options include:

- Baud rate

- Number of data bits

- Parity bit, if present

- Number of stop bits

- Handshake protocol

For our demo we will choose the fast and convenient configuration:115200, 8, N, 1, CTS/RTS—that is:

- 115,200 baud

- 8 data bits

- No parity

- 1 stop bit

- Hardware handshake using the CTS and RTS lines

UART Configuration

Use the New Project Setup checklist to create a new project called **Serial** and a new source file, similarly called **serial.c**. We will start by adding a few useful I/O definitions to help us control the hardware handshake lines:

```
/*
** Asynchronous Serial Communication
** UART2 RS232 asynchronous communication demonstration code
*/

#include <p32xxxx.h>

// I/O definitions for the Explorer16
#define CTS      _RF12                 // Clear To Send, input
#define RTS      _RF13                 // Request To Send, output
#define TRTS     TRISFbits.TRISF13     // Tris control for RTS pin
```

The hardware handshake is especially necessary in communicating with a Windows terminal application, since Windows is a multitasking operating system and its applications can sometimes experience long delays that would otherwise cause significant loss of data. We will use one I/O pin as an input (RF12 on the Explorer 16 board) to sense when the terminal is ready to receive a new character (Clear To Send), and one I/O pin as an output (RF13 on the Explorer 16 board) to advise the terminal when our application is ready to receive a character (Request To Send).

To set the baud rate, we get to play with the Baud Rate Generator (U2BREG), a 16-bit counter that feeds on the peripheral bus clock. From the device datasheet we learn that in the normal mode of operation (BREGH=0) it operates off a 1:16 divider versus a high-speed mode (BREGH=1) where its clock operates off a 1:4 divider. A simple formula, published on the datasheet, allows us to calculate the ideal setting for our configuration (see Equation 9.1).

Equation 9.1. UART Baud Rate with UxBREG = 1.

$$Baud\ Rate = \frac{F_{PB}}{4 \cdot (U \times BRG + 1)}$$

$$U \times BRG = \frac{F_{PB}}{4 \cdot Baud\ Rate} - 1$$

In our case, Equation 9.1 translates to the following expression:
U2BREG = (36,000,000/4/115,200) − 1 = 77.125

To decide how to best round out the result, we need a 16-bit integer after all. We will use the reverse formula to calculate the actual baud rate and determine the percentage error:
Error = ((Fpb/4/(U2BREG+1)) − baud rate)/baud rate %

Rounding up to a value of 77, we obtain an actual baud rate of 115,384 baud with an error of just 0.2 percent—well within acceptable tolerance. However, with a value of 78 we obtain 113,924 baud, a larger 1.1 percent error but still within the acceptable tolerance range for a standard RS232 port (±2 percent).

We can therefore define the constant BRATE as:

```
#define BRATE       77      // 115,200 Bd (BREGH=1)
```

Two more constants will help us define the initialization values for the UART2 main control registers called U2MODE and U2STA (see Figure 9.2).

U-0	U-0	U-0	U-0	U-0	U-0	U-0	U-0
—	—	—	—	—	—	—	—
bit 31							bit 24

U-0	U-0	U-0	U-0	U-0	U-0	U-0	U-0
—	—	—	—	—	—	—	—
bit 23							bit 16

R/W-0	R/W-0	R/W-0	R/W-0	R/W-0	R/W-0	R/W-0	R/W-0
ON	FRZ	SIDL	IREN	RTSMD	—	UEN<1:0>	
bit 15							bit 8

R/W-0	R/W-0	R/W-0	R/W-0	R/W-0	R/W-0	R/W-0	R/W-0
WAKE	LPBACK	ABAUD	RXINV	BRGH	PDSEL<1:0>		STSEL
bit 7							bit 0

Figure 9.2: The UxMODE control registers.

The initialization value for U2MODE will include the BREGH bit, the number of stop bits, and the parity bit settings.

```
#define U_ENABLE 0x8008 // enable,BREGH=1, 1 stop, no parity
```

The initialization for U2STA will enable the transmitter and clear the error flags (see Figure 9.3):

```
#define U_TX  0x0400  // enable tx, clear all flags
```

U-0	U-0	U-0	U-0	U-0	U-0	U-0	R/W-0
—	—	—	—	—	—	—	ADM_EN

bit 31 — bit 24

R/W-0	R/W-0	R/W-0	R/W-0	R/W-0	R/W-0	R/W-0	R/W-0
ADDR<7:0>							

bit 23 — bit 16

R/W-0	R/W-0	R/W-0	R/W-0	R/W-0	R/W-0	R-0	R-1
UTXISEL<1:0>		TXINV	RXEN	TXBRK	TXEN	TXBF	TRMT

bit 15 — bit 8

R/W-0	R/W-0	R/W-0	R-1	R-0	R-0	R/C-0	R-0
URXISEL<1:0>		ADDEN	RIDLE	PERR	FERR	OERR	RXDA

bit 7 — bit 0

Figure 9.3: The UxSTA control registers.

Using the constants defined above, let's initialize the UART2module, the baud rate generator, and the I/O pins used for the handshake:

```
void initU2( void)
{
    U2BRG = BRATE;          // initialize the baud rate generator
    U2MODE = U_ENABLE;      // initialize the UART module
    U2STA = U_TX;           // enable the Transmitter
    TRTS = 0;               // make RTS an output pin
    RTS = 1;                // set RTS default status (not ready)
} // initU2
```

Sending and Receiving Data

Sending a character to the serial port is a three-step procedure:

1. Make sure that the terminal (PC running Windows HyperTerminal) is ready. Check the Clear to Send (CTS) line. CTS is an active low signal; that is, while it is high, we better wait patiently.

2. Make sure that the UART is not still busy sending some previous data. PIC32 UARTs have a four-level-deep FIFO buffer, so all we need to do is wait until at least the top level frees up; in other words, we need to check for the transmit buffer full flag UTXBF to be clear.

3. Finally, transfer the new character to the UART transmit buffer (FIFO).

All of the above can be nicely packaged in one short function that we will call putU2(), respecting a rule that wants all C language I/O libraries (stdio.h) to use the put- prefix to offer a series of character output functions such as putchar(), putc(), fputc() and so on:

```c
int putU2( int c)
{
  while ( CTS);                 // wait for !CTS, clear to send
  while ( U2STAbits.UTXBF);     // wait while Tx buffer full
  U2TXREG = c;
  return c;
} // putU2
```

To receive a character from the serial port, we will follow a very similar sequence:

1. Alert the terminal that we are ready to receive by asserting the RTS signal (active low).

2. Patiently wait for something to arrive in the receive buffer, checking the URXDA flag inside the UART2 status register U2STA.

3. Fetch the character from the receive buffer (FIFO).

Again, all of the above steps can be nicely packaged in one last function:

```c
char getU2( void)
{
  RTS=0;                        // assert Request To Send !RTS
  while ( !U2STAbits.URXDA);    // wait for a new char to arrive
  RTS=1;
  return U2RXREG;               // read char from receive buffer
} // getU2
```

Testing the Serial Communication Routines

To test our serial port control routines, we can now write a small program that will initialize the serial port, send a prompt, and let us type on the terminal keyboard while echoing each character back to the terminal screen:

```
main()
{

  char c;

  // 1. init the UART2 serial port
  initU2();

  // 2. prompt
  putU2( '>');

  // 3. main loop
  while ( 1)
  {

     // 3.1 wait for a character
     c = getU2();

     // 3.2 echo the character
     putU2( c);

  } // main loop
} // main
```

1. Build the project first, then follow the standard checklist to activate the Debugger and to program the Explorer 16.

2. Connect the serial cable to the PC (directly or via a Serial-to-USB converter) and configure HyperTerminal for the same communication parameters: 115200, n, 8, 1, RTS/CTS on the available COM port.

3. Click the HyperTerminal **Connect** button to start the terminal emulation.

4. Select **Run** from the Debugger menu to execute the demonstration program.

Note

I recommend, for now, that you do not attempt to single-step or use breakpoints or the RunToCursor function when using the UART! See the "Tips & Tricks" section at the end of the chapter for a detailed explanation.

If HyperTerminal is already set to provide an echo for each character sent, you will see double—literally! To disable this functionality, first hit the **Disconnect** button on HyperTerminal. Then select **File | Properties** to open the Properties dialog box, and select the **Settings Pane** tab (see Figure 9.4). This will be a good opportunity to set a couple more options that will come in handy in the rest of the exploration.

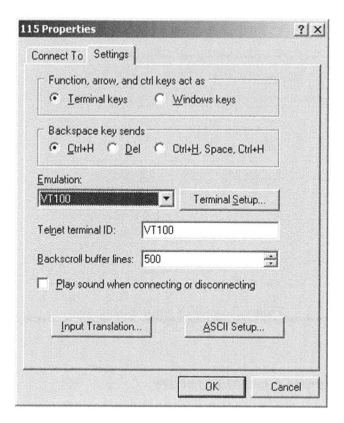

Figure 9.4: The HyperTerminal Properties dialog box Settings pane.

5. Select the **VT100 terminal** emulation mode so that a number of commands (activated by special "escape" strings) will become available and will give us more control over the cursor position on the terminal screen.

6. Select **ASCII Setup** to complete the configuration. In particular, make sure that the **Echo typed characters locally** function is *not* checked (this will immediately improve your . . . vision). See Figure 9.5.

7. Also check the **Append line feeds to incoming line ends** option. This will make sure that every time an ASCII carriage return (\r) character is received, an additional line feed (\n) character is inserted automatically.

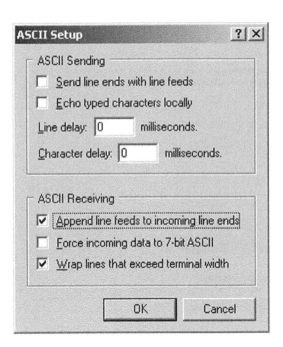

Figure 9.5: ASCII Setup dialog box.

Building a Simple Console Library

To transform our demo project in a proper terminal console library that could become handy in future projects, we need only a couple more functions that will complete the puzzle: a function to print an entire (zero-terminated) string and a function to input a full text line. Printing a string is, as you can imagine, the simple part:

```
int puts ( char *s)
{
  while ( *s)              // loop until *s == '\0', end of string
    putU2 ( *s++);         // send char and point to the next one
  putU2 ( '\r');           // terminate with a cr / line feed
} // puts
```

It is just a loop that keeps calling the putU2 () function to send, one after the other, each character in the string to the serial port.

Reading a text string from the terminal (console) into a string buffer can be equally simple, but we have to make sure that the size of the buffer is not exceeded (should the user type a really long string), and we have to convert the carriage return character at the end of the line in a proper \0 character for the string termination:

```
char *getsn ( char *s, int len)
{
  char *p = s;            // copy the buffer pointer
  do{
    *s = getU2 ();        // wait for a new character
    if ( *s=='\r')        // end of line, end loop
      break;

    s++;                  // increment buffer pointer
    len--;
  } while ( len>1 );      // until buffer full

  *s='\0';                // null terminate the string

  return p;               // return buffer pointer
} // getsn
```

In practice, the function as presented would prove very hard to use. There is no echo of what is being typed, and the user has no room for error. Make only the smallest typo and the entire line must be retyped. If you're like me, you make a lot of typos all the time, and the most battered key on your keyboard is the Backspace key. A better version of the getsn() function must include character echo and at least provisions for the Backspace key to perform basic editing. It really takes only a couple more lines of code. The echo is quickly added after each character is received. The Backspace character (identified by the ASCII code 0x8) is decoded to move the buffer pointer one character backward (as long as we are not at the beginning of the line already). We must also output a specific sequence of characters to visually remove the previous character from the terminal screen:

```
char *getsn( char *s, int len)
{

  char *p = s;            // copy the buffer pointer
  int cc = 0;             // character count
  do{

    *s = getU2();         // wait for a new character
    putU2( *s);           // echo character

    if (( *s == BACKSPACE)&&( s>p))
    {

      putU2( ' ');        // overwrite the last character
      putU2( BACKSPACE);
      len++;
      s--;                // back the pointer
      continue;
    }
    if ( *s=='\n')        // line feed, ignore it
      continue;
    if ( *s=='\r')        // end of line, end loop
      break;

    s++;                  // increment buffer pointer
    len--;
  } while ( len>1 );      // until buffer full

  *s = '\0';              // null terminate the string

  return p;               // return buffer pointer
} // getsn
```

Put all the functions in a separate file that we will call **conU2.c**. Then create a small header file **conU2.h**, to decide which functions (prototypes) and which constants to publish and make visible to the outside world:

```
/*
** CONU2.h
** console I/O library for Explorer16 board
*/

// I/O definitions for the Explorer16
#define CTS     _RF12    // Cleart To Send, in, HW handshake
#define RTS     _RF13    // Request To Send, out, HW handshake
#define BACKSPACE 0x8    // ASCII backspace character code

// init the serial port UART2, 115200, 8, N, 1, CTS/RTS
void initU2( void);

// send a character to the serial port
int putU2( int c);

// wait for a new character to arrive to the serial port
char getU2( void);

// send a null terminated string to the serial port
int puts( char *s);

// receive a null terminated string in a buffer of len char
char * getsn( char *s, int n);
```

Testing a VT100 Terminal

Since we have enabled the VT100 terminal emulation mode (see the previous HyperTerminal settings), we now have a few commands available to better control the terminal screen and cursor position, such as:

- `clrscr()`, to clear the terminal screen

- `home()`, to move the cursor to the home position in the upper-left corner of the screen

These commands are performed by sending so-called "escape sequences," defined in the ECMA-48 standard (also ISO/IEC 6429 and ANSI X3.64), also referred to as ANSI escape codes. They all start with the characters ESC (ASCII 0x1b) and the character [(left square bracket).

```
// useful macros for VT100 terminal emulation
#define clrscr() putsU2( "\x1b[2J")
#define home()   putsU2( "\x1b[1,1H")
```

To test the console library we can now write a small program that will:

1. Initialize the serial port

2. Clear the terminal screen

3. Send a welcome message/banner

4. Send a prompt character

5. Read a full line of text

6. Print the text on a new line

Save the following code in a new file that we will call **CONU2test.c**:

```
/*
** CONU2 Test
** UART2 RS232 asynchronous communication demonstration code
*/
// configuration bit settings, Fcy=72MHz, Fpb=36MHz
#pragma config POSCMOD=XT, FNOSC=PRIPLL
#pragma config FPLLIDIV=DIV_2, FPLLMUL=MUL_18, FPLLODIV=DIV_1
#pragma config FPBDIV=DIV_2, FWDTEN=OFF, CP=OFF, BWP=OFF

#include <p32xxxx.h>
#include "CONU2.h"

#define BUF_SIZE 128

main()
{
  char s[BUF_SIZE];

  // 1. init the console serial port
  initU2();
```

```
  // 2. text prompt
  clrscr();
  home();
  puts( "Exploring the PIC32!");

  // 3. main loop
  while ( 1)
  {

    // 3.1 read a full line of text
    getsn( s, sizeof(s));
    // 3.2 send a string to the serial port
    puts( s);
  } // main loop
}// main
```

1. Create a new project using the New Project checklist, and add all three files conU2.h, conU2.c, and conU2test.c to the project and build all.

2. Use the appropriate debugger checklist to connect and program the Explorer 16 board.

3. Test the editing capabilities of the new console library you just completed.

The Serial Port as a Debugging Tool

Once you have a small library of functions to send and receive data to a console through the serial port, you have a new powerful debugging tool available. You can strategically position calls to print functions to present the content of critical variables and other diagnostic information on the terminal. You can easily format the output to the most convenient format for you to read. You can add input functions to set parameters that can help better test your code, or you can use the input function to simply pause the execution and give you time to read the diagnostic output when required. This is one of the oldest debugging tools, effectively used since the first computer was invented and connected to a teletypewriter.

The Matrix Project

To finish today's exploration on a more entertaining note, let's develop a new demo project that we will call the *matrix*. The intent is that of testing the speed of the serial port and the PC terminal emulation by sending large quantities of text to the terminal and clocking its performance. The only problem is that we don't (yet) have access to a large storage device from which to extract some meaningful content to send to the terminal. So the next best option is that of "generating" some content using a pseudo-random number generator. The stdlib.h library offers a convenient rand() function that returns a positive integer between 0 and RAND_MAX (which, in the MPLAB C32 implementation, can be verified to be equal to the largest signed 32-bit integer available).

Using the "reminder of" operator (denoted by the % symbol in C language), we can reduce its output to any smaller integer range and, in our example, produce a subset of values that corresponds to ASCII printable characters only. The following statement, for example, will produce only characters in the range from 33 to 127:

```
putU2( 33+(rand()%94));
```

To generate a more appealing and entertaining output, especially if you happened to watch the movie *The Matrix*, we will present the (random) content by columns instead of rows. We will use the pseudo-random number generator to change the content and the "length" of each column as we periodically redraw the entire screen:

```
/*
** The UART Matrix
**
*/
#include <p32xxxx.h>
#include <stdlib.h>

#include "CONU2.h"

#define COL     40
#define ROW     23
#define DELAY   3000

main()
{
   int v[40];      // length of each column
   int i, j, k;

   // 1. initializations
   T1CON = 0x8030; // TMR1 on, prescale 256, int clock (Tpb)
   initU2();       // initialize UART (115200, 8N1, CTS/RTS)
   clrscr();       // clear the terminal (VT100 emulation)
```

```
// 2. randomize the sequence
getU2();          // wait for a character input
srand( TMR1);    // use the current timer value as seed
// 3. init each column length
for( j = 0; j<COL; j++)
        v[j]=rand()%ROW;
// 4. main loop
while( 1)
{
  home();

  // 4.1 refresh the entire screen, one row at a time
  for( i=0; i<ROW; i++)
  {

    // 4.1.1 refresh one column at a time
    for( j=0; j<COL; j++)
    {

      // update each column
      if ( i < v[j])
        putU2( 33 + (rand()%94));
      else
        putU2( ' ');

      // additional column spacing
      putU2( ' ');
    } // for j
    // 4.1.2 empty string, advance to next line
    puts("");
  } // for i

  // 4.2 randomly increase or reduce each column length
  for( j=0; j<COL; j++)
  {
    switch ( rand()%3)
    {
      case 0: // increase length
              v[j]++;
              if (v[j]>ROW)
                v[j]=ROW;
          break;
```

```
        case 1: // decrease length
          v[j]--;
          if (v[j]<1)
              v[j]=1;
          break;

      default:// unchanged
          break;
      } // switch
    } // for j
    } // main loop
} // main
```

Forget the performance—watching this code run is fun. It is too fast anyway; in fact, you will have to add a small delay loop (inside the `for` loop in 4.1) to make it more pleasing to the eye:

```
// 4.1.0 delay to slow down the screen update
TMR1 = 0;
while( TMR1≤DELAY);
```

> **Note**
>
> Remember to take the blue pill next time!

Debriefing

In this lesson we developed a small console I/O library while reviewing the basic functionality of the UART module for operation as an RS232 serial port. We connected the Explorer 16 board to a VT100 terminal (emulated by Windows HyperTerminal). We will take advantage of this library in the next few lessons to provide us with a new debugging tool and possibly as a user interface for more advanced projects.

Notes for the C Experts

I am sure that, at this point, you are wondering about the possibility of using the more advanced library functions defined in the stdio.h library, such as `printf()`, to direct the `stdout` output stream to the UART2 peripheral. Not only is this possible, but you can consider it done!

In addition, the stdio.h library defines two helper functions, `_mon_putc()` and `_mon_getc()`, that can be used to customize the behavior of the standard library. They are declared with the attribute `weak`, which means that the MPLAB C32 linker won't complain about you trying to redefine them. In fact, you are supposed to redefine them in order to implement new functionalities, such as using the SPI port as your input/output stream or redirecting the output to an LCD display and so on.

> **Note**
>
> Remember that whether you customize the stdio.h functions or not, you are always responsible for the proper interface initialization. So before the first call to `printf()`, make sure the UART2 or your communication peripheral of choice is enabled and the baud rate is set correctly.

Notes for the PIC® Microcontroller Experts

Sooner or later, every embedded control designer will have to come to terms with the USB bus. If, for now, a small "dongle" (converting the serial port to a USB port) can be a reasonable solution, eventually you will find opportunities and designs that will actually benefit from the superior performance and compatibility of the USB bus. Several 8- and 16-bit PIC microcontroller models already incorporate a USB Serial Interface Engine (SIE) as a standard communication interface. Microchip offers a free USB software stack with drivers and ready-to-use solutions for the most common classes of application.

One of them, known as the Communication Device Class (CDC), makes the USB connection look completely transparent to the PC application so that even HyperTerminal cannot tell the difference. Most important, you will not need to write and/or install any special Windows drivers. When writing the application in C, you won't even notice the difference, if not for the absence of a need to specify any communication parameter. In USB there is no baud rate to set, no parity to calculate, no port number to select (incorrectly), and the communication speed is so much higher . . .

Tips & Tricks

As we mentioned during one of the early exercises presented in this lesson, single-stepping through a routine that enables and uses the UART to transmit and receive data from the HyperTerminal program is a bad idea. You will be frustrated seeing the

HyperTerminal program misbehave and/or simply lock up and ignore any data sent to it without any apparent reason.

To understand the problems, you need to know more about how the MPLAB ICD2 in circuit debugger operates. After executing each instruction when in single-step mode or upon encountering a breakpoint, the ICD2 debugger not only stops the CPU execution but also "freezes" all the peripherals. It freezes them, as in dead-cold-ice all of a sudden; not a single clock pulse is transmitted through their digital veins. When this happens to a UART peripheral that is busy in the middle of a transmission, the output serial line (TX) is also frozen in the current state. If a bit was being shifted out in that precise instant, and specifically if it was a 1, the TX line will be held in the "break" state (low) indeterminately. The HyperTerminal program, on the other side, would sense this permanent "break" condition and interpret it as a line error. It will assume that the connection is lost and it will disconnect. Since HyperTerminal is a pretty "basic" program, it will not bother letting you know what is going on; it will not send a beep, not an error message, nothing—it will just lock up!

If you are aware of the potential problem, this is not a big deal. When you restart your program with the ICD2, you will only have to remember to click the HyperTerminal Disconnect button first and then the Connect button again. All operations will resume normally.

Exercises

1. Write a console library with buffered I/O (using interrupts) to minimize the impact on program execution (and debugging).

2. Develop a simple command-line interpreter that recognizes a small defined set of keywords to assist in debugging by inspecting and modifying the value of RAM memory locations and/or providing hexadecimal memory dumps of the Flash memory.

Books

Axelson, J., *Serial Port Complete*, second edition (Lakeview Research, Madison, WI, 2007). This new edition was published just in time for me to include it here. The author is most famous for her *USB Complete*" book (see below), considered *the* reference book for all embedded-control programmers. Over time she has developed and maintained a whole series completely dedicated to serial and parallel communication interfaces.

Axelson, J., *USB Complete*, third edition (Lakeview Research, Madison, WI, 2005). By the time you read this book, most probably new models of the PIC32MX family will have been announced offering USB communication capabilities. So, I thought you might appreciate this recommendation. Jan Axelson's book has reached the third edition already. She has continued to add material at every step and still managed to keep things very simple.

Eady, F., *Implementing 802.11 with Microcontrollers: Wireless Networking for Embedded Systems Designers* (Newnes, Burlington, MA, 2005). Fred brings his humor and experience in embedded programming to make even wireless networking seem easy.

Links

http://*en.wikipedia.org/wiki/ANSI_escape_code*. This is a link to the complete table of ANSI escape codes as implemented by the VT100 HyperTerminal emulation.

www.cs.utk.edu/~shuford/terminal/dec.html. This is a real dive into a piece of the history of computers. I used these terminals; does this make me look old?

Glass = Bliss

The Plan

I would be surprised if you told me that on your desk next to your PC there was still a large and bulky CRT computer monitor. In a matter of a few years the entire personal computer industry has shifted to the new technology: flat LCD panels of ever larger size and higher resolution. In the embedded-control world, something similar has happened. LED seven-segment displays are so 1990s! Small LCD displays have become ubiquitous and, besides consuming a fraction of the power of their LED counterparts, they provide alphanumeric output (i.e., they support text) and, ever more often, graphics as well. But wait, maybe there is already another generation of organic LED displays (OLEDs) just around the corner and ready to demand revenge.

In this lesson, we will learn how to interface with a small and inexpensive LCD alphanumeric display module. This project will be a good excuse for us to learn and use the Parallel Master Port (PMP), a flexible parallel interface available on all PIC32MX microcontrollers.

Preparation

In addition to the usual software tools, including the MPLAB® IDE, the MPLAB C32 compiler, and the MPLAB SIM simulator, this lesson will require only the use of the Explorer 16 demonstration board and your In-Circuit Debugger of choice (PIC32 Starter kit, ICD2, REAL ICE, or the like).

The Exploration

The Explorer 16 board can accommodate three different types of dot-matrix, alphanumeric LCD display modules and one type of graphic LCD display module. By default, it comes with a simple "2-rows by 16-character" display and a 3V alphanumeric LCD module (most often a Tianma TM162JCAWG1) compatible with the industry-standard HD44780 controllers. These LCD modules are complete display systems composed of the LCD glass, column, and row multiplexing drivers; power supply circuitry; and an intelligent controller, all assembled together in so-called Chip On Glass (COG) technology. Thanks to this high level of integration, the circuitry required to control the dot-matrix display is greatly simplified. Instead of the hundreds of pins required by the column-and-row drivers to directly control each pixel, we can interface to the module with a simple 8-bit parallel bus using just 11 I/Os.

On alphanumeric modules (see Figure 10.1) in particular, we can directly place ASCII character codes into the LCD module controller RAM buffer (known as the Display Data RAM buffer, or DDRAM). The output image is produced by an integrated character generator (a table) using a 5×7 grid of pixels to represent each character. The table (see Figure 10.2) typically contains an extended ASCII character set in the sense that it has been somewhat merged with a small subset of Japanese Kata Kana characters as well some symbols of common use. While the character generator table is mostly implemented in the display controller ROM, various display models offer the possibility to extend the character set by modifying/creating new characters (from 2 to 8) accessing a second small internal RAM buffer (the Character Generator RAM buffer, or CGRAM).

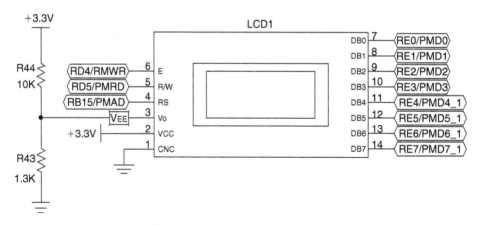

Figure 10.1: Default alphanumeric LCD module connections.

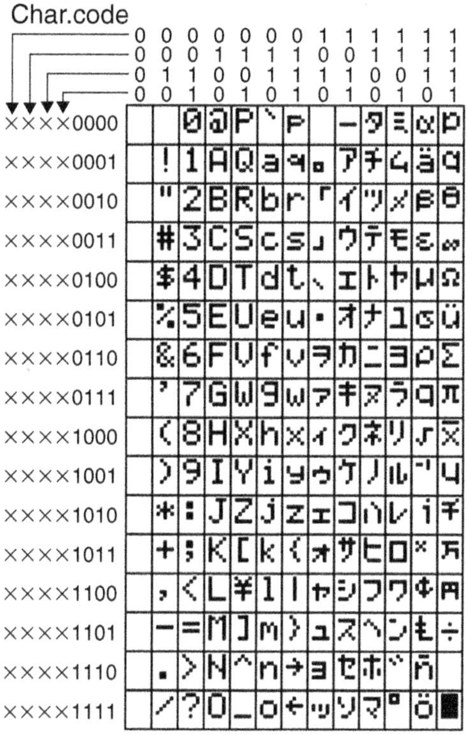

Char.code

Figure 10.2: Character generator table used by HD44780-compatible LCD display controllers.

HD44780 Controller Compatibility

As mentioned, the 2 × 16 LCD module used in the Explorer 16 board is one among a vast selection of LCD display modules available on the market in configurations ranging from 1 to 4 lines of 8, 16, 20, 32, and up to 40 characters each and that are compatible with the original HD44780 chipset, today considered an industry standard.

The HD44780 compatibility means that the integrated controller contains just two separately addressable 8-bit registers: one for ASCII data and one for commands/status. The standard sets of commands shown in Tables 10.1 and 10.2 can be used to set up and control the display.

Thanks to this commonality, any code we will develop to drive the LCD on the Explorer 16 board will be immediately available for use with any of the other HD44780-compatible alphanumeric LCD display modules.

Table 10.1: The HD44780 instruction set.

Instruction	Code										Description	Execution Time
	RS	R/W	DB7	DB6	DB5	DB4	DB3	DB2	DB1	DB0		
Clear display	0	0	0	0	0	0	0	0	0	1	Clears display and returns cursor to the home position (address 0).	1.64 mS
Cursor home	0	0	0	0	0	0	0	0	1	*	Returns cursor to home position (address 0). Also returns display being shifted to the original position. DDRAM contents remain unchanged.	1.64 mS
Entry mode set	0	0	0	0	0	0	0	1	I/D	S	Sets cursor move direction (I/D), specifies to shift the display (S). These operations are performed during data read/write.	40 uS
Display on/off control	0	0	0	0	0	0	1	D	C	B	Sets on/off of all display (D), cursor on/off (C), and blink of cursor position character (B).	
Cursor/ display shift	0	0	0	0	0	1	S/C	R/L	*	*	Sets cursor move or display shift (S/C), shift direction (R/L). DDRAM contents remain unchanged.	40 uS
Function set	0	0	0	0	1	DL	N	F	*	*	Sets interface data length (DL), number of display lines (N), and character font (F).	
Set CGRAM address	0	0	0	1	CGRAM address						Sets the CGRAM address. CGRAM data is sent and received after this setting.	40 uS
Set DDRAM address	0	0	1	DDRAM address							Sets the DDRAM address. DDRAM data is sent and received after this setting	40 uS
Read busy flag and address counter	0	1	BF	CGRAM/DDRAM address							Reads busy flag (BF), indicating internal operation is being performed, and reads CGRAM or DDRAM address counter contents (depending on previous instruction).	0 uS
Write to CGRAM or DDRAM	1	0	write data								Writes data to CGRAM or DDRAM.	40 uS
Read from CGRAM or DDRAM	1	1	read data								Reads data from CGRAM or DDRAM.	40 uS

Table 10.2: HD44780 command bits.

Bit Name	Setting/Status	
I/D	0 = Decrement cursor position	1 = Increment cursor position
S	0 = No display shift	1 = Display shift
D	0 = Display off	1 = Display on
C	0 = Cursor off	1 = Cursor on
B	0 = Cursor blink off	1 = Cursor blink on
S/C	0 = Move cursor	1 = Shift display
R/L	0 = Shift left	1 = Shift right
DL	0 = 4-bit interface	1 = 8-bit interface
N	0 = 1/8 or 1/11 Duty (1 line)	1 = 1/16 Duty (2 lines)
F	0 = 5 × 7 dots	1 = 5 × 10 dots
BF	0 = Can accept instruction	1 = Internal operation in progress

The Parallel Master Port

The simplicity of the 8-bit bus shared by all these display modules is remarkable. Beside the eight bidirectional data lines (which, by enabling a special "nibble" mode, could be reduced to just four for further I/O saving), there is:

- An Enable strobe line (E)

- A Read/Write selection line (R/W)

- An address line (RS) for the register selection

It would be simple enough to control the 11 I/Os by accessing manually (bit banging) the individual PORTE and PORTD pins to implement each bus sequence, but we will take this opportunity instead to explore the capabilities of a new peripheral introduced with the PIC24 architecture and enhanced in the PIC32 architecture: the Parallel Master Port (PMP). This addressable parallel port was designed to ease access to a large number of external parallel devices of common use, ranging from analog-to-digital converters, RAM buffers, ISA bus compatible interfaces, LCD display modules, and even hard disk drives and Compact Flash cards.

You can think of the PMP as a sort of flexible I/O bus added to the PIC32 architecture that relieves the microcontroller of the mundane task of managing slow external peripherals. The PMP offers:

- Eight- or 16-bit bidirectional data path

- Up to 64 k of addressing space (16 address lines)

- Six additional strobe/control lines, including:

  1. Enable

  2. Address latch

  3. Read and write (separate or combined)

  4. Chip Select (2x)

The PMP can also be configured to operate in slave mode to attach, as an addressable peripheral, to a larger microprocessor/microcontroller system.

Both bus read and bus write sequences are fully programmable so that not only can the polarity and choice of control signals be configured to match the target bus, the timing can also be finely tuned to adapt to the speed of the peripherals to which we want to interface.

Configuring the PMP for LCD Module Control

As in all other PIC32 peripherals, there is a set of control registers dedicated to the PMP configuration. The first and most important one is PMCON. You will recognize the familiar sequence of control bits common to all the modules xxCON registers (see Figure 10.3).

The list of control registers that we will need to initialize is a bit longer this time and also includes PMMODE, PMADDR, PMSTAT, and PMAEN. They are packed with powerful options and they all require your careful consideration. Instead of proceeding through a lengthy review of each and every one of them, I will list only the key choices required specifically by the LCD module interface:

- PMP enabled

- Fully demultiplexed interface (separate data and address lines will be used)

- Enable strobe signal (on pin RD4)

- Read signal (on pin RD5)

U-0	U-0	U-0	U-0	U-0	U-0	U-0	U-0
—	—	—	—	—	—	—	—
bit 31							bit 24

U-0	U-0	U-0	U-0	U-0	U-0	U-0	U-0
—	—	—	—	—	—	—	—
bit 23							bit 16

R/W-0	R/W-0	R/W-0	R/W-0	R/W-0	U-0	R/W-0	R/W-0
ON	FRN	SIDL	ADRMUX1	ADRMUX0	—	PTWREN	PTRDEN
bit 15							bit 8

R/W-0	R/W-0	R/W-0	R/W-0	R/W-0	U-0	R/W-0	R/W-0
CSF1	CSF0	ALP	CS2P	CS1P	—	WRSP	RDSP
bit 7							bit 0

Figure 10.3: PMCON control register.

- Enable strobe active high

- Read active high, write active low

- Master mode with read and write signals on the same pin (RD5)

- Eight-bit bus interface (using PORTE pins)

- Only one address bit is required, so we will choose the minimum configuration, including PMA0 (on pin RB15) and PMA1 (unused)

Also, considering that the typical LCD module is an extremely slow device, we will better select the most generous timing, adding the maximum number of wait states allowed at each phase of a read or write sequence:

- 4 × Tpb wait for data set up before read/write

- 15 × Tpb wait between R/W and enable

- 4 × Tpb wait data set up after enable

A Small Library of Functions to Access an LCD Display

Create a new project called **Liquid** using the New Project checklist and a new source file **liquid.c** to start creating a small LCD interface library.

We will start writing the LCD initialization routine first. It is natural to start with the initialization of the PMP port key control registers:

```
void LCDinit(void)
{

  // PMP initialization
  PMCON = 0x83BF;         // Enable the PMP, long waits
  PMMODE = 0x3FF;         // Master Mode 1
  PMPEN = 0x0001;         // PMA0 enabled
```

After these steps, we are able to communicate with the LCD module for the first time, and we can follow a standard LCD initialization sequence as recommended by the manufacturer. The initialization sequence must be timed precisely (see the HD44780 instruction set for details) and cannot be initiated before at least 30 ms have been granted to the LCD module to proceed with its own internal initialization (power on reset) sequence. For simplicity and safety, we will hardcode a delay in the LCD module initialization function, and we will use the Timer1 module to obtain simple but precise timing loops for all subsequent steps:

```
// init TMR1
  T1CON = 0x8030;         // Enabled,1:256 Fpb, 1 tick ~ 6us

  // wait for >30ms
  TMR1 = 0; while(TMR1<6000); // 6000 x 6us = 36ms
```

For our convenience, we will also define a couple of constants that will hopefully help us make the following code more readable:

```
#define LCDDATA 1        // RS = 1 ; access data register
#define LCDCMD 0         // RS = 0 ; access command register
#define PMDATA PMDIN1    // PMP data buffer
```

To send each command to the LCD module, we will select the command register (setting the address PMA0 = RS = 0) first. (see Figure 10.4).

Then we will start a PMP write sequence by depositing the desired command byte in the PMP data output buffer:

```
PMADDR = LCDCMD;         // command register (ADDR = 0)
PMDATA = 0x38;           // set: 8-bit interface, 2 lines, 5x7
```

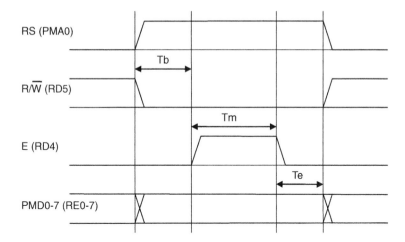

Figure 10.4: PMP-to-LCD display 8-bit interface write command sequence.

The PMP will perform the complete bus write sequence as follows:

1. The address will be published on the PMP address bus (PMA0).

2. The content of PMDATA will be published on the PMP data bus (PMD0-PMD7).

3. The R/W signal will be asserted low (RD5).

4. After $4 \times$ Tpb (Tb) the strobe signal E will be asserted high.

5. After $15 \times$ Tpb (Tm) the Enable strobe will be de-asserted.

6. After $4 \times$ Tpb (Te) the data will be removed from the bus.

Notice how this sequence is quite long as it extends for $20 \times$ Tpb or more than 0.5 us after the PIC32 has initiated it. In other words, the PMP will still be busy executing part of this sequence while the PIC32 will have already executed at least another 40 instructions or more. Since we are going to wait for a considerably longer amount of time anyway ($>$40 us) to allow the LCD module to execute the command, we will not worry about the time the PMP requires to complete the command; we'll just have to wait patiently.

```
TMR1 = 0; while( TMR1<8);    // 8 x 6us = 48us
```

We will then proceed similarly with the remaining steps of the LCD module initialization sequence:

```
PMDATA = 0x0c;                   // ON, no cursor, no blink
TMR1 = 0; while( TMR1<8);        // 8 x 6us = 48us
PMDATA = 0x01;                   // clear display
TMR1 = 0; while( TMR1<300);      // 300 x 6us = 1.8ms

PMDATA = 0x06;                   // increment cursor, no shift
TMR1 = 0; while( TMR1<300);      // 300 x 6us = 1.8ms
```

After the LCD module initialization, things will get a little easier and the timing loops will no longer be necessary, because we will be able to use the LCD module Read Busy Flag command. This will tell us whether the integrated LCD module controller has completed the last command and is ready to receive and process a new one. To read the LCD status register containing the LCD busy flag, we will need to instruct the PMP to execute a bus read sequence. This is a two-step process: First, we initiate the read sequence by reading (and discarding) the contents of the PMP data buffer (PMPDIN) a first time. When the PMP sequence is completed, the data buffer will contain the actual value read from the bus, and we will read its contents from the PMP data buffer again. But how can we tell when the PMP read sequence is complete?

Simple: We can check the PMP busy flag (PMMODEbits.BUSY) in the PMMODE control register (see Figure 10.5).

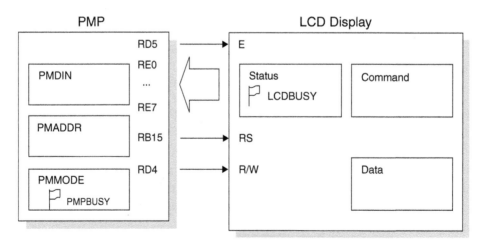

Figure 10.5: PMP-to-LCD connection block diagram.

In summary, to check the LCD module busy flag, we will need to check the PMP busy flag first to make sure that any previous command is completed, issue a read command, wait for the PMP busy flag again, and only at this point will we gain access to the actual LCD module status register contents, including the LCD busy flag.

By passing the register address as a parameter to the read function, we will obtain a more generic function that will be able to read the LCD status register or the data register, as in the following code:

```
char readLCD( int addr)
{

    int dummy;
    while( PMMODEbits.BUSY);    // wait for PMP to be available
    PMADDR = addr;              // select the command address
    dummy = PMDATA;            // init read cycle, dummy read
    while( PMMODEbits.BUSY);    // wait for PMP to be available
    return( PMDATA);           // read the status register

} // readLCD
```

The LCD module status register contains two pieces of information: the LCD busy flag and the LCD RAM pointer current value. We can use two simple macros, busyLCD() and addrLCD(), to split the two pieces and a third one, getLCD(), to access the data register:

```
#define busyLCD() readLCD( LCDCMD) & 0x80
#define addrLCD() readLCD( LCDCMD) & 0x7F
#define getLCD() readLCD( LCDDATA)
```

Using the busyLCD() function we can create a function to write data or commands to the LCD module:

```
void writeLCD( int addr, char c)
{
    while( busyLCD());
    while( PMMODEbits.BUSY);    // wait for PMP to be available
    PMADDR = addr;
    PMDATA = c;
} // writeLCD
```

A few additional macros will help complete the library:

- putLCD() will send ASCII data to the LCD module:

```
#define putLCD( d)   LCDwrite( LCDDATA, (d))
```

- cmdLCD() will send generic commands to the LCD module:

```
#define cmdLCD( c)   writeLCD( LCDCMD, (c))
```

- homeLCD() will reposition the cursor on the first character of the first row:

```
#define homeLCD()   writeLCD( LCDCMD, 2)
```

- clrLCD() will clear the entire contents of the display:

```
#define clrLCD()   writeLCD( LCDCMD, 1)
```

And finally, for our convenience, we might want to add putsLCD(), a function that will send an entire null terminated string to the display module:

```
void putsLCD( char *s)
{
  while( *s)
    putLCD( *s++);
}//putsLCD
```

Let's put all of our work together, adding a short main function:

```
main( void)
{

  // initializations
  initLCD();

  // put a title on the first line
  putsLCD( "Exploring        ");

  // put the cursor on the second line (addr 0x40)
  cmdLCD( 0x80 | 0x40);
  putsLCD( "      the PIC32");

  // main loop, empty for now
  while ( 1)
  {
  }
} // main
```

If all went well, after building the project and programming the Explorer 16 board with the debugger of choice, you will now have the great satisfaction of seeing the title string showing, split between the two rows of the LCD display.

Building an LCD Library and Using the PMP Library

The exact same functionality can be obtained using the specific PMP peripheral library by including the pmp.h library or simply including plib.h. Four functions in particular provide us with all the tools we need to control the PMP and dialog with the LCD display:

- mPMPOpen(), which helps us configure the parallel master port

- PMPSetAddress(), which allows us to set the address register

- PMPMasterWrite(), which initiates a basic write sequence

- mPMPMasterReadByte(), which initiates a basic read sequence and returns a byte value

Since we are at it, we will not only rewrite the code to use the more descriptive macros and definitions offered by the library, we will also rearrange the code a little so to transform it into a practical little library of its own to be used in the near future in other projects with the Explorer 16 demonstration board.

Let's start by creating a new project that we will call **LCD library**. Then let's create a new source file called **LCDlib.c**. Here is the new initLCD() function as expressed using the PMP library functions and macros:

```
void initLCD( void)
{
    // PMP initialization
    mPMPOpen( PMP_ON | PMP_READ_WRITE_EN | 3,

              PMP_DATA_BUS_8 | PMP_MODE_MASTER1 |
              PMP_WAIT_BEG_4 | PMP_WAIT_MID_15 |
              PMP_WAIT_END_4,
              0x0001,            // only PMA0 enabled
              PMP_INT_OFF);      // no interrupts used

    // wait for >30ms
    Delayms( 30);
```

```
//initiate the HD44780 display 8-bit init sequence
    PMPSetAddress( LCDCMD);       // select command register
    PMPMasterWrite( 0x38);        // 8-bit int, 2 lines, 5x7
    Delayms( 1);                  //>48us

    PMPMasterWrite( 0x0c);        // ON, no cursor, no blink
    Delayms( 1);                  //>48us

    PMPMasterWrite( 0x01);        // clear display
    Delayms( 2);                  //>1.6ms

    PMPMasterWrite( 0x06);        // increment cursor, no shift
    Delayms( 2);                  //>1.6ms
} // initLCD
```

Notice how I exaggerated the timing delays in the initialization sequence in order to use a single delay function that operates in basic increments of 1 millisecond called Delayms(). We will see shortly how and where to define it.

Here are the other core functions that will populate our simple LCD library:

```
char readLCD( int addr)
{
    PMPSetAddress( addr);             // select register
    mPMPMasterReadByte();             // initiate read sequence
    return mPMPMasterReadByte();      // read actual data
} // readLCD

void writeLCD( int addr, char c)
{
    while( busyLCD());
    PMPSetAddress( addr);             // select register
    PMPMasterWrite( c);               // initiate write sequence
} // writeLCD
```

If you found in the previous project (Liquid) that setting the cursor on the second line of the display was a bit awkward, you will agree that adding a little smarts to the putsLCD() function could be helpful. In particular, it would be nice to allow the routine to interpret a few special characters, like the *line end*, *tab*, and the *new line*, similarly to the way a serial port and/or a console are expected to.

```
void putsLCD( char *s)
{
    char c;
while( *s)
{
    switch (*s)
    {
    case '\n':                // point to second line
      setLCDC( 0x40);
      break;
    case '\r':                // home, point to first line
      setLCDC( 0);
      break;
    case '\t':                // advance next tab (8) positions
      c = addrLCD();
      while( c & 7)
      {
          putLCD( ' ');
          c++;
      }
      if ( c > 15)            // if necessary move to second line
          setLCDC( 0x40);
      break;
    default:                  // print character
      putLCD( *s);
      break;
    } //switch
    s++;
  } //while
} //putsLCD
```

This way, printing a string containing (or terminating) with the character \n (new line) will set the cursor to the beginning of the second line of the LCD display. A \r character (line end) will place the cursor back to the beginning of the first line, and \t character (tab) will produce the expected result.

A standard header and a few #include statements will complete the module:

```
/*
** LCDlib.c
*/
#include <p32xxxx.h>
#include <plib.h>
#include <explore.h>
#include <LCD.h>

#define PMDATA PMDIN
```

Save the LCDlib.c code file we just completed and then start a new source file in the MPLAB editor window. This will be the include file **LCD.h**, which will complete the library by publishing all the macros and function prototypes required:

```
/*
**
** LCD.h
**
*/
#define HLCD     16      // LCD width=16 characters
#define VLCD     2       // LCD height=2 rows

#define LCDDATA 1        // address of data register
#define LCDCMD 0         // address of command register

void initLCD( void);
void writeLCD( int addr, char c);
char readLCD( int addr);

#define putLCD( d)  writeLCD( LCDDATA, (d))
#define cmdLCD( c)  writeLCD( LCDCMD, (c))

#define clrLCD()    writeLCD( LCDCMD, 1)
#define homeLCD()   writeLCD( LCDCMD, 2)

#define setLCDG( a) writeLCD( LCDCMD, (a & 0x3F) | 0x40)
#define setLCDC( a) writeLCD( LCDCMD, (a & 0x7F) | 0x80)

#define busyLCD()   ( readLCD( LCDCMD) & 0x80)
#define addrLCD()   ( readLCD( LCDCMD) & 0x7F)
#define getLCD()    readLCD( LCDDATA)

void putsLCD( char *s);
```

Finally, to test the newly created LCD library, let's write a small new test program that we will call **LCDlib test.c**:

```
/*
** LCDlib test
**
*/
// configuration bit settings, Fcy=72 MHz, Fpb=36 MHz
#pragma config POSCMOD=XT, FNOSC=PRIPLL
#pragma config FPLLIDIV=DIV_2, FPLLMUL=MUL_18, FPLLODIV=DIV_1
#pragma config FPBDIV=DIV_2, FWDTEN=OFF, CP=OFF, BWP=OFF

#include <p32xxxx.h>

#include <LCD.h>

main()
{
    initLCD();

    clrLCD();
    putsLCD( "Exploring \nthe \tPIC32");

    while( 1);
}
```

THE EXPLORER.C LIBRARY

To help us initialize the PIC32 for maximum performance (see Day 7), vectored interrupts (see Day 5) and use the features offered by the Explorer16 board (such as the LED bar, see Day 1–3), at this point, we should start aggregating in a new small library a couple of handy functions. We will keep adding gradually new functions to it in the next few chapters but here is its first incarnation:

```
/*
** Explore.c
**
*/

#include <p32xxxx.h>
#include <plib.h>
#include <explore.h>
```

```
void initEX16( void)
{
  // 1. disable the JTAG port to make the LED bar
  // available if not using the Starter Kit
#ifndef PIC32_STARTER_KIT
  mJTAGPortEnable( 0);
#endif

  // 2. Sysytem config performance
  SYSTEMConfigPerformance( FCY);

  // 7. allow vectored interrupts
  INTEnableSystemMultiVectoredInt();   // Interrupt vectoring

  // 8. PORTA output LEDs0..6, make RA7 an input button
  LATA = 0;
  TRISA = 0xFF80;

} // initEX16
//
void _general_exception_handler( unsigned c, unsigned s)
{
  while (1);
} // exception handler
//
/*
** Simple Delay functions
**
** uses:    Timer1
** Notes:   Blocking function
*/

void Delayms( unsigned t)
{
  T1CON = 0x8000;      // enable TMR1, Tpb, 1:1
  while (t--)
  {  // t x 1ms loop
      TMR1 = 0;
      while (TMR1 < FPB/1000);
    }
} // Delayms
```

The corresponding include file: explore.h will gather as well some useful definitions and the first two functions' prototypes:

```
/*
** Explore.h
**
*/
#define FALSE    0
#define TRUE     !FALSE
#define FCY      72000000L
#define FPB      36000000L

// uncomment the following line if using the PIC32 Starter Kit
//#define PIC32_STARTER_KIT

// function prototypes
void initEX16( void);
void Delayms( unsigned);
```

Creating the include *and* lib *Directories*

To keep our files in order and our projects clean and tidy, we should apply a little discipline here and start grouping all the simple libraries we created so far two subdirectories:

- *include*, where we will put all the .h files created for the simple libraries we worked on so far, including:
    1. explore.h
    2. LCD.h
    3. conU2.h
    4. SEE.h
- *lib*, where we will put all the corresponding .c modules, including:
    1. explore.c
    2. LCDlib.c
    3. conU2.c
    4. SEE.c

From now on we will refer automatically to these modules by adding the *include* directory to the *include search path* of each new project. The sequence of steps required will be the following:

1. Open the **Build Options** dialog box (see Figure 10.6) by choosing **Project | BuildOptions. . . | Project**.

2. In the "Show Directories for" box, select **Include Search Path**.

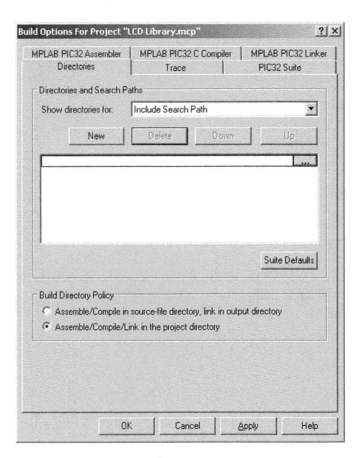

Figure 10.6: Build options for project dialog box.

3. Click the **New** button to create a new empty entry.

Figure 10.7: Browse for folder dialog box.

4. Select the . . . button on the rightmost edge to open the Browse dialog box (see Figure 10.7).

5. Select our new *include* directory.

6. Click **OK** to close the dialog box.

7. Click **OK** to accept the new setting.

8. Save the project by selecting **Project | SaveProject**,

With these settings, we will be able to refer to the LCD.h file with the *default include* statement, as in:

```
#include <LCD.h>
```

without needing to add details of the path required to reach the directory where the file is actually stored.

Note

Notice the use of the angled brackets (<>) as opposed to the double quotes (" ") syntax. The difference between the two notations lies in where the compiler will look for the file to be included. The double quotes method we used in all previous projects tells the compiler to look for a file inside the current project directory. The angled brackets, on the other hand, tell the compiler to look for the file inside a series of directories known as the *include search path* that typically contains all the compiler-specific (MPLAB C32) library directories defined during the installation of the program on our computer but also all the additional directories we listed in the Include Search Path dialog box.

Advanced LCD Control

If you felt that the preceding discussion was not too complex or perhaps not rewarding enough, here we have some more interesting stuff and a new challenge for you to consider.

When we introduced the HD44780 compatible alphanumeric LCD modules, we mentioned how the display content was generated by the LCD module controller using a table, the character generator, located in ROM. But we also mentioned the possibility to extend the character set with *user-defined* symbols using an additional RAM buffer (known as the CGRAM). Writing to the CGRAM, it is possible to create from two to eight new character patterns, depending on the LCD display model. Of course, if we had 32 user-defined characters, we could almost turn the entire alphanumeric display into a complete graphical display. Unfortunately, the most popular and inexpensive LCD modules, in particular the ones used on the Explorer 16 board, have only space for two user-defined characters. Still, there are a number of interesting things we can do with those. In the following, for example, we use just one of the two user-defined characters to illustrate how to develop a simple progress bar effect.

We will need a function to set the LCD module RAM buffer pointer to the beginning of the CGRAM area using the Set CGRAM Address command, or better a macro that uses the writeLCD() function:

```
#define setLCDG( a) writeLCD( LCDCMD, (a & 0x3F) | 0x40)
```

Once the buffer pointer is set on the CGRAM and specifically at the beginning of the buffer (setLCDG(0)), we can use the putLCD() function to place 8 bytes of data in the buffer. Each byte of data will contribute 5 bits (LSb) to the construction of the eight rows composing the new character pattern. After repositioning the buffer pointer into the DDRAM area (using the macro `setLCDC( 0)`), we can use the newly defined character with the ASCII code `0x00`.

Notice that by convention, although the first line of the display corresponds to addresses from `0` to `15` of the DDRAM buffer, the second line is always found at addresses from `0x40` to `0x4f` independently of the display width—the number of characters that compose each line of the actual display.

Progress Bar Project

It is time to start our last project for the day. We'll call it **Progress**. Let's proceed with the usual New Project checklist, and remember at the end to add the *include* directory in the *include search path*.

A new source file, **ProgressBar.c**, can be immediately created by inserting the standard template and *include* statements list:

```
/*
**
** Progress Bar
**
*/
// configuration bit settings, Fcy = 72 MHz, Fpb = 36 MHz
#pragma config POSCMOD = XT, FNOSC = PRIPLL
#pragma config FPLLIDIV = DIV_2, FPLLMUL = MUL_18, FPLLODIV = DIV_1
#pragma config FPBDIV = DIV_2, FWDTEN = OFF, CP = OFF, BWP = OFF
#include <p32xxxx.h>
#include <explore.h>
#include <LCD.h>
```

We could draw a blocky progress bar using just a string of (up to) 16 "brick" characters that can be obtained from the LCD font table by selecting the code 0xff, giving a solid 5×8 black pixels pattern. But to obtain a finer resolution and smoother motion, we can exploit instead the user-defined character feature we just learned to use. The trick is to build most of the progress bar with (5×8) bricks and then define a single new character of the required thickness for the tip (see Figure 10.8).

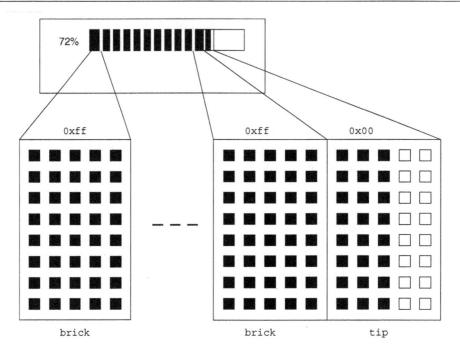

Figure 10.8: Drawing a progress bar.

Here is the code required to define a progress bar tip of given thickness:

```
void newBarTip( int i, int width)
{
    char bar;
    int pos;

    // save cursor position
    while( busyLCD());
    pos = addrLCD();

    // generate a new character at position i
    // set the data pointer to the LCD CGRAM buffer
    setLCDG( i*8);

    // as a horizontal bar (0-4)x thick moving left to right
    // 7 pixel tall
    if ( width > 4)
      width = 0;
```

```
    else
      width = 4 - width;

    for( bar=0xff; width > 0; width--)
      bar<<=1;                  // bar >>= 1; if right to left

    // fill each row (8) with the same pattern
    putLCD( bar);
    putLCD( bar);
    putLCD( bar);
    putLCD( bar);
    putLCD( bar);
    putLCD( bar);
    putLCD( bar);
    putLCD( bar);
    // restore cursor position
    setLCDC( pos);
} // newBarTip
```

Given this essential building block, drawing an actual progress bar requires only a few more lines of code:

```
void drawProgressBar( int index, int imax, int size)
{   // index is the current progress value
    // imax is the maximum value
    // size is the number of character positions available
    int i;

    // scale the input values in the available space
    int width=index * (size*5)/imax;

    // generate a character to represent the tip
    newBarTip( TIP, width % 5); // user defined character 0

    // draw a bar of solid blocks
    for ( i=width/5; i>0; i--)
        putLCD( BRICK); // filled block character

    // draw the tip of the bar
    putLCD( TIP); // use character 0

} // drawProgressBar
```

As you can see, to make the drawProgressBar() function really friendly, I included a little scaling of the input values so that the bar itself can be made to fit the desired number of spaces on the LCD display and the progress level is made relative to a given maximum value passed as a parameter. To put it to the test, we'll define a loop where a counter value (index) is cycling slowly through a range of values from 0 to 99. Each value is shown in the first three characters of the first line of the display. The rest of the line is filled with the progress bar.

```c
main( void)
{
    int index;
    char s[8];

    // LCD initialization
    initLCD();

    index = 0;

    // main loop
    while( 1)
    {

      clrLCD();

      sprintf( s, "%2d", index);
      putsLCD( s); putLCD( '%');

      // draw bar
      drawProgressBar( index, 100, HLCD-3);

      // advance and keep index in boundary
      index++;
      if ( index > 99)
      index=0;

      // slow down the action
      Delayms( 100);

    } // main loop
} // main
```

Notice that it is important to slow the execution of the main loop by inserting a small delay; otherwise the refresh of the display is so rapid that all we get to see is a sort of ghostly faint image. Remember, LCD displays are slow little things; be patient with them!

Finally, before you start building the project, remember to add all the required library modules we used. You will need to select the project window and right-click the source files to **Add file.** Browse to the *lib* directory we created today and select both the **explore.c** module (that will give us the Delayms() function) and the **LCDlib.c** module.

Now build the project, program the Explorer 16 board with the debugger of your choice, and observe the code running and drawing a progress bar that moves smoothly from left to right to fill the entire top line of the LCD display. This is true (glass) bliss!

Debriefing

Today we learned how to use the Parallel Master Port to interface to an alphanumeric LCD display module, just one of many common devices that require an 8-bit parallel interface. Since the LCD display modules are relatively slow peripherals, it might seem that there has been little or no significant advantage in using the PMP instead of a traditional bit-banged I/O solution. In reality, even when accessing such simple and slow peripherals, the use of the PMP can provide two important benefits:

- The timing, sequence, and multiplexing of the control signals are always guaranteed to match the configuration parameters, eliminating the risk of dangerous bus collisions and/or unreliable operation as a consequence of coding errors or unexpected execution and timing conditions (interrupts, bugs, and so on).

- The MCU is completely free from tending at the external (peripheral) bus, allowing simultaneous execution of any number of higher-priority tasks without disruption of the interface timing.

Notes for the PIC24 Experts

The PMP module of the PIC32 is mostly identical to the PIC24 peripheral, yet some important enhancements have been included in its design. Here are the major differences that will affect your code while porting an application to the PIC32:

1. The PMCON register control bits layout has been updated to resemble more closely the layout of most other peripherals so that the module ON, FRZ, and IDL bits are now located in the standard position (bit 15, bit 14, bit 13).

2. The PMBE output signal has been removed.

3. The PMPTTL control bit is now found in the PMCON register to select Schmitt trigger or TTL input levels. It used to be part of the PADCFG1 register on the PIC24.

4. In the PMMODE register, the IRQM=11 and IRQM=10 selections have been modified.

5. The PMPEN register is now renamed PMAEN. This has been similarly updated on the latest revision of the PIC24 datasheets as well.

6. A single PMDIN and a single PMDOUT registers (now 32 bits wide) give simultaneous access to all data buffers.

Tips & Tricks

Though basic alphanumeric displays are pretty much standardized around the HD44780 controller interface and command set, things are very different when it comes to graphic displays. A variety of controllers are being currently offered with very different capabilities. The most common controllers for small LCD displays are probably the New Japan Radio (NJU6679) used in many monochrome displays (up to 128 × 128) and using a parallel interfaces very similar to the HD44780. But the new trend is represented by the serially interfaced EPSON (S1D15G10) controllers used in many inexpensive color LCD displays, often referred to as "Nokia knock-offs" because their low price is mostly driven by the large volumes of production supposedly achieved on the latest generations of multimedia phones. OLED displays are also going the way of the serial interfaces (SPIs). Finally, when the resolution of the display grows beyond the QVGA (320*240), you can no longer rely on finding a complete controller chip on glass, and you have to start producing a complex synchronized waveform while continuously refreshing the screen. A QVGA or more advanced display peripheral module becomes a necessity.

Exercises

1. As suggested in the previous explorations using asynchronous serial interfaces, it is possible to redirect the output of the stdio.h library routines, such as printf(), to the LCD display. Redefine the _mon_putc() Function (see the MPLAB C32 C Library Guide for details) to send characters to the LCD via the parallel master port interface.

2. LCD displays are typically very slow devices. A lot of processing power is wasted while the PIC32 is waiting for the LCD display to perform a command. Using a buffering mechanism and timer interrupts implement a background LCD display interface. (A basic example of such a mechanism is provided in the LCD. c code provided with the Explorer 16 demonstration board for the PIC24 and dsPIC platforms).

Books

Bentham, Jeremy, *TCP/IP Lean: Web Servers for Embedded Systems* (CMP Books, Lawrence, KS).This book will take you one level of complexity higher, showing you how the TCP/IP protocols, the foundation of the Internet, can be easily implemented in a "few" lines of C code. The author knows how to keep things "lean" as necessary in every embedded-control application.

Links

www.microchip.com/graphics. Microchip is offering graphic libraries capable of supporting the most popular LCD display controllers for the 16-bit and 32-bit architectures. Check the availability of free and third-party supported libraries on the Web Graphic Design Center. www.microchip.com/stellent/idcplg?IdcService=SS_GET_PAGE&nodeId=1824&appnote=en011993. This is a link to Microchip Application Note 833, a free TCP/IP stack for all PICmicros. *www.microchip. com/stellent/idcplg?IdcService=SS_GET_PAGE&nodeId=1824&appnote=en012 108.* Application Note 870 describes a Simple Network Management Protocol for Microchip TCP/IP stack-based applications.

It's an Analog World

The Plan

We live in an analog world. Temperature, humidity, and pressure but also voltages and currents are analog. If we want our embedded-control applications to interact with the outside world, we need to learn to interpret analog information and convert it to digital so that a microcontroller can elaborate it and possibly produce an analog output again. The analog-to-digital converter module is one of the key interfaces to the "real" world. The PIC32MX family was designed with embedded-control applications in mind and therefore is ideally prepared to deal with the analog nature of this world. A fast analog-to-digital converter (ADC), capable of 500,000 conversions per second, is available on all models with an input multiplexer that allows you to monitor a number of analog inputs quickly and with high resolution. In this lesson we will learn how to use the 10-bit ADC module available on the PIC32MX family to perform two simple measurements on the Explorer 16 board: reading a voltage input from a potentiometer first and a voltage input from a temperature sensor later.

Preparation

In addition to the usual software tools, including the MPLAB® IDE, the MPLAB C32 compiler, and the MPLAB SIM simulator, this lesson will require the use of the Explorer 16 demonstration board and the In-Circuit Debugger of your choice.

The Exploration

The first step in using the ADC, like any other peripheral module inside the PIC32, is to familiarize yourself with the module building blocks and the key control registers.

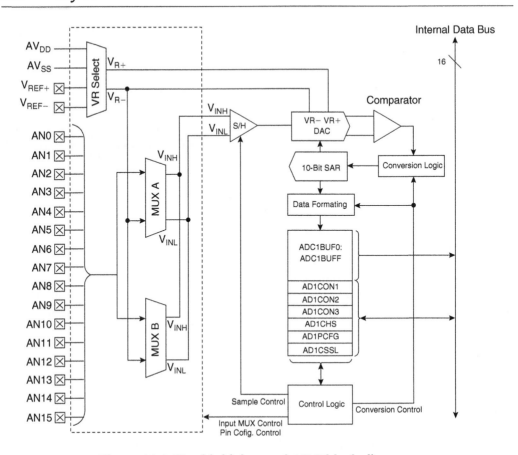

Figure 11.1: Ten-bit high-speed ADC block diagram.

Yes, this means reading the datasheet once more, and even the Explorer 16 User Guide, to find out the schematics.

We can start by looking at the ADC module block diagram (see Figure 11.1).

This is a pretty sophisticated structure that offers many interesting capabilities:

- Up to 16 input pins can be used to receive the analog inputs.

- Two input multiplexers can be used to select different input analog channels and different reference sources each.

- The output of the 10-bit converter can be formatted for integer or fixed-point arithmetic, signed or unsigned, 16-bit and 32-bit output.

- The control logic allows for many possible automated conversion sequences to synchronize the process to the activity of other related modules and inputs.

- The conversion output is stored in a 32-bit-wide, 16-words-deep buffer that can be configured for sequential scanning or simple FIFO buffering.

All these capabilities require a number of control registers to be properly configured, and I understand how, especially at the beginning, the number of options available and decisions to take could make you a bit dizzy. So we will start by taking the shortest and simplest approach with the simplest example application: reading the position of the R6 potentiometer on the Explorer 16 board.

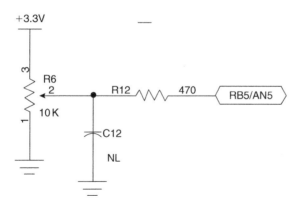

Figure 11.2: Detail of the Explorer 16 demonstration board R6 potentiometer.

The 10k Ohm potentiometer is directly connected to the power supply rails so that its output can span the entire range of values from 3.3 V to the ground reference. It is connected to the RB5 pin that corresponds to the analog input AN5 of the ADC input multiplexer.

After creating a new project using the appropriate checklist, we can create a new source file **pot.c**, including the usual header file and adding the definition of a couple useful constants. The first one, POT, defines the input channel assigned to the potentiometer; the second one, `AINPUTS`, is a mask that will help us define which inputs should be treated as analog and which ones as digital:

```
/*
** It's an analog world
** Converting the analog signal from a potentiometer
*/
// configuration bit settings, Fcy=72 MHz, Fpb=36 MHz
#pragma config POSCMOD=XT, FNOSC=PRIPLL
#pragma config FPLLIDIV=DIV_2, FPLLMUL=MUL_18, FPLLODIV=DIV_1
#pragma config FPBDIV=DIV_2, FWDTEN=OFF, CP=OFF, BWP=OFF
#include <p32xxxx.h>

#define POT 5 // 10 k potentiometer on AN5 input
#define AINPUTS 0xffef // Analog inputs POT, TSENS
```

The actual initialization of all the ADC control registers can be best performed by a short function, `initADC()`, that will produce the desired initial configuration:

- `AD1PCFG` will be passed the mask selecting the analog input channels: 0 s will mark the analog inputs, 1 s will configure the respective pins as digital inputs.

- `AD1CON1` will set the conversion to start automatically, triggered by the completion of the auto-timed sampling phase. Also, the output will be formatted for a simple unsigned, right-aligned (integer) value.

- `AD1CSSL` will be cleared because no scanning function will be used (only one input).

- `AD1CON2` will select the use of MUXA and will connect the ADC reference inputs to the analog input rails AVdd and AVss pins.

- `AD1CON3` will select the conversion clock source and divider.

- Finally we set `ADON`, and the entire ADC peripheral will be activated and ready for use.

```
void initADC( int amask)
{
  AD1PCFG = amask;      // select analog input pins
  AD1CON1 = 0;     // manual conversion sequence control
  AD1CSSL = 0;     // no scanning required
  AD1CON2 = 0;     // use MUXA, AVss/AVdd used as Vref+/-
```

```
    AD1CON3=0x1F02;          // Tad=2+1) x 2 x Tpb=6x27 ns>75 ns
    AD1CON1bits.ADON=1;      // turn on the ADC
} //initADC
```

By keeping `amask` as a parameter to the initialization routine, we make it flexible so it's able to accept different (multiple) input channels in future applications.

Note

As for all other peripheral modules found inside the PIC32, a corresponding peripheral library (adc.h) offers a set of functions and macros that are supposed to simplify or at least make the code that accesses the ADC module more readable. Because of the great flexibility of the ADC module, it is my very personal opinion that it is best if you familiarize yourself first with the low-level details of its operation by directly accessing the few control registers rather than seeking early refuge in the peripheral library.

The First Conversion

The actual analog-to-digital conversion is a two-step process. First we need to take a sample of the input voltage signal; then we can disconnect the input and perform the actual conversion of the sampled voltage to a digital value. The two distinct phases are controlled by two separate control bits in the `AD1CON1` register: `SAMP` and `DONE`. The timing of the two phases is important to provide the necessary accuracy of the measurement:

- During the sampling phase, the external signal is connected to an internal capacitor that needs to be charged up to the input voltage. Enough time must be provided for the capacitor to track the input voltage, and this time is mainly proportional to the impedance of the input signal source (in our case, known to be 10 k Ohm) as well as the internal capacitor value. In general, the longer the sampling time, the better the result, compatible with the input signal frequency (not an issue in our case).

- The conversion phase timing depends on the selected ADC clock source. This is derived by the peripheral bus clock signal via a divider or, alternatively, by a dedicated RC oscillator. The RC option, although appealing for its simplicity, is a good choice when a conversion needs to be performed when the PIC32 is in a low-power mode, when the peripheral clock can be turned off. The oscillator clock divider on the other end is a better option in more general cases since it provides synchronous operation with the peripheral bus and therefore a better rejection of the internal noise. The conversion clock should be the fastest possible, compatibly with the specifications of the ADC module.

Here is a basic conversion routine:

```
int readADC( int ch)
{
  AD1CHSbits.CH0SA = ch;        // 1. select analog input
  AD1CON1bits.SAMP = 1;         // 2. start sampling
  T1CON = 0xs8000; TMR1 = 0;    // 3. wait for sampling time
  while (TMR1 < 100);           //
  AD1CON1bits.SAMP = 0;         // 4. start the conversion
  while (!AD1CON1bits.DONE);    // 5. wait conversion complete
  return ADC1BUF0;              // 6. read result
} // readADC
```

Automating Sampling Timing

As you can see, using this basic method, we have been responsible for providing the exact timing of the sampling phase, dedicating a timer to this task and performing two waiting loops. But on the PIC32 ADC module, the sampling phase can be self-timed up to a maximum of 32 $\times$ Tad periods. Whether we can use this feature or not will depend ultimately on the product of the source impedance and the ADC input capacitance. By setting the SSRC bits in the AD1CON1 register to the 0x7 configuration, we can enable an automatic start of conversion upon termination of the self-timed sampling period. The sampling period itself is selected by the AD1CON3 register SAM bits. Here is a new and improved example that uses the self-timed sampling and conversion trigger:

```
void initADC( int amask)
{
  AD1PCFG = amask;          // select analog input pins
  AD1CON1 = 0x00E0;         // automatic conversion after sampling
  AD1CSSL = 0;              // no scanning required
  AD1CON2 = 0;              // use MUXA, use AVdd & AVss as Vref+/-
  AD1CON3 = 0x1F3F;         // Tsamp = 32 x Tad;
  AD1CON1bits.ADON = 1;     // turn on the ADC
} //initADC
```

Notice how making the conversion-start, triggered automatically by the completion of the self-timed sampling phase, gives us two advantages:

- Proper timing of the sampling phase is guaranteed without requiring us to use any timed delay loop and/or other resource.

- One command (start of the sample phase) suffices to complete the entire sampling and conversion sequence.

With the ADC so configured, starting a conversion and reading the output is a trivial matter:

- AD1CHS selects the input channel for MUXA.

- Setting the SAMP bit in AD1CON1 starts the timed-sampling phase, immediately followed by the conversion.

- The DONE bit will be set in the AD1CON1 register as soon as the entire sequence is completed and a result is ready.

- Reading the ADC1BUF0 register will immediately return the desired conversion result.

```
int readADC( int ch)
{
  AD1CHSbits.CH0SA = ch;       // 1. select input channel
  AD1CON1bits.SAMP = 1;        // 2. start sampling
  while (!AD1CON1bits.DONE);   // 3. wait conversion complete
  return ADC1BUF0;             // 4. read conversion result
} // readADC
```

Developing a Demo

All that remains to do at this point is to figure out an entertaining way to put the converted value to use on the Explorer 16 demo board. The LEDs connected to PORTA are an intriguing choice, but those of you using a PIC32 Starter Kit would not be able to enjoy the experience, since most of the PORTA pins would be tied up by the JTAG port. Instead we will use the LCD library developed in the previous chapter to display a blocky bar graph. Yes, we could use the nice and smooth progress bar developed in the previous chapter (Day 10) but I don't want you to get distracted by the details. Here is the main routine we will use to test our analog-to-digital conversion functions:

```
main ()
{
  int i, a;

  // initializations
  initADC( AINPUTS); // initialize the ADC
  initLCD();// initialize the LCD display

  // main loop
  while( 1)
  {
    a = readADC( POT); // select the POT input and convert

    // reduce the 10-bit result to a 4 bit value (0..15)
    // (divide by 64 or shift right 6 times
    a >> = 6;

    // draw a bar on the display
    clrLCD();
    for ( i=0; i<=a; i++)
      putLCD( 0xFF);

    // slow down to avoid flickering
    Delayms( 200);
  } // main loop
} // main
```

After the call to the ADC initialization routine, we can initialize the LCD display module. Then in the main loop we perform the conversion on AN5 and we reformat the output to fit our special display requirements. As configured, the 10-bit conversion output will be returned as a right-aligned integer in a range of values between 0 and 1023. By dividing that value by 64 (or, in other words, shifting it right six times) we can reduce the range to a 0 to 15 value. Printing the resulting number of "bricks" gives a blocky bar whose length is proportional to the position of the potentiometer.

Remember to add an **#include** <> statement for the LCD.h library and add to the project source files list both the **explore.c** and **LCDlib.c** modules we placed in the *lib* directory.

Build the project and, following the usual In Circuit Debugging checklist, program the Explorer 16 board. If all goes well, you will be able to play with the potentiometer, moving it from side to side while observing a bar of 16 blocks moving from left to right correspondingly.

Creating Our Own Mini ADC Library

We will use over and over the two simple routines that initialize the ADC module and perform a single self-timed conversion. Let's separate them into a standalone small library called **ADClib.c** that we will add to our new collection inside the *lib* directory.

```
/*
** ADClib.c
**
*/
#include <p32xxxx.h>
#include <ADC.h>

// initialize the ADC for single conversion, select input pins
void initADC( int amask)
{
  AD1PCFG = amask;            // select analog input pins
  AD1CON1 = 0x00E0;           // auto convert after end of sampling
  AD1CSSL = 0;                // no scanning required
  AD1CON2 = 0;                // use MUXA, AVss/AVdd used as Vref+/-
  AD1CON3 = 0x1F3F;           // max sample time = 31Tad
  AD1CON1SET = 0x8000;        // turn on the ADC
} //initADC

int readADC( int ch)
{
  AD1CHSbits.CH0SA = ch;      // select analog input channel
  AD1CON1bits.SAMP = 1;       // start sampling
  while (!AD1CON1bits.DONE);  // wait to complete conversion
  return ADC1BUF0;            // read the conversion result
} // readADC
```

Similarly we can isolate the include file LCD.h that offers the basic set of definitions and prototypes required to access the library functions. We will save it in the *include* directory.

```
/*
** ADC.h
**
*/
#define POT 5            //10k potentiometer on AN5 input
#define TSENS 4          // TC1047 Temperature sensor on AN4
#define AINPUTS 0xffcf  // Analog inputs for POT and TSENS

// initialize the ADC for single conversion, select input pins
void initADC( int amask) ;
int readADC( int ch);
```

Simple enough. We are ready to proceed with more fun and games!

Fun and Games

Okay, I'll admit it, the previous project was not too exciting. After all, we have been using a 32-bit machine operating at 72 MHz, capable of performing a 10-bit analog-to-digital conversion several hundred thousands of times per second, only to discard all but 4 bits of the conversion result and watch a blocky bar moving on an LCD display. How about making it a bit more challenging and playful instead? How about developing a monodimensional Pac-Man game, or should we call it the "Pot-Man" game?

If you remember the old Pac-Man game—please don't tell me you never heard of it, but if you really have to, check the link to a Wikipedia entry at the end of this chapter—there is a hungry little "thing," the Pac, that roams a two-dimensional labyrinth in a desperate search for food. Now, with a little fantasy, we can imagine a monodimensional reduction of the game, where the Pac is represented by a single < or > character, depending on the direction of movement. It is limited to a left/right movement on a line of the LCD display as it is controlled by the potentiometer position. Bits of food are represented by a * character and are placed randomly, one at a time, on the same line. As soon as the Pac reaches a piece of food, it gulps it and moves on, and a new piece is placed in a different location.

Once more, the pseudo-random number generator function rand() (defined in stdlib.h) will be very helpful here. All games need a certain degree of unpredictability, and pseud-random number generators are the way computer games provide it in a world of logic and otherwise infinite repetition.

We can start by modifying the previous project code or typing away from scratch a brand-new **Pot-Man.c** file. A new project needs to be created, and I suggest we call it simply **POT**. Just a few more lines of code are truly needed to perform the simple animation:

```c
/*
** Pot-Man.c
**
*/
// configuration bit settings, Fcy=72 MHz, Fpb=36 MHz
#pragma config POSCMOD=XT, FNOSC=PRIPLL
#pragma config FPLLIDIV=DIV_2, FPLLMUL=MUL_18, FPLLODIV=DIV_1
#pragma config FPBDIV=DIV_2, FWDTEN=OFF, CP=OFF, BWP=OFF

#include <p32xxxx.h>
#include <explore.h>
#include <LCD.h>
#include <ADC.h>

main ()
{
    int a, r, p, n;
    // 1. initializations
    initLCD();
    initADC( AINPUTS);

    // 2. use the first reading to randomize
    srand( readADC( POT));

    // 3. init the hungry Pac
    p = '<';

    // 4. generate the first random food bit position
    r = rand() % 16;

    // main loop
    while( 1)
    {
        // 5. select the POT input and convert
        a = readADC( POT);

        // 6. reduce the 10-bit result to a 4 bit value (0..15)
        // (divide by 64 or shift right 6 times
        a>> = 6;
```

```
// 7. turn the Pac in the direction of movement
if ( a < n) // moving to the left
   p = '>';
if ( a > n) // moving to the right
   p = '<';

// 8. when the Pac eats the food, generate more food
while (a == r )
   r=rand() % 16;

// 9. update display
clrLCD();
setLCDC( a); putLCD( p);
setLCDC( r); putLCD( '*');

// 10. provide timing and relative position
Delayms( 200); // limit game speed
n=a;             // memorize previous position
} // main loop
} // main
```

- In 1, we perform the usual initialization of the ADC module and the LCD display.

- In 2, we read the potentiometer value for the first time and we use its position as the *seed* value for the pseudo-random number generator. This makes the game experience truly unique each time, provided the potentiometer is not always found in the leftmost or rightmost position. That would provide a seed value of 0 or 1023, respectively, every time and therefore would make the game quite repetitive because the pseudo-random sequence would proceed through exactly the same steps at any game restart.

- In 3, we assign a first arbitrary direction to the Pac.

- In 4, we determine a first random position for the first bit of food.

- In 5, we are already inside the main loop checking for the latest position of the potentiometer cursor.

- In 6, we reduce the integer 10-bit value to the four most significant bits to obtain a value between 0 and 15.

- In 7, we compare the new position with the previous loop position to determine which way the mouth of the Pac should be facing. If the ADC reading has reduced, it means we moved the potentiometer counter-clockwise. Hence we will make the Pac turn to the left. Vice versa, if the ADC reading has increased compared to the previous loop value, the potentiometer must have been turned clockwise, and we'd better turn the Pac to the right.

- In 8, we compare the new position of the Pac—the ADC reading—with the food position and, if the two coincide (the Pac got his lunch), a new random food position is immediately calculated. The operation needs to be repeated in a while loop because each time a new random value (r) is calculated, there is a chance (exactly 1/16 if our pseudo-random generator is a good one) that the new value could be just the same. In other words, we could be creating a new "food nibblet" right in the Pac's mouth. Now we don't want that—it would not be very sporting, don't you agree?

- Finally, in 9, we get to clean the display content and then place the two symbols for the Pac and the food piece in their respective positions.

- In 10, we close the loop with a short delay and save the Pac's position for the next loop to compare.

Don't forget to include in the project the LCDlib.c, ADClib.c, and Explore.c files found in the *lib* directory. Build the project and program it onto the Explorer 16 board. You will have to admit it: Analog-to-digital conversions are so much more entertaining now!

Sensing Temperature

Moving on to more serious things, there is a temperature sensor mounted on the Explorer 16 board, and it happens to be a Microchip TC1047A integrated temperature-sensing device with a nice linear voltage output. This device is very small, it is offered in a SOT-23 (three-pin, surface-mount) package. The power consumption is limited to 35uA (typ.) while the power supply can cover the entire range from 2.5 V to 5.5 V. The output voltage is independent from the power supply and is an extremely linear function of the temperature (typically within 0.5 degree C) with a slope of exactly 10 mV/C. The offset is adjusted to provide an absolute temperature indication according to the formula shown in Figure 11.3.

We can apply our newly acquired abilities to convert the voltage output to digital information using, once more, the ADC of the PIC32. The temperature sensor is directly connected to the AN4 analog input channel as per the Explorer 16 board schematic (see Figure 11.4).

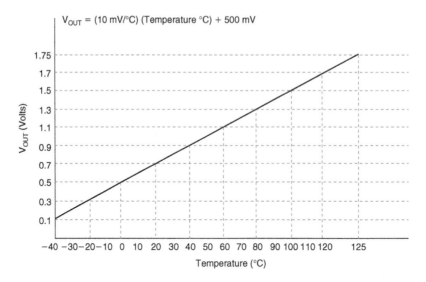

Figure 11.3: TC1047 output voltage vs. temperature characteristics.

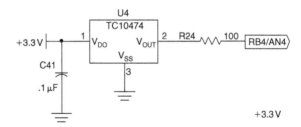

Figure 11.4: Detail of the Explorer 16 demonstration board TC1047A temperature sensor.

We can reuse the ADC library developed for the previous exercise and put it in a new project called **TEMP**, saving the previous source file as **Temp.c**.

Let's start modifying the code to include a new constant definition: TSENS for the ADC input channel assigned to the temperature sensor.

```
/*
** Temp.c
** Converting the analog signal from a TC1047 Temp Sensor
*/
```

```
// configuration bit settings, Fcy=72 MHz, Fpb=36 MHz
#pragma config POSCMOD=XT, FNOSC=PRIPLL
#pragma config FPLLIDIV=DIV_2, FPLLMUL=MUL_18, FPLLODIV=DIV_1
#pragma config FPBDIV=DIV_2, FWDTEN=OFF, CP=OFF, BWP=OFF

#include <p32xxxx.h>
#include <explore.h>
#include <LCD.h>
#include <ADC.h>
```

As you can see, nothing else needed to change with regard to the ADC configuration or activation of the conversion sequence. Presenting the result on the LCD display might be a little tricky, though. Temperature sensors provide a certain level of noise, and to give a more stable reading it is common to perform a little filtering. Taking groups of 10 samples over a period of a second, for example, and performing an average will give us a cleaner value to work with.

```
a=0;
  for ( j=0; j < 10; j++)
    a+=readADC( TSENS);     // add up 10 readings
  i=a/10;     // divide by 10 to average
```

Referring to the formula in Figure 11.3, we can now calculate the absolute temperature value as measured by the TC1047 on the Explorer 16 board. In fact, resolving for the temperature in degrees C, we obtain:

$$T = \frac{Vout - 500\,mV}{10\,mV/C}$$

where:

$$Vout = ADC\ reading * ADC\ resolution\ (mV/bit)$$

Since we have configured the PIC32 ADC module to use as an internal voltage reference the AVdd line connected to Vdd (3.3 V), and knowing it operates as a 10-bit, we derive that the ADC resolution is 3.3 mV/bit. Hence the temperature can be expressed as:

$$T = \frac{(3.3 * i) - 500}{10};$$

We could easily print the resulting absolute temperature on the LCD display, but it would not be fun, would it? How about providing instead a relative temperature indication using a single character (cursor) position as an index, or even better, how about using the temperature as a way to control the monodimensional Pac-Man game we developed in the previous project? We could heat the sensor by breathing hot air onto the sensor to move it to the right or blowing cold air on it to move it to left.

From a practical point of view, it seems easy to implement. We can sample the initial temperature value just before the main loop and then use it as a reference to determine an offset for the Pac position relative to the center of the display. In the main loop we will update the cursor position, moving it to the right as the sensed temperature increases or to the left as the sensed temperature decreases. Here is the complete code for the new Temp-Man game, or should we call it the Breathalyzer game instead?

```c
main ()
{
  int a, i, j, n, r, p;
  // 1. initializations
  initADC( AINPUTS); // initialize the ADC
  initLCD();

  // 2. use the first reading to randomize
  srand( readADC( TSENS));
  // generate the first random position
  r = rand() % 16;
  p = '<';

  // 3. compute the average value for the initial reference
  a = 0;
  for ( j=0; j<10; j++)
  {
    a+=readADC( TSENS); // read the temperature
    Delayms( 100);
  }
  i=a/10; // average

  // main loop
  while( 1)
  {
    // 4. take the average value over 1 second
    a = 0;
    for ( j=0; j<10; j++)
```

```
{
  a += readADC( TSENS); // read the temperature
  Delayms( 100);
}
a /= 10;        // average result

// 5. compare initial reading, move the Pac
a=7+(a-i);

// 6. keep the result in the value range 0..15
if ( a > 15)
  a = 15;
if ( a < 0)
  a = 0;

// 7. turn the Pac in the direction of movement
if ( a < n)// moving to the left
  p = '>';
if ( a > n)// moving to the right
  p = '<';

// 8. as soon as the Pac eats the food, generate new
while (a == r )
  r = rand() % 16;

// 9. update display
clrLCD();
setLCDC( r); putLCD( '*');
setLCDC( a); putLCD( p);

// 10. remember previous postion
n = a;
} // main loop
} // main
```

You will notice how most of the code has remained absolutely identical to our previous project/game. The only notable differences are found in the following sections:

- In 3 and in 4, we use a simple average of 10 values taken over a period of a second instead of a single instantaneous reading.

- In 5, we compute the temperature difference and use it as an offset with respect to the center position (7).

- In 6, we check for boundaries. Once the difference becomes negative and more than 4 bits wide, the display must simply indicate the leftmost position. When the difference is positive and more than 4 bits wide, the rightmost position must be used.

- In 10, we don't need further delays because the temperature reading and averaging already provide already a natural pace to the game.

Build the project with the usual checklists, remembering to *include* all the libraries required. Program it to the Explorer 16 board using the In-Circuit debugger of choice and give it a try.

The first problem you will encounter will be to identify the minuscule temperature sensor on the board. (Hint: It is close to the lower-left corner of the processor module and it looks just like any surface-mount transistor). The second immediate problem will be to find the right way to breathe on the board to produce warm or cold air as required to move the Pac. It is more complex than it might appear. In fact, personally, I found the cooling part to be the hardest; some friends are suggesting that this might be a problem related to my current position. If you work in marketing, they say, it's just hot air!

Debriefing

In this lesson we have just started scratching the surface and exploring the possibilities provided by the ADC module of the PIC32. We have used one simple configuration of the many possible and only a few of the advanced features available. We have tested our newly acquired capabilities with two types of analog input available on the Explorer 16 board, and hopefully we had some fun in the process.

Notes for the PIC24 Experts

The ADC module of the PIC32 is mostly identical to the PIC24 peripheral, yet some important enhancements have been included in its design. Here are the major differences that will affect your code while porting an application to the PIC32:

1. In the AD1CON1 register, the conversion format options are now extended to a 32-bit fractional word.

2. The CLRASAM control bit has been added to the AD1CON1 register to allow the conversion sequence to be stopped after the first interrupt.

3. In the AD1CON2 register a new autocalibration mode has been added to reduce the ADC offset. The OFFCAL control bit has been added to enter the calibration mode.

4. The AD1CHS register control bits are now in the upper half of the register 32-bit word. There is also a single CH0NB0 control bit for the selection of the negative input of the second input multiplexer.

Tips & Tricks

If the sampling time required is longer than the maximum available option (32×Tad), you can try to extend Tad first or, a better option, swap things around and enable the automatic sampling start (at the end of the conversion). This way the sampling circuit is always open, charging, whenever the conversion is not occurring. Manually clearing the SAMP bit will trigger the actual conversion start. Further, having Timer3 periodically clearing the SAMP control bit for you (one of the options for the SSRC bits in AD1CON1) and enabling the ADC end of conversion interrupt will provide the widest choice of sampling periods possible for the least amount of MCU overhead possible. No waiting loops, only a periodic interrupt when the results are available and ready to be fetched.

Further, not all applications require a complete conversion of analog input values. The PIC32MX family offers also analog comparator modules (two), with dedicated input multiplexers. They can assist in those applications in which we need a fast response to an analog input as it crosses a threshold. No need to set up the ADC, select a channel, and perform a conversion; the comparison is done continuously. An interrupt (or an output signal) is produced immediately as the reference voltage is reached.

Speaking of reference voltages, yet another module, called the Comparator Reference, effectively representing a small digital-to-analog converter of sorts, can generate up to 32 reference voltages to be used with the comparator modules or independently.

Exercises

1. Use the ADC FIFO buffer to collect conversion results and set up Timer3 for automatic conversion and the interrupt mechanism so that a call is performed only once the buffer is full and temperature values are ready to be averaged.

2. Experiment with interfacing other types of analog sensors (using the prototyping area of the Explorer 16 board) such as pressure sensors, humidity sensors, and

even accelerometers. Two- and/or three-axis solid-state accelerometers are getting very inexpensive and readily available. All it takes to interface to them is a few analog input pins and a fast 10-bit ADC module.

Books

Baker, Bonnie, *A Baker's Dozen: Real Analog Solutions for Digital Designers* (Newnes, Burlington, MA). For proper care and feeding of an analog-to-digital converter, look no further than this cookbook.

Links

www.microchip.com/filterlab. Download the free FilterLab software from the Microchip Web site; it will help you quickly and efficiently design antialiasing filters for your analog inputs.

www.microchip.com/stellent/idcplg?IdcService=SS_GET_PAGE&nodeId=2102&param=en021419&pageId=79&pageId=79. Temperature sensors are available in many flavors and a choice of interface options, including direct I^2C or SPI digital output.

Expansion

Congratulations, you have endured five more days of hard fieldwork. You have learned to use some of the key hardware peripheral modules of the PIC32MX, and you have put them to use on the Explorer 16 demo board.

In the third part of this book we will start developing new projects that will require you to master several peripheral modules at once. Since the complexity of the examples will grow a bit more, not only is it recommended you have an actual demonstration board (the Explorer 16) at hand, but you'll also need the ability to perform small modifications and utilizing the prototyping area to add new functionality to the demonstration board. Simple schematics and component part numbers will be offered in the following chapters as required. On the companion Web site, *www.ExploringPIC32.com*, you will find additional expansion boards and prototyping options to help you enjoy even the most advanced projects.

Capturing User Inputs

The Plan

If analog inputs are the essence of the interface between an embedded-control application and the outside world, digital inputs are, sadly, the true foundation of the user interface. As wrong as this might seem, for a long time now we humans have been trained to reduce our interaction with them, the machines, to buttons and switches. Probably this is because the alternative, using speech, gestures, and vision, requires such a leap in the complexity of the interface that we have rather learned to accept the limitation and reduced ourselves to communicate essentially through ones and zeros. Perhaps this explains the attention and enthusiasm that some recent innovations are producing as pioneered by video games and mobile phone manufacturers; think of the Wii accelerometer-filled wand and the iPhone multitouch sensing screen, for example.

Today we will explore various methods to capture "traditional" user inputs by detecting the activation of buttons and simple mechanical switches, reading the inputs from rotary encoders, and eventually interfacing to computer keyboards. This will give us the motivation to investigate a few alternative methods and evaluate their trade-offs. We'll implement software state machines, practice using interrupts, and possibly learn to use a few new peripherals. It's going to be a long day, so be rested and ready to start at dawn!

Preparation

In addition to the usual software tools, including the MPLAB® IDE, the MPLAB C32 compiler, and the MPLAB SIM simulator, this lesson will require the use of the Explorer 16 demonstration board and an In-Circuit Debugger of your choice. You will also need a soldering iron and a few components ready at hand to expand the board capabilities using the prototyping area or a small expansion board. You can check on the companion Web site (www.exploringPIC32.com) for the availability of expansion boards that will help you with the experiments.

Buttons and Mechanical Switches

Reading the input from a button, a mechanical switch, is one of the most common activities for an embedded-control application. After all, a single bit of information needs to be retrieved from a port pin configured as a digital input. But the great speed of a microcontroller and the mechanical (elastic) properties of the switch require that we pay some attention to the problem.

In Figure 12.1 you can see the connection of one of the four buttons present on the Explorer 16 demonstration board. At idle, the switch offers an open circuit and the input pin is kept at a logic high level by a pull-up resistor. When the button is pressed, the contact is closed and the input pin is brought to a logic low level. If we could consider the switch as an ideal component, the transition between the two states would be immediate and unambiguous, but the reality is a little different. As represented in Figure 12.2, when the button is pressed and the mechanical contact is established, we obtain all but a clean transition. The elasticity of the materials, the surface oxidation of the contacts, and a number of other factors make it so that there can be a whole series of transitions,

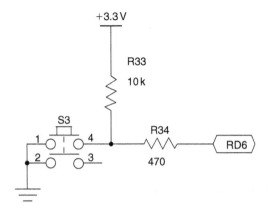

Figure 12.1: Explorer 16 button schematic detail.

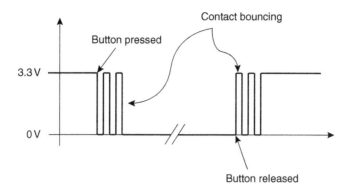

Figure 12.2: Electrical response of a mechanical switch.

increasing in number and spaced with the aging and general wear of the device. This phenomenon, generally referred to as *contact bouncing*, can continue in the worst cases for several hundred microseconds if not for milliseconds.

When the button is released, a similar bouncing effect can be detected as the pressure between the two contact surfaces is removed and the circuit is opened.

For a PIC32 operating at a high clock frequency, the timescale of the event is enormous. A tight loop polling the status of the input line could detect each and every bounce and count them as numerous distinct activations and releases of the button. In fact, as a first experiment, we could devise a short piece of code to do just that so we can access the "quality" of the buttons available on the Explorer 16 board.

Let's create a new project called **Buttons**, and let's add a first new source file to it that we'll call **bounce.c**:

```
/*
** bounce.c
**
*/
// configuration bit settings, Fcy=72MHz, Fpb=36MHz
#pragma config POSCMOD=XT, FNOSC=PRIPLL
#pragma config FPLLIDIV=DIV_2, FPLLMUL=MUL_18, FPLLODIV=DIV_1
#pragma config FPBDIV=DIV_2, FWDTEN=OFF, CP=OFF, BWP=OFF
#include <p32xxxx.h>
```

```
main( void)
{
  int count;      // the bounces counter

  count = 0;
  // main loop
  while( 1)
  {
    // wait for the button to be pressed
    while (_RD6);

    // count one more button press
    count++;

    // wait for the button to be released
    while ( ! _RD6);

  } // main loop

} // main
```

After initializing an integer counter, we directly enter the main loop, where we wait for the leftmost button on the board (marked S3 and connected to the RD6 input pin) to be pressed (transition to logic level low). As soon as we detect the button pressure, we increment the counter and proceed to the next loop, where we wait for the button to be released, only to continue in the main loop and start from the top.

Build the project immediately and program the code on the Explorer 16 board using your in circuit debugger of choice. To perform our first experiment, you can now run the code and slowly press the **S3** button for a predetermined number of times: let's say 20! Stop the execution and inspect the current value of the variable count. You can simply move your mouse over the variable in the editor window to see a small popup message appear (if the MPLAB option is enabled), or you can open the **Watch** window and add the variable **count** to it. (I suggest you set its visualization properties to **Decimal**.)

In my personal experimentation, after 20 button pushes I obtained a value of count varying generally between 21 and 25. As car manufacturers say: "Your mileage might vary"! This is actually a very good result, indicating that most of the time there have been no bounces at all. It's a testament to good-quality contacts, but it also reflects the fact that the board button has been used very little so far. If we are to design applications that use buttons and mechanical switches, we have to plan for the worst and consider a substantial degradation of performance over the life of the product.

Button Input Packing

Planning for a general solution to apply to all four buttons available on the Explorer 16 and extensible to an entire array of similar buttons if necessary, we will start developing a simple function that will collect all the inputs and present them conveniently encoded in a single integer code. Save the previous source file (**Save As**) with the new name **Buttons.c** and add it to the project (replacing bounce.c):

```
int readK( void)
{ // returns 0..F if keys pressed, 0 = none
  int c = 0;

  if ( !_RD6) // leftmost button
    c |=8;
  if ( !_RD7)
    c |=4;
  if ( !_RA7)
    c |=2;
  if ( !_RD13) // rightmost button
    c |=1;

  return c;
} // readK
```

In fact, the designers of the of the Explorer 16 board have "fragmented" the input pins, corresponding to the four buttons, between two ports in noncontiguous positions, probably in an attempt to ease the board layout rather than to please us, the software developers.

The function readK() as proposed collects the four inputs and packs them contiguously in a single integer returned as the function value. Figure 12.3 illustrates the resulting encoding.

The position of the buttons is now reflected in the relative position of each bit in the function return value, with the MSb (bit 3) corresponding to the leftmost button status. Also, the logic of each input is inverted so that a pressed button is represented by a 1. As a result, when called in the idle condition, no button pressed, the function returns 0, and when all the buttons are pressed, the function returns a value 0x0f.

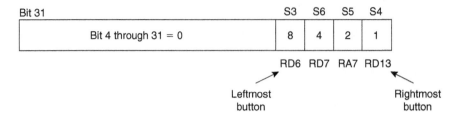

Figure 12.3: readK() button encoding.

Notice that we have performed no *debouncing* yet. All readK() does is grab a picture of the status of the inputs and present them in a different convenient format. Should we have a matrix of buttons arranged in a 3 × 4, 4 × 4, or larger keypad, it would be easy to modify the function while maintaining the output format and leaving untouched the rest of the code we will develop from here.

We can quickly modify the main() function to visualize the output on the LCD display using the LCD.h library we developed in the previous chapters:

```
main( void)
{
  char s[16];
  int b;

  initLCD();        // init LCD display

  // main loop
  while( 1)
  {
    clrLCD();
    putsLCD( "Press any button\n");
    b = readK();
    sprintf( s, "Code = %X", b);
    putsLCD( s);
    Delayms( 100);
  } // main loop
} // main
```

Build the project after adding the **LCDlib.c** module to the list of the project sources and program the Explorer 16 board with your In-Circuit Debugger of choice.

As you run the simple demo, you will see that as soon as a button is pressed, a new code is immediately displayed. Multiple buttons can be pressed simultaneously, producing all possible codes from 0x01 to 0x0f.

For our convenience, we will add the readK() function to our *explore.c* library module. In fact, if you are working with the code provided with the CD-ROM that accompanies this book, you will notice that the function is already there but under another name, readKEY(), so as not to create any conflict with the previous and following examples.

Button Inputs Debouncing

It is time now to start working on the actual debouncing. The basic technique used to filter out the spurious commutations of the mechanical switch consists of adding a small delay after the first input commutation is detected and subsequently verifying that the output has reached a stable condition. When the button is released, a new short delay is inserted before verifying once more that the output has reached the idle condition.

Here is the code for the new function getK() that performs the four steps listed previously and some more:

```
int getK( void)
{ // wait for a key pressed and debounce
  int i=0, r=0, j=0;
  int c;

  // 1. wait for a key pressed for at least .1sec
  do{
    Delayms( 10);
    if ( (c = readKEY()))
    {
      if ( c>r)              // if more than one button pressed
        r = c;               // take the new code
      i++;
    }
    else
      i=0;
} while ( i<10);
```

In 1, we have a do..while loop that, at regular intervals 10ms apart, uses the function readKEY() to check on the inputs status. The loop is designed to terminate only after

10 iterations (for a total of 100ms) during which there has been no bouncing. During that time, though, the user might have pressed more buttons. The function accommodates for one or more buttons to be "added" over time rather than assuming they will all be pressed together with absolute synchronicity. The variable r will contain the "most complete" button code.

```
// 2. wait for key released for at least .1 sec
i =0;
do {
   Delayms( 10);
   if ( (c = readKEY()))
   {
      if (c>r)      // if more then one button pressed
         r = c;     // take the new code
      i=0;
      j++;          // keep counting
   }
   else
      i++;
} while ( i<10);
```

In 2, the situation is reversed as buttons are released. The do.. while is designed to wait for all buttons to be released until the inputs stabilize in the idle condition for at least 100ms.

```
// 3. check if a button was pushed longer than 500ms
if ( j>50)
   r+=0x80;       // add a flag in bit 7 of the code
```

In 3, we are actually making use of an additional counter represented by the variable j that had been added to the second loop. Its role is that of detecting when the button-pressed condition is prolonged beyond a certain threshold. In this case it's set to 500ms. When this happens, an additional flag (bit 7) is added to the return code. This can be handy to provide additional functionalities to an interface without adding more hardware (buttons) to the Explorer 16 board. So, for example, pressing the leftmost button for a short amount of time produces the code 0x08. Pressing the same button for more than half a second will return the code 0x88 instead.

```
// 4. return code
   return r;
} // getK
```

It is only in 4 that the button code encoded in the variable r is returned to the calling program.

To test the new functionality and verify that we have eliminated all button bouncing, we can now replace the main() function with the following code and save the resulting file as **Buttons2.c**:

```
main( void)
{
  char s[16];
  int b;

  initLCD();        // init LCD display
  putsLCD( "Press any button\n");

  // main loop
  while( 1)
  {
    b = getK();
    sprintf( s, "Code = %X", b);
    clrLCD();
    putsLCD( s);
  } // main loop
} // main
```

Remember to include the **LCDlib.c** and **explore.c** modules found in the *lib* directory to the project.

Replace **Buttons2.c** in the project source list in place of buttons.c and build the project. After programming the Explorer 16 board with your in-circuit debugger of choice, run the code and observe the results on the LCD display.

First you will notice that contrary to what happened in the previous demo, new codes are displayed only after buttons are released. The function getK() is in fact a *blocking function*. It waits for the user inputs and returns only when a new return code is ready.

Play with various combinations of buttons, pressing two or three of them more or less simultaneously, and observe how the order of press and release does not affect the outcome, simplifying the user input. Try long and short button combinations. You can modify the threshold or even introduce secondary thresholds for very long button presses.

Once more, because of its usefulness, I suggest we add the getK() function to our explore.c library module. If you are using the code from the CD-ROM attached to this book, you will find it already there with the name changed in getKEY() to avoid conflicts with the examples in this chapter.

Rotary Encoders

Another type of input device based on mechanical switches (sometimes replaced by optical sensors) and very common in many embedded-control applications is the *rotary encoder*. In the past we have seen the use of a potentiometer attached to the PIC32 ADC module to provide user input (and control the position of the Pac-Man), but rotary encoders are pure digital devices offering a higher degree of freedom. Their main advantage is that they offer no limitation to the movement in any of the rotation directions. Some encoders provide information on their *absolute* position; others of simpler design and lower cost, known as *incremental encoders*, provide only a relative indication of movement.

In embedded applications, absolute rotary encoders can be used to identify the position (angle) of a motor/actuator shaft. Incremental encoders are used to detect direction of motion and speed of motors but also for user interfaces as a rapid input tool to select an entry in a menu system on a display panel: think of the omnipresent input knob on car navigators and digital radios. Another good example of a user interface application of an incremental encoder is a (ball) mouse, assuming you can still find one nowadays. They used to contain two (optical) rotary encoders to detect relative motion in two dimensions. In fact, if you think of it, your computer has no idea "where" the mouse is at any given point in time, but it knows exactly how far you moved it and in which direction. Don't look at modern "optical" mice, though; the technology they are based on is completely different.

To experiment with a simple and inexpensive rotary encoder (I used an ICW model from Bourns), I suggest you test your prototyping skills by soldering only a couple of resistors (10 K Ohm) onto the Explorer 16 board prototyping area and connecting just three wires between the encoder and the PIC32 I/O pins, as illustrated in Figure 12.4.

When so connected, the encoder provides two output waveforms (shown in Figure 12.5) that can be easily interpreted by the PIC32. Notice that the motion of the encoder is in steps between detent positions. At each step the encoder produces two commutations, one on each mechanical switch corresponding to an input pin. The order of the two

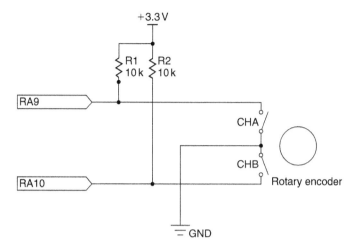

Figure 12.4: Rotary encoder interface detail.

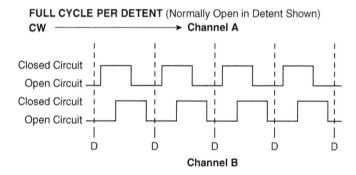

Figure 12.5: Rotary encoder output waveforms detail.

commutations tells us about the direction of rotation. Since the two waveforms are identical but appear to be out of phase by a 90-degree angle, these simple encoders are often referred to as *quadrature encoders*.

At rest, both switches are open and the corresponding input pins are pulled up at a logic level high. When rotating clockwise, the CHA switch is closed first, bringing the RA9 input pin to a logic low, then the CHB switch is closed, bringing the RA10 pin to a logic low level. When rotating counter-clockwise, the sequence is inverted. As the encoder reaches the next detent position, both switches are opened again.

Here is a simple program that can be used to demonstrate how to interface to a rotary encoder to track the position of a rotating knob and display a relative counter on the LCD display:

```c
/*
** Rotary.c
**
*/
// configuration bit settings, Fcy=72MHz, Fpb=36MHz
#pragma config POSCMOD=XT, FNOSC=PRIPLL
#pragma config FPLLIDIV=DIV_2, FPLLMUL=MUL_18, FPLLODIV=DIV_1
#pragma config FPBDIV=DIV_2, FWDTEN=OFF, CP=OFF, BWP=OFF

#include <p32xxxx.h>
#include <explore.h>
#include <LCD.h>

#define ENCHA   _RA9            // channel A
#define ENCHB   _RA10           // channel B

main( void)
{
    int i = 0;
    char s[16];

    initLCD();

    // main loop
    while( 1)
    {
        while( ENCHA);          // detect CHA falling edge
        Delayms( 5);            // debounce
        i += ENCHB ? 1 : -1;
        while( !ENCHA);         // wait for CHA rising edge
        Delayms( 5);            // debounce

        // display relative counter value
        clrLCD();
        sprintf( s, "%d", i);
        putsLCD( s);
    } // main loop

} // main
```

The idea behind the code in the main loop is based on a simple observation: by focusing only on one input commutations—say, ENCHA—we can detect motion. By observing the status of the second input ENCHB immediately after the activation of the first channel, we can determine the direction of movement. This can be seen in Figure 12.5 as you move your eyes from left to right (corresponding to a clockwise rotation); when the CHA switch is closed (represented as a rising edge), the CHB switch is still open (low). But if you read the same figure from right to left (corresponding to a counter-clockwise rotation of the encoder), when CHA is closed (rising edge), CHB is already closed (high).

Since we have not forgotten the lesson about switch bouncing, we have also added a pair of calls to a delay routine, to make sure that we don't read multiple commutations when there is really just one. The length of the delays was decided based on information provided by the encoder manufacturer on the device datasheet. The ICW encoders' contacts are in fact rated for a maximum of 5ms bounces when operated at a rotation speed of 15 RPM.

Create a new project called **Rotary**. Save the preceding code as **rotary.c** and remember to add our default *include* directory, as well as the **LCDlib.c** and **explore.c** source files found in the *lib* directory, to the list of project source files.

Build and program the Explorer 16, modified for the application, to run the short demo.

If all went well, you will see a counter displayed in decimal format being continuously updated on the LCD display as you turn the encoder knob. The counter is a signed (32-bit) integer and as such it can swing between positive and negative values, depending on how much and how long you turn clockwise and counter-clockwise.

Interrupt-Driven Rotary Encoder Input

The main problem with the simple demonstration code we have just developed is in its assumption that the entire attention of the microcontroller can be devoted to the task at hand: detecting the commutations on the CHA and CHB input pins. This is perhaps an acceptable use of resources when the application is waiting for user input and there are no other tasks that need to be handled by the microcontroller. But if there are and, as often is the case, they happen to be of higher priority and importance than our application, we cannot afford the luxury to use a *blocking* input algorithm. We need to make the encoder input a background task.

As we saw in Day 5, the simplest way to obtain a sort of multitasking capability in embedded-control applications is to use the PIC32 interrupt mechanisms. A background task becomes a small state machine that follows a simple set of rules. In our case, transforming the algorithm developed in the previous demonstration into a state machine and drawing its diagram (see Figure 12.6), we learn that only two states are required:

- An idle state (R_IDLE), when the CHA encoder input is not active

- An active state (R_DETECT), when the CHA encoder input is active

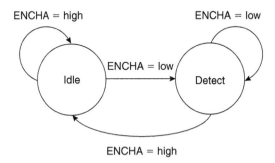

Figure 12.6: Rotary encoder state machine diagram.

The transitions between the two states are simply expressed in Table 12.1.

Table 12.1: Rotary encoder state machine transition.

State	Conditions	Effect
R_IDLE	ENCHA active (low)	If ENCHB is active, the direction of rotation is counterclockwise (d = −1) Transition to R_DETECT state
	ENCHA inactive (high)	Set default direction clockwise (d = 1) Remain in current state (wait)
R_DETECT	ENCHA inactive (high)	Update counter Transition to R_IDLE state
	ENCHA active (low)	Remain in current state (wait)

By binding the execution of the state machine to a periodic interrupt produced by one of the timers (Timer2, for example) we can ensure that the task will be performed continuously and, with the proper choice of timing, obtain a natural debouncing in the process.

We can create a new source file that we will call **Rotary2.c**, starting with the usual template and the following few declarations:

```
/*
** Rotary2.c
**
*/
// configuration bit settings, Fcy=72MHz, Fpb=36MHz
#pragma config POSCMOD=XT, FNOSC=PRIPLL
#pragma config FPLLIDIV=DIV_2, FPLLMUL=MUL_18, FPLLODIV=DIV_1
#pragma config FPBDIV=DIV_2, FWDTEN=OFF, CP=OFF, BWP=OFF

#include <p32xxxx.h>
#include <plib.h>
#include <explore.h>
#include <LCD.h>

#define ENCHA _RA9        // encoder channel A
#define ENCHB _RA10       // encoder channel B
#define TPMS (FPB/1000)   // PB clock ticks per ms

// state machine definitions
#define R_IDLE      0
#define R_DETECT    1

volatile int RCount;
char RState;
```

Notice that RCount, the variable used to maintain the relative movement counter, is declared as a volatile to inform the compiler that its value could change unpredictably at the hands of the interrupt service routine (state machine). This will ensure that the compiler won't try to optimize access to it in the main() function by making wrong assumptions, since the variable is never written to in the main loop.

Choosing to use the vectored interrupt mechanism of the PIC32 for efficiency, we can code the interrupt service routine as follows:

```
void __ISR( _TIMER_2_VECTOR, ipl1) T2Interrupt( void)
{
  static char d;
```

```
  switch ( RState) {
    default:
    case R_IDLE:          // waiting for CHA rise
      if ( ! ENCHA)
      {
        RState = R_DETECT;
        if ( ! ENCHB)
          d = -1;
      }
      else
        d = 1;
      break;

    case R_DETECT:        // waitin for CHA fall
      if ( ENCHA)
      {
        RState = R_IDLE;
        RCount += d;
      }
      break;
  } // switch

  mT2ClearIntFlag();
} // T2 Interrupt
```

Finally, a small initialization routine is necessary to set up the initial conditions required for the Timer2 peripheral (with a 5 ms period) and the state machine to operate correctly:

```
void initR( void)
{
  // init state machine
  RCount = 0;             // init counter
  RState = 0;             // init state machine

  // init Timer2
  T2CON = 0x8020;         // enable Timer2, Fpb/4
  PR2 = 5*TPMS/4;         // 5ms period
  mT2SetIntPriority( 1);
  mT2ClearIntFlag();
  mT2IntEnable( 1);
} // init R
```

The Timer2 interrupt can be set to a level 1 priority, the lowest, since the absolute timing is not relevant here. Even when rotating the encoder very fast (120 RPM max, according to the device datasheet), the commutations are going to happen on a timescale that is an order of magnitude larger (20 ms). Any other task present in your application can, in fact, be assumed to have a higher priority.

Finally, here is a new main() function designed to put our rotary encoder routines to the test by periodically (10 times a second) checking the value of RCount and displaying its current value on the LCD display:

```
main( void)
{
  int  i = 0;
  char s[16];

  initEX16();            // init and enable interrupts
  initLCD();             // init LCD module
  initR();               // init Rotary Encoder

  // main loop
  while( 1)
  {
    Delayms( 100);       // place holder for a complex app.

    clrLCD();
    sprintf( s, "RCount = %d", RCount);
    putsLCD( s);

  } // main loop

} // main
```

Notice the call to the initEX16() function that, if you remember from Day 10, besides performing the fine tuning of the PIC32 for performance, enables the vectored interrupt mode.

Notice also that where the Delayms(100) call is made, in the main() function, you could actually replace the core of a complex application that will now be able to operate continuously without being "blocked" by the encoder detection routines.

Keyboards

If a few buttons, a keypad, or a rotary encoder offer the possibility to inexpensively accept user input to an embedded-control application, they pale compared to the convenience of a real computer keyboard.

With the advent of the USB bus, computers have finally been freed of a number of "legacy" interfaces that had been in use for decades, since the introduction of the first IBM PC. The PS/2 mouse and keyboard interface is one of them. The result of this transition is that a large number of the "old" keyboards are now flooding the surplus market, and even new PS/2 keyboards are selling for very low prices. This creates the opportunity to give our future PIC32 projects a powerful input capability in return for very little complexity and cost.

Note

Interfacing to a USB keyboard is a completely different deal. You will need a USB host interface, with all the hardware and software complexity that it implies. New PIC32 models with USB host peripherals will address these needs, but a discussion of their use and the command of the USB protocol required are well beyond the scope of this book.

PS/2 Physical Interface

The PS/2 interface uses a five-pin DIN (see Figure 12.7) or a six-pin mini-DIN connector. The first was common on the original IBM PC-XT and AT series but has not been in use for a while. The smaller six-pin version has been more common in recent years. Once the different pin-outs are taken into consideration, you will notice that the two are electrically identical.

The host must provide a 5 V power supply. The current consumption will vary with the keyboard model and year, but you can expect values between 50 and 100 mA. (The original specifications used to call for up to 275 mA max.)

The data and clock lines are both open-collector with pull-up resistors (1–10 k ohm) to allow for two-way communication. In the normal mode of operation, it is the keyboard that drives both lines to send data to the personal computer. When it is necessary, though, the computer can take control to configure the keyboard and to change the status LEDs (Caps Lock and Num Lock).

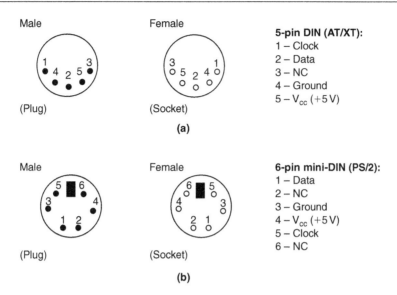

Figure 12.7: (a) Electrical interface (5-pin DIN) and (b) Physical interface (6-pin DIN).

The PS/2 Communication Protocol

At idle, both the data and clock lines are held high by the pull-ups (located inside the keyboard). In this condition the keyboard is enabled and can start sending data as soon as a key has been pressed. If the host holds the clock line low for more than 100us, any further keyboard transmissions are suspended. If the host holds the data line low and then releases the clock line, this is interpreted as a request to send a command (see Figure 12.8).

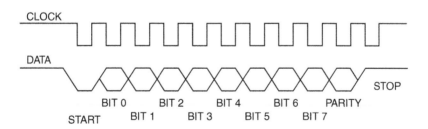

Figure 12.8: Keyboard-to-host communication waveform.

The protocol is a curious mix of synchronous and asynchronous communication protocols we have seen in previous chapters. It is synchronous since a clock line is provided, but it is similar to an asynchronous protocol because a start, a stop, and a parity bit are used to bracket the actual 8-bit data packet. Unfortunately, the baud rate used is not a standard value and can change from unit to unit over time, with temperature and the phase of the moon. In fact, typical values range from 10 to 16 kbits per second. Data changes during the clock high state. Data is valid when the clock line is low. Whether data is flowing from the host to the keyboard or vice versa, it is the keyboard that always generates the clock signal.

> **Note**
>
> The USB bus reverses the roles as it makes each peripheral a synchronous slave of the host. This simplifies things enormously for a non real-time, nonpreemptive multitasking operating system like Windows. The serial port and the parallel port were similarly asynchronous interfaces and, probably for the same reason, both became legacy with the introduction of the USB bus specification.

Interfacing the PIC32 to the PS/2

The unique peculiarities of the protocol make interfacing to a PS/2 keyboard an interesting challenge, since neither the PIC32 SPI interface nor the UART interface can be used. In fact, the SPI interface does not accept 11-bit words (8-bit or 16-bit words are the closest options), whereas the PIC32 UART requires the periodic transmission of special break characters to make use of the powerful auto baud-rate detection capabilities. Also notice that the PS/2 protocol is based on 5 V level signals. This requires care in choosing which pins can be directly connected to the PIC32. In fact, only the 5 V-tolerant digital input pins can be used, which excludes the I/O pins that are multiplexed with the ADC input multiplexer.

Input Capture

The first idea that comes to mind is to implement in software a PS/2 serial interface peripheral using the input capture peripheral (see Figure 12.9).

Five input capture modules are available on the PIC32MX360F512L, connected to the IC1-IC5 pins multiplexed on PORTD pins 8, 9, 10, 11, and 12, respectively.

Each input capture module is controlled by a single corresponding control register ICxCON and works in combination with one of two timers, either Timer2 or Timer3.

One of several possible events can trigger the input capture:

- Rising edge
- Falling edge

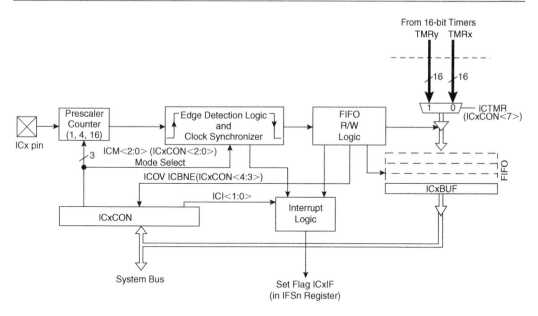

Figure 12.9: Input capture module block diagram.

- Rising and falling edge

- Fourth rising edge

- Sixteenth rising edge

The current value of the selected timer is recorded and stored in a FIFO buffer to be retrieved by reading the corresponding ICxBUF register. In addition to the capture event, an interrupt can be generated after a programmable number of events (each time, every second, every third or every fourth).

To put the input capture peripheral to use and receive the data stream from a PS/2 keyboard, we can connect the IC1 input (RD8) to the clock line and configure the peripheral to generate an interrupt on each and every falling edge of the clock (see Figure 12.10).

After creating a new project that we will call **IC** and following our usual template, we can start adding the following initialization code to a new source file we'll call **PS2IC.c**:

```
#define PS2DAT  _RG12      // PS2 Data input pin
#define PS2CLK  _RD8       // PS2 Clock input pin (IC1)
```

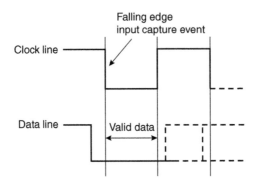

Figure 12.10: PS/2 interface bit timing and the input capture trigger event.

```
void initKBD( void)
{
  // init I/Os
  _TRISD8 = 1;     // make RD8, IC1 an input pin, PS2 clock
  _TRISG12 = 1;    // make RG12 an input pin, PS2 data

  // clear the kbd flag
  KBDReady = 0;

  // init input capture
  IC1CON = 0x8082;          // TMR2, int every cap, fall'n edge
  mIC1ClearIntFlag();       // clear the interrupt flag
  mIC1SetIntPriority( 1);
  mIC1IntEnable( 1);        // enable the IC1 interrupt

  // init Timer2
  mT2ClearIntFlag();        // clear the timer interrupt flag
  mT2SetIntPriority( 1);
  mT2IntEnable( 1);         // enable (TMR2 is not active yet)
} // init KBD
```

We will also need to create an interrupt service routine for the IC1 interrupt vector. This routine will have to operate as a state machine and perform in a sequence the following steps:

1. Verify the presence of a start bit (data line low).

2. Shift in 8 bits of data and compute a parity.

3. Verify a valid parity bit.

4. Verify the presence of a stop bit (data line high).

If any of the above checks fails, the state machine must reset and return to the start condition. When a valid byte of data is received, we will store it in a buffer—think of it as a mailbox—and a flag will be raised so that the main program or any other "consumer" routine will know a valid key code has been received and is ready to be retrieved. To fetch the code, it will suffice to copy it from the mailbox first and then clear the flag (see Figure 12.11).

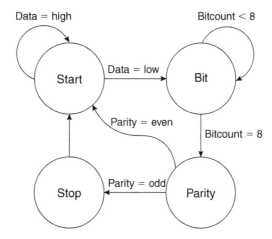

Figure 12.11: The PS/2 receive state machine diagram.

The state machine requires only four states and a counter. All the transitions can be summarized in Table 12.2.

Theoretically I suppose we should consider this an 11-state machine, counting each time the bit state is entered with a different bitcount value as a distinct state. But the four-state model works best for an efficient C language implementation. Let's define a few constants and variables that we will use to maintain the state machine:

```
// definition of the keyboard PS/2 state machine
#define PS2START      0
#define PS2BIT        1
#define PS2PARITY     2
#define PS2STOP       3
```

Table 12.2: PS/2 receive state machine transitions.

State	Conditions	Effect
Start	Data = low	Init bitcount Init parity Transition to bit state
Bit	Bitcount < 8	Shift in key code, LSB first (shift right) Update parity Increment bitcount
	Bitcount = 8	Transition to parity state
Parity	Parity = even	Error; transition back to start
	Parity = odd	Transition to stop
Stop	Data = low	Error; transition back to start
	Data = high	Save the key code in buffer Set flag Transition to start

```
#define TPS     (FPB/1000000)    // timer ticks per uS
#define TMAX    500*TPS          // 500uS time out limit

// PS2 KBD state machine and buffer
int PS2State;
unsigned char KBDBuf;
int KCount, KParity;

// mailbox
volatile int KBDReady;
volatile unsigned char KBDCode;
```

The interrupt service routine for the input capture IC1 module can finally be implemented using a simple switch statement:

```
void __ISR( _INPUT_CAPTURE_1_VECTOR, ipl1) IC1Interrupt( void)
{ // input capture interrupt service routine
  int d;

  // 1. reset timer on every edge
  TMR2 = 0;
```

```
switch( PS2State){
default:
case PS2START:
  if ( ! PS2DAT)                // verify start bit
  {
    KCount = 8;                 // init bit counter
    KParity = 0;                // init parity check
    PR2 = TMAX;                 // init timer period
    T2CON = 0x8000;             // enable TMR2, 1:1
    PS2State = PS2BIT;
  }
  break;

case PS2BIT:
  KBDBuf >>=1;                  // shift in data bit
  if ( PS2DAT)
    KBDBuf += 0x80;
  KParity ^= KBDBuf;            // update parity
  if ( --KCount == 0)           // if all bit read, move on
    PS2State = PS2PARITY;
  break;

case PS2PARITY:
  if ( PS2DAT)                  // verify parity bit
    KParity ^= 0x80;
  if ( KParity & 0x80)          // if parity odd, continue
    PS2State = PS2STOP;
  else
    PS2State = PS2START;
  break;

case PS2STOP:
  if ( PS2DAT)                  // verify stop bit
  {
    KBDCode = KBDBuf;           // save code in mail box
    KBDReady = 1;               // set flag, code available
    T2CON = 0;                  // stop the timer
  }
  PS2State = PS2START;
  break;

} // switch state machine
```

```
    // clear interrupt flag
    d = IC1BUF;                    // discard capture
    mIC1ClearIntFlag();

} // IC1 Interrupt
```

Testing Using a Stimulus Scripts

The small perforated prototyping area can be used to attach a PS/2 mini-DIN connector to the Explorer 16 demonstration board, the only alternative being the development of a custom daughter board (PICTail) for the expansion connectors. Before committing to designing such a board, though, we would like to make sure that the chosen pin-out and code is going to work. The MPLAB SIM software simulator will once more be our tool of choice.

In previous chapters we have used the software simulator in conjunction with the Watch window, the StopWatch, and the Logic Analyzer to verify that our programs were generating the proper timings and outputs, but this time we will need to simulate inputs as well. To this end, MPLAB SIM offers a considerable number of options and resources— so many in fact that the system might seem a bit intimidating. First, the simulator offers two types of input stimuli:

- Asynchronous ones, typically triggered manually by the user

- Synchronous ones, triggered automatically by the simulator after a scripted amount of time (expressed in processor cycles or seconds)

The scripts containing the descriptions of the synchronous stimuli (which can be quite complex) are prepared using the Stimulus window (see Figure 12.12). You must have the MPLAB SIM selected as your active debugging tool (**Debugger | Select Tool | MPLAB SIM**) to open the Stimulus window by selecting **Stimulus | New Workbook** from the Debugger menu. To prepare the simplest type of stimulus script, one that assigns values to specific input pins (but also entire registers) at given points in time, you can select the first tab, **Pin/Register Actions**.

After selecting the unit of measurement of choice, microseconds in our case, click the **first row** of the table that occupies most of the dialog box window space (where it says "click here to Add Signals"). This will allow you to add columns to the table. Add **one column** for every pin for which you want to simulate inputs. In our example, that would be RG12 for the PS/2 Data line and IC1 for the Input Capture pin that we want connected to the PS2 Clock line. At this point we can start typing in the stimulus timing table. To simulate a generic PS/2 keyboard transmission, we need to produce a 10 kHz clock

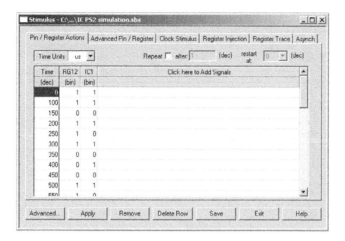

Figure 12.12: The Stimulus window.

signal for 11 cycles, as represented in the PS/2 keyboard waveform in Figure 12.6. This requires an event to be inserted in the timing table each 50us. As an example, Table 12.3 illustrates the trigger events I recommend you add to the Stimulus window timing table to simulate the transmission of key code 0x79.

Once the timing table is filled, you can save the current content for future use with the **Save** button. The file generated will be an ASCII file with the .SBS extension. In theory you could edit this file manually with an MPLAB IDE editor or any basic ASCII editor, but you are strongly discouraged from doing so. The format is more rigid than meets the eye and you might end up trashing it. If you were wondering why the term "workbook" is used for what looks like a simple table, you are invited to explore the other panes (accessible by clicking the tabs at the top of the dialog box) of the Stimulus window. You will see that what we are using in this example is just one of the many methods available, representing a minuscule portion of the capabilities of the MPLAB SIM simulator. A workbook file can contain a number of different types of stimuli produced by any (or multiple) of those panes.

```
Segment of the Stimulus workbook file
## SCL Builder Setup File: Do not edit!!

## VERSION: 3.60.00.00
## FORMAT:  v2.00.01
## DEVICE:  PIC32MX360F512L

## PINREGACTIONS
us
```

Table 12.3: SCL Generator timing example for basic.

Time (us)	RG12	IC1	Comment
0	1	1	Idle state, both lines are pulled up
100	1	1	
150	0	0	First falling edge, start bit (0)
200	1	1	
250	1	0	Bit 0, k ey code LSb (1)
300	0	1	
350	0	0	Bit 1 (0)
400	0	1	
450	0	0	Bit 2 (0)
500	1	1	
550	1	0	Bit 3 (1)
600	1	1	
650	1	0	Bit 4 (1)
700	1	1	
750	1	0	Bit 5 (1)
800	1	1	
850	1	0	Bit 6 (1)
900	0	1	
950	0	0	Bit 7, key code MSb (0)
1000	0	1	
1050	0	0	Parity bit (0)
1100	1	1	
1150	1	0	Stop bit (1)
1200	1	1	Idle

```
No Repeat
RG12
IC1
--
0
1
1
--
100
1
1
--
150
0
0
```

Before we get to use the generated stimulus file, we have to complete the project with a few final touches. Let's prepare an include file to publish the accessible function: initKBD(), the flag KBDReady, and the buffer for the received key code KBDCode:

```
/*
**
** PS2IC.h
**
** PS/2 keyboard input library using input capture
*/
extern volatile int KBDReady;
extern volatile unsigned char KBDCode;

void initKBD( void);
```

Note that there is no reason to publish any other detail of the inner workings of the PS/2 receiver implementation. This will give us freedom to try a few different methods later without changing the interface. Save this file as **PS2IC.h** and include it in the project.

Let's also create a new file, **PS2ICTest.c**, that will contain the usual template with the main() routine and will use the PS2IC.c module to test its functionality:

```
/*
** PS2ICTest.c
**
*/
```

```
// configuration bit settings, Fcy=72MHz, Fpb=36MHz
#pragma config POSCMOD=XT, FNOSC=PRIPLL
#pragma config FPLLIDIV=DIV_2, FPLLMUL=MUL_18, FPLLODIV=DIV_1
#pragma config FPBDIV=DIV_2, FWDTEN=OFF, CP=OFF, BWP=OFF

#include <p32xxxx.h>
#include <explore.h>
#include "PS2IC.h"

main()
{
  int Key;
  initEX16();              // init and enable interrupts
  initKBD();               // initialization routine
  while ( 1)
  {
    if ( KBDReady)         // wait for the flag
    {
      Key = KBDCode;       // fetch the key code
      KBDReady = 0;        // clear the flag
    }
  } // main loop
} //main
```

The `initEX16()` function takes care of the fine tuning of the PIC32 for performance but also enables the vectored interrupts mode. The call to the `initKBD()` function takes care of the PS/2 state machine initialization, sets the chosen input pins, and configures the interrupts for the Input Capture module. The main loop will wait for the interrupt routine to raise the `KBDready` flag, indicating that a key code is available; it will fetch the key code and copy it in the local variable `Key`. Finally, it will clear the `KBDReady` flag, ready to receive a new character.

Now remember to add the file to the project and build all. Instead of immediately launching the simulation, select the **Stimulus** window once more, and click the **Apply** button.

Note

Keep the Stimulus window open (in the background). Resist the temptation to click the Exit button, as that would close the workbook and leave us without stimuli.

Click the **Reset** button (or select **Debugger | Reset**) and watch for the first stimulus to arrive as the microsecond 0 trigger is fired. Remember, both lines RG12 and IC1 are supposed to be set high according to our timetable. A message will confirm this in the Output window (see Figure 12.13).

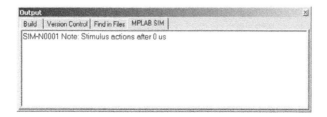

Figure 12.13: In the Output window (MPLAB SIM pane), a stimulus action has been triggered.

It is your choice now to proceed by single-stepping or animating through the program to verify its correct execution. My suggestion is that you place a **breakpoint** inside the main loop on the instruction copying KBDCode to the Key variable. Open the **Watch** window and add **Key** from the Symbol list, then **RUN**.

After a few seconds, the execution should terminate at the breakpoint, and the content of Key should reflect the data we sent through the simulated PS/2 stimulus script: 0x79!

The Simulator Profiler

If you were curious about how fast the simulation of a PIC32 could run on your computer, there is an interesting feature available to you in the MPLAB SIM Debugger menu: the profile. Select the Profile submenu (**Debugger | Profile**) and click **Reset Profile** (see Figure 12.14).

This will clear the simulator profile counters and timers. Then click the **Reset** button and repeat the simulation (**Debugger | Run**) until it encounters the breakpoint again. This time select **Debugger | Profiler | Display Profile** to display the latest statistics from MPLAB SIM (see Figure 12.15).

A relatively long report will be available in the output window (MPLAB SIM pane) detailing how many times each instruction was used by the processor during the simulation and, at the very bottom, offering an assessment of the absolute "simulation"

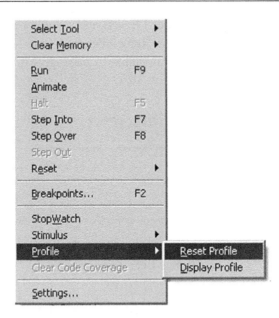

Figure 12.14: The Simulator Profile submenu.

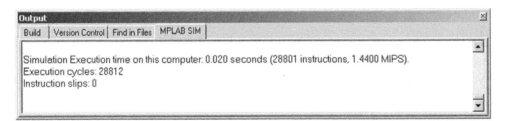

Figure 12.15: Simulator Profile output.

speed. In my case, that turned out to be 1.4 MIPS. A respectable result after all, although nothing to write home about. Contrary to the simulation of other PIC® microcontrollers, where these numbers would have compared well with the actual processor real-time performance, compared to the PIC32 the software simulation (on my laptop) ran at just 1/50th of the actual silicon speed!

Change Notification

Though the input capture technique worked all right, there are other options that we might be curious to explore to interface efficiently with a PS/2 keyboard. In particular there is another interesting peripheral available on the PIC32 that could offer an alternative method to implement a PS/2 interface: the Change Notification (CN) module. There are as many as 22 I/O pins connected with this module, and this can give us some freedom in choosing the ideal input pins for the PS/2 interface while making sure they don't conflict with other functions required in our project or already in use on the Explorer 16 board.

Only three control registers are associated with the CN module. The CNCON register contains the basic control bits to enable the module, and the CNEN register contains the enable bits for each of the CN input pins. Note that only one interrupt vector is available for the entire CN module; therefore it will be the responsibility of the interrupt service routine to determine which one has actually changed if more than one is enabled. Finally, the CNPUE register controls the individual activation of internal pull-up resistors available for each input pin (see Figure 12.16).

Virtual Address	Name		Bit 31/23/15/7	Bit 30/22/14/6	Bit 29/21/13/5	Bit 28/20/12/4	Bit 27/19/11/3	Bit 26/18/10/2	Bit 25/17/9/1	Bit 24/16/8/0
BF88_61C0	CNCON	31:24	—	—	—	—	—	—	—	—
		23:16	—	—	—	—	—	—	—	—
		15:8	ON	FRZ	SIDL	—	—	—	—	—
		7:0	—	—	—	—	—	—	—	—
BF88_61C4	CNCONCLR	31:0	Write clears selected bits in CNCON, Read yields undefined							
BF88_61C8	CNCONSET	31:0	Write sets selected bits in CNCON, Read yields undefined							
BF88_61CC	CNCONINV	31:0	Write inverts selected bits in CNCON, Read yields undefined							
BF88_61D0	CNEN	31:24	—	—	—	—	—	—	—	—
		23:16	—	—	CNEN 21[1]	CNEN 20[1]	CNEN 19[1]	CNEN 18	CNEN 17	CNEN 16
		15:8	CNEN[15:8]							
		7:0	CNEN[7:0]							
BF88_61D4	CNENCLR	31:0	Write clears selected bits in CNEN, Read yields undefined							
BF88_61D8	CNENSET	31:0	Write sets selected bits in CNEN, Read yields undefined							
BF88_61DC	CNENINV	31:0	Write inverts selected bits in CNEN, Read yields undefined							
BF88_61E0	CNPUE	31:24	—	—	—	—	—	—	—	—
		23:16	—	—	CNPUE 21[1]	CNPUE 20[1]	CNPUE 19[1]	CNPUE 18	CNPUE 17	CNPUE 16
		15:8	CNEN[15:8]							
		7:0	CNEN[7:0]							
BF88_61E4	CNPUECLR	31:0	Write clears selected bits in CNPUE, Read yields undefined							
BF88_61E8	CNPUESET	31:0	Write sets selected bits in CNPUE, Read yields undefined							
BF88_61EC	CNPUEINV	31:0	Write inverts selected bits in CNPUE, Read yields undefined							
Note 1: CNEN and CNPUE bit(s) are not implemented on 64-pin variants and read as '0'										

Figure 12.16: The CN control registers table.

In practice, all we need to support the PS/2 interface is just one of the CN inputs connected to the PS2 clock line. The PIC32 weak pull-up will not be necessary in this case since it is already provided by the keyboard. There are 22 pins to choose from, and we will find a CN input that is not shared with the ADC (remember, we need a 5 V tolerant input) and is not overlapping with some other peripheral used on the Explorer 16 board. This takes a little studying between the device datasheet and the Explorer 16 user guide. But once the input pin is chosen, say, CN11 (multiplexed with pin RG9, the SS line of the SPI2 module and the PMP module Address line PMA2), a new initialization routine can be written in just a couple of lines (see Figure 12.17):

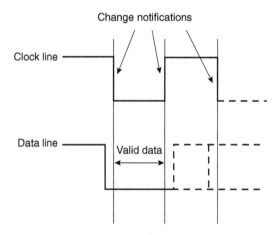

Figure 12.17: PS/2 interface bit timing Change Notification event detail.

```
#define PS2DAT  _RG12       // PS2 Data input pin
#define PS2CLK  _RG9        // PS2 Clock input pin (CN11)

void initKBD( void)
{
  // init I/Os
  _TRISG9 = 1;              // make RG9 an input pin
  _TRISG12 = 1;            // make RG12 an input pin

  // clear the flag
  KBDReady = 0;
```

```
// configure Change Notification system
CNENbits.CNEN11 = 1;      // enable PS2CLK (CN11)
CNCONbits.ON = 1;         // turn on Change Notification
mCNSetIntPriority( 1);    // set interrupt priority >0
mCNClearIntFlag();        // clear the interrupt flag
mCNIntEnable( 1);         // enable interrupt
} // init KBD
```

As per the interrupt service routine, we can use exactly the same state machine used in the previous example, adding only a couple of lines of code to make sure that we are looking at a falling edge of the clock line.

In fact, using the input capture module, we could choose to receive an interrupt only on the desired clock edge, whereas the change notification module will generate an interrupt both on falling and rising edges. A simple check of the status of the clock line immediately after entering the interrupt service routine will help us tell the two edges apart:

```
void __ISR( _CHANGE_NOTICE_VECTOR, ipl1) CNInterrupt( void)
{ // change notification interrupt service routine
  // 1. make sure it was a falling edge
  if ( PS2CLK == 0)
  {
    switch( PS2State){
    default:
    case PS2START:            // verify start bit
      if ( ! PS2DAT)
      {
        KCount = 8;           // init bit counter
        KParity = 0;          // init parity check
        PS2State = PS2BIT;
      }
      break;
    case PS2BIT:
      KBDBuf >>=1;            // shift in data bit
      if ( PS2DAT)
        KBDBuf += 0x80;
      KParity ^= KBDBuf;      // update parity
      if ( --KCount == 0)     // if all bit read, move on
        PS2State = PS2PARITY;
      break;
```

```
    case PS2PARITY:
      if ( PS2DAT)              // verify parity
        KParity ^= 0x80;
      if ( KParity & 0x80)      // if parity odd, continue
        PS2State = PS2STOP;
      else
        PS2State = PS2START;
      break;

    case PS2STOP:
      if ( PS2DAT)              // verify stop bit
      {
        KBDCode = KBDBuf;       // save code in mail box
        KBDReady = 1;           // set flag, code available
      }
      PS2State = PS2START;
      break;

    } // switch state machine
  } // if falling edge

  // clear interrupt flag
  mCNClearIntFlag();
} // CN Interrupt
```

Add the constants and variables declarations already used in the previous example:

```
// definition of the keyboard PS/2 state machine
#define PS2START    0
#define PS2BIT      1
#define PS2PARITY   2
#define PS2STOP     3

// PS2 KBD state machine and buffer
int PS2State;
unsigned char KBDBuf;
int KCount, KParity;

// mailbox
volatile int KBDReady;
volatile unsigned char KBDCode;
```

Package it all together in a file that we will call **PS2CN.c**.

The include file PS2CN.h will be practically identical to the previous example, since we are going to offer the same interface:

```
/*
**
** PS2CN.h
**
** PS/2 keyboard input module using Change Notification
*/

extern volatile int KBDReady;
extern volatile unsigned char KBDCode;

void initKBD( void);
```

Create a new project called **PS2CN** and add both the .c and the .h files to the project.

Finally, create a main module to test this new technique. One more time, it will be mostly identical to the previous project:

```
/*
** PS2CNTest.c
**
*/
// configuration bit settings, Fcy=72MHz, Fpb=36MHz
#pragma config POSCMOD=XT, FNOSC=PRIPLL
#pragma config FPLLIDIV=DIV_2, FPLLMUL=MUL_18, FPLLODIV=DIV_1
#pragma config FPBDIV=DIV_2, FWDTEN=OFF, CP=OFF, BWP=OFF

#include <p32xxxx.h>
#include <explore.h>
#include "PS2CN.h"

main()
{
  initEX16();                    // init and enable interrupts
  initKBD();                     // kbd initialization

  while ( 1)
  {
    if ( KBDReady)               // wait for the flag
    {
```

```
        PORTA = KBDCode;   // fetch the key code
        KBDReady = 0;      // clear the flag
    }
  } // main loop
} //main
```

Save the project, then build the project (**Project | BuildAll**) to compile and link all the modules. To test the change notification technique, we will use once more MPLAB SIM stimulus generation capabilities. Once more we will repeat most of the steps performed in the previous project. Starting with the Stimulus window (**Debugger | Stimulus | New Workbook**), we will create a new workbook. Inside the window, create **two columns**, one for the same PS2 Data line connected to RG12, but the PS2 Clock line will be connected to the CN11 Change Notification module input this time. Add the same sequence of stimuli as presented in Table 12.3, replacing the IC1 **input** column with the CN11 column. Save the workbook as **PS2CN.sbs** and then click the **Apply** button to activate the stimulus script.

We are ready now to execute the code and test the proper functioning of the new PS/2 interface. Open the **Watch** window and add Key from the symbols list. Then set a **breakpoint** inside the main loop on the line where KBDCode is copied to the Key variable. Finally, perform a reset (**Debugger | Reset**) and verify that the first event is triggered (setting both PS/2 input lines high at time 0 us). Run the code (**Debugger | RUN**) and, if all goes well, you will see the processor stop at the breakpoint after less than a second, and you will see the contents of Key to be updated to reflect the key code 0x79. Success again!

Evaluating Cost

Changing from the Input Capture to the Change Notification method was almost too easy. The two peripherals are extremely potent and, although designed for different purposes, when applied to the task at hand they performed almost identically. In the embedded world, though, you should constantly ask yourself if you could solve the problem with fewer resources even when, as in this case, there seems to be abundance.

Let's evaluate the real cost of each solution by counting the resources used and their relative scarcity. In using the Input Capture, we have in fact used one of five IC modules available in the PIC32MX360F512L model. This peripheral is designed to operate in conjunction with a timer (Timer2 or Timer3), although we are not using the timing information in our application but only the interrupt mechanism associated with the input edge trigger. When using the Change Notification, we are using only one of 22 possible inputs, but we are also taking control of the sole interrupt vector available to this peripheral. In other words, should we need any other input pin to be controlled by the change notification peripheral, we will have to share the interrupt vector, adding latency and complexity to the solution. I would call this a tie.

I/O Polling

There is one more method that we could explore to interface to a PS/2 keyboard. It is the most basic one and it implies the use of a timer, set for a periodic interrupt, and any 5V tolerant I/O pin of the microcontroller. In a way, this method is the most flexible from a configuration and layout point of view. It is also the most generic since any microcontroller model, even the smallest and most inexpensive, will offer at least one timer module suitable for our purpose. The theory of operation is pretty simple. At regular intervals an interrupt will be generated, set by the value of the period register associated with the chosen timer (see Figure 12.18).

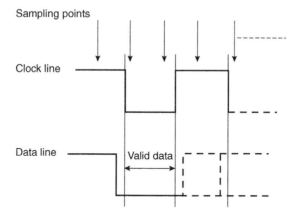

Figure 12.18: PS/2 interface bit timing I/O polling sampling points.

We will use Timer4 this time, just because we never used it before, and its associated period register PR4. The interrupt service routine T4Interrupt() will sample the status of the PS/2 Clock line and it will determine whether a falling edge has occurred on the PS/2 Clock line over the previous period. When a falling edge is detected, the PS/2 Data line status will be considered to receive the key code. To determine how frequently we should perform the sampling and therefore identify the optimal value of the PR4 register, we should look at the shortest amount of time allowed between two edges on the PS/2 clock line. This is determined by the maximum bit rate specified for the PS/2 interface that, according to the documentation in our possession, corresponds to about 16 k bit/s. At that rate, the clock signal can be represented by a square wave with an approximately 50-percent duty cycle and a period of approximately 62.5 us. In other words, the clock line will stay low for little more than 30 us each time a data bit is presented on the PS/2 Data line, and it will stay high for approximately the same amount of time, during which the next bit will be shifted out.

By setting PR4 to a value that will make the interrupt period shorter than 30 us (say 25 us), we can guarantee that the clock line will always be sampled at least once between two consecutive edges. The keyboard transmission bit rate, though, could be as slow as 10 k bit/s, giving a maximum distance between edges of about 50 us. In that case we would be sampling the clock and data lines twice and possibly up to three times between each clock edge. In other words, we will have to build a new state machine to detect the actual occurrence of a falling edge and to properly keep track of the PS/2 clock signal (see Figure 12.19).

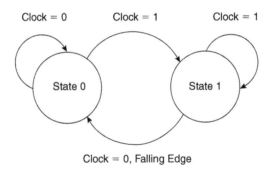

Figure 12.19: Clock-polling state machine graph.

The state machine requires only two states, and all the transitions can be summarized in the Table 12.4.

When a falling edge is detected, we can still use the same state machine developed in the previous projects to read the data line. It is important to note that in this case

Table 12.4: Clock-polling state machine transitions.

State	Conditions	Effect
State0	Clock = 0	Remain in State0
	Clock = 1	Rising Edge, Transition to State1
State1	Clock = 1	Remain in State1
	Clock = 0	Falling edge detected Execute the Data state machine Transition to State0

the value of the data line is not guaranteed to be sampled right after the actual falling edge of the clock line has occurred but instead could be considerably delayed. To avoid the possibility of reading the data line outside the valid period, it is imperative to simultaneously sample both the clock and the data line. This will be performed by copying the value of the two inputs in two local variables (d and k) at the very beginning of the interrupt service routine. In our example, we will choose to use RG12 (again) for the data line and RG13 for the clock line. Here is the skeleton implementation of the Clock-polling state machine illustrated previously:

```
#define PS2CLK  _RG13       // PS2 Clock output
#define PS2DAT  _RG12       // PS2 Data input pin

// PS2 KBD state machine and buffer
int PS2State;
unsigned char KBDBuf;

// mailbox
volatile int KBDReady;
volatile unsigned char KBDCode;

void __ISR( _TIMER_4_VECTOR, ipl1) T4Interrupt( void)
{
  int d, k;

  // sample the inputs clock and data at the same time
  d = PS2DAT;
  k = PS2CLK;
```

```
// keyboard state machine
if ( KState)
{ // previous time clock was high KState 1
  if ( !k)                // PS2CLK == 0
  { // falling edge detected,
    KState = 0;           // transition to State0

          <<<< insert data state machine here >>>>

  } // falling edge
  else
  { // clock still high, remain in State1

  } // clock still high
} // state 1

else
{ // state 0
  if ( k) // PS2CLK == 1
  { // rising edge, transition to State1
    KState = 1;

  } // rising edge
  else
  { // clocl still low, remain in State0

  } // clock still low
} // state 0

// clear the interrupt flag
mT4ClearIntFlag();
} // T4 Interrupt
```

Thanks to the periodic nature of the polling mechanism we just developed, we can add a new feature to the PS2 interface to make it more robust with minimal effort. First, we can add a counter to idle loops of both states of the clock state machine. This way we will be able to create a timeout to be able to detect and correct error conditions, should the PS/2 keyboard be disconnected during a transmission or if the receive routine should lose synchronization for any reason.

The new transition table (Table 12.5) is quickly updated to include the timeout counter `KTimer`.

Table 12.5: Clock-polling (with timeout) state machine transition table.

State	Conditions	Effect
State0	Clock = 0	Remain in State0 Decrement KTimer If KTimer = 0, error Reset the data state machine
	Clock = 1	Rising Edge, Transition to State1
State1	Clock = 1	Remain in State1 Decrement KTimer If KTimer = 0, error Reset the data state machine
	Clock = 0	Falling edge detected Execute the Data state machine Transition to State0 Restart KTimer

The new transition table adds only a few instructions to our interrupt service routine:

```
void __ISR( _TIMER_4_VECTOR, ipl1) T4Interrupt( void)
{
  int d, k;

  // sample the inputs clock and data at the same time
  d = PS2DAT;
  k = PS2CLK;

  // keyboard state machine
  if ( KState)
  { // previous time clock was high KState 1
    if ( !k)                             // PS2CLK = 0
    { // falling edge detected,
      KState = 0;                        // transition to State0
      KTimer = KMAX;                     // restart the counter

            <<<< insert data state machine here >>>>

    } // falling edge
    else
    { // clock still high, remain in State1
```

```
    KTimer--;
    if ( KTimer == 0)                    // Timeout
      PS2State = PS2START;               // Reset data SM
  } // clock still high
} // Kstate 1
else
{ // Kstate 0
  if ( k)                                // PS2CLK == 1
  { // rising edge, transition to State1
    KState = 1;
  } // rising edge
  else
  { // clocl still low, remain in State0
    KTimer--;
    if ( KTimer = 0)                     // Timeout
      PS2State = PS2START;               // Reset data SM
  } // clock still low
} // Kstate 0

// clear the interrupt flag
mT4ClearIntFlag();
} // T4 Interrupt
```

Testing the I/O Polling Method

Let's now insert the Data state machine from the previous projects, modified to operate on the value sampled in d and k at the interrupt service routine entry. It fits entirely in a single switch statement:

```
switch( PS2State){
  default:
  case PS2START:
    if ( !d) // PS2DAT == 0
    {
      KCount = 8;                        // init bit counter
      KParity = 0;                       // init parity check
      PS2State = PS2BIT;
    }
    break;
```

```
case PS2BIT:
  KBDBuf >>=1;              // shift in data bit
  if ( d)                  // PS2DAT == 1
    KBDBuf += 0x80;
  KParity ^= KBDBuf;       // calculate parity
  if ( --KCount == 0)      // all bit read
    PS2State = PS2PARITY;
  break;

case PS2PARITY:
  if ( d)                  // PS2DAT == 1
    KParity ^= 0x80;
  if ( KParity & 0x80)     // parity odd, continue
    PS2State = PS2STOP;
  else
    PS2State = PS2START;
  break;

case PS2STOP:
  if ( d)                  // PS2DAT == 1
  {
    KBDCode = KBDBuf;      // write in the buffer
    KBDReady = 1;
  }
  PS2State = PS2START;
  break;

} // switch
```

Let's complete this third module with a proper initialization routine:

```
void initKBD( void)
{
  // init I/Os
  ODCGbits.ODCG13 = 1;     // make RG13 open drain (PS2clk)
  _TRISG13 = 1;            // make RG13 an input pin (for now)
  _TRISG12 = 1;            // make RG12 an input pin

  // clear the kbd flag
  KBDReady = 0;
```

```
  // configure Timer4
  PR4 = 25*TPS - 1;        // 25 us
  T4CON = 0x8000;          // T4 on, prescaler 1:1
  mT4SetIntPriority( 1);   // lower priority
  mT4ClearIntFlag();       // clear interrupt flag
  mT4IntEnable( 1);        // enable interrupt
} // init KBD
```

This is quite straightforward.

Let's save it all in a module we can call **PS2T4.c.** Let's create a new include file, too:

```
/*
**
** PS2T4.h
**
** PS/2 keyboard input library using T4 polling
*/

extern volatile int KBDReady;
extern volatile unsigned char KBDCode;

void initKBD( void);
```

It is practically identical to all previous modules include files, and the main test module will not be much different either:

```
/*
** PS2T4 Test
**
*/
// configuration bit settings, Fcy=72MHz, Fpb=36MHz
#pragma config POSCMOD=XT, FNOSC=PRIPLL
#pragma config FPLLIDIV=DIV_2, FPLLMUL=MUL_18, FPLLODIV=DIV_1
#pragma config FPBDIV=DIV_2, FWDTEN=OFF, CP=OFF, BWP=OFF

#include <p32xxxx.h>
#include <explore.h>
#include "PS2T4.h"

main()
{
  initEX16();              // init and configure interrupts
  initKBD();               // initialization routine
```

```
while ( 1)
{
  if ( KBDReady)          // wait for the flag
  {
    PORTA = KBDCode;      // fetch the key code
    KBDReady = 0;         // clear the flag
  }
} // main loop
} //main
```

Create a new project **T4** and add all three files to it. Build all and follow the same series of steps used in the previous two examples to generate a stimulus script. Remember that this time the stimulus for the Clock line must be provided on the RG13 pin. Open the **Watch** window and add **PORTA** and **KBDCode**. Finally set a **breakpoint** to the line after the assignment to PORTA and execute **Debug | Run**. If all goes well, even this time you should be able to see PORTA updated in the Watch window and showing a new value of 0x79. Success again!

Cost and Efficiency Considerations

Comparing the cost of this solution to the previous two, we realize that the I/O polling approach is the one that gives us the most freedom in choosing the input pins and uses only one resource, a timer, and one interrupt vector. The periodic interrupt can also be seamlessly shared with other tasks to form a common time base if they all can be reduced to multiples of the polling period. The time-out feature is an extra bonus; to implement it in the previous techniques, we would have had to use a separate timer and another interrupt service routine in addition to the Input Capture or Change Notification modules and interrupts.

Looking at the efficiency, the Input Capture and the Change Notification methods appear to have an advantage because an interrupt is generated only when an edge is detected. Actually, as we have seen, the Input Capture is the best method from this point of view, since we can select precisely the one type of edge we are interested in—that is, the falling edge of the PS/2 Clock line.

The I/O polling method appears to require the longest interrupt routine, but the number of lines does not reflect the actual *weight* of the interrupt service routine. In fact, when we look closer, of the two nested state machines that compose the I/O polling interrupt service routine, only a few instructions are executed at every call, resulting in a very short execution time and minimal overhead.

To verify the actual software overhead imposed by the interrupt service routines, we can perform one simple test on each one of the three implementations of the PS/2 interface. I will use only the last one as an example. We can allocate one of the I/O pins (one of the LED outputs on PORTA not used by the JTAG port would be a logical choice) to help us visualize when the microcontroller is inside an interrupt service routine. We can set the pin on entry and reset it right before exit:

```
void __ISR(..) T4Interrupt( void)
{
  _RA2 = 1;        // flag up, inside the ISR

  <<< Interrupt service routine here >>

  _RA2 = 0;        // flag down, back to the main
}
```

Using MPLAB SIM simulator Logic Analyzer view, we can visualize it on our computer screen. Follow the Logic Analyzer checklist so you will remember to enable the **Trace** buffer, and set the correct **simulation** speed. Select the **RA0** channel and rebuild the project.

To test the first two methods (IC and CN), you will need to open the **Stimulus** window and apply the **scripts** to simulate the inputs. Without them there will be no interrupts at all. When testing the I/O polling routine, you won't necessarily need it; the Timer4 interrupt keeps coming anyway and, after all, we are interested in seeing how much time is wasted by the continuous polling when no keyboard input is provided.

Let MPLAB SIM run for a few seconds, then stop the simulation and switch back to the **Logic Analyzer** window. You will have to zoom in quite a bit to get an accurate picture (see Figure 12.20).

Activate the **cursors** ▶◀ and drag them to measure the number of cycles between two consecutive rising edges of RA2, marking two successive entries in the interrupt service routine. Since we selected a 25 us period, you should read 900 cycles between calls (25 us * 36 cycles/us @72 MHz).

Measuring the number of cycles between a rising edge and a falling edge of RA2 instead will tell us, with good approximation, how much time we are spending inside the interrupt service routine; 36 cycles is what I found. The ratio between the two quantities will give us an indication of the computing power absorbed by the PS/2 interface. In our case that turns out to be just 4 percent.

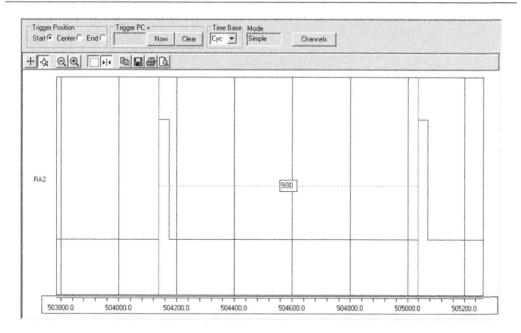

Figure 12.20: Logic Analyzer view, measuring the I/O polling period.

Keyboard Buffering

Independently from the solution you will choose of the three we have explored so far, there are a few more details we need to take care of before we can claim to have completed the interface to the PS/2 keyboard. First, we need to add a buffering mechanism between the PS/2 interface routines and the "consumer" or the main application. So far, in fact, we have provided a simple mailbox mechanism that can store only the last key code received. If you investigate further how the PS/2 keyboard protocol works, you will discover that when a single key is pressed and released, a minimum of three (and a maximum of five) key codes are sent to the host. If you consider Shift, Ctrl, and Alt key combinations, things get a little more complicated and you realize immediately that the single-byte mailbox is not going to be sufficient. My suggestion is to use at least a 16-byte first-in/first-out (FIFO) buffer. The input to the buffer can be easily integrated with the receiver interrupt service routines so that when a new key code is received it is immediately inserted in the FIFO.

The buffer can be declared as an array of characters, and two pointers will keep track of the *head* and *tail* of the buffer in a circular scheme (see Figure 12.21).

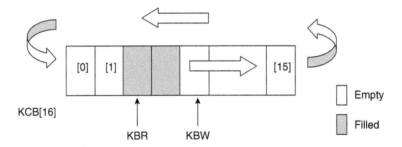

Figure 12.21: Circular buffer (FIFO).

```
// circular buffer
unsigned char KCB[ KB_SIZE];

// head and tail or write and read pointers
volatile int KBR, KBW;
```

Following a few simple rules, we can keep track of the buffer content:

- The write pointer KBW (or head) marks the first empty location that will receive the next key code.

- The read pointer KBR (or tail) marks the first filled location.

- When the buffer is empty, KBR and KBW are pointing at the same location.

- When the buffer is full, KBW points to the location before KBR.

- After reading or writing a character to/from the buffer, the corresponding pointer is incremented.

- Upon reaching the end of the array, each pointer will wrap around to the first element of the array.

Insert the following snippet of code into the initialization routine:

```
// init the circular buffer pointers

  KBR = 0;
  KBW = 0;
```

Then update the state machine STOP state:

```
case PS2STOP:
  if ( PS2IN & DATMASK)      // verify stop bit
  {
    KCB[ KBW] = KBDBuf;      // write in the buffer
    // check if buffer full
    if ( (KBW+1)%KB_SIZE != KBR)
      KBW++;                   // else increment ptr
    KBW %= KB_SIZE;          // wrap around
  }
  PS2State = PS2START;
  break;
```

Notice the use of the % operator to obtain the reminder of the division by the buffer size. This allows us to keep the pointers wrapping around the circular buffer.

A few considerations are required for fetching key codes from the FIFO buffer. In particular, if we choose the input capture or the change notification methods, we will need to make a new function available (getKeyCode()) to replace the mailbox/flag mechanism. The function will return FALSE if there are no key codes available in the buffer and TRUE if there is at least one key code in the buffer, and the code is returned via a pointer:

```
int getKeyCode( char *c)
{
  if ( KBR == KBW)          // buffer empty
    return FALSE;

  // else buffer contains at least one key code
  *c = KCB[ KBR++];         // extract the first key code
  KBR %= KB_SIZE;           // wrap around the pointer

  return TRUE;
} // getKeyCode
```

Notice that the extraction routine modifies only the read pointer; therefore it is safe to perform this operation when the interrupts are enabled. Should an interrupt occur during the extraction, there are two possible scenarios:

- The buffer was empty, a new key code will be added, but the getKeyCode() function will "notice" the available character only at the next call.

- The buffer was not empty, and the interrupt routine will add a new character to the buffer tail, if there is enough room.

In both cases there are no particular concerns of conflicts or dangerous consequences.

But if we choose the polling technique, the timer interrupt is constantly active and we can use it to perform one more task for us. The idea is to maintain the simple mailbox and flag mechanism for delivering key codes as the interface to the receive routine and have the interrupt constantly checking the mailbox, ready to replenish it with the content from the FIFO. This way we can confine the entire FIFO management to the interrupt service routine, making the buffering completely transparent and maintaining the simplicity of the mailbox delivery interface. The new and complete interrupt service routine for the polling I/O mechanism is presented here:

```
void __ISR( _TIMER_4_VECTOR, ipl1) T4Interrupt( void)
{
  int d, k;

//_RA2 =1;

  // 1. check if buffer available
  if ( !KBDReady && ( KBR!=KBW))
  {
    KBDCode = KCB[ KBR++];
    KBR %= KB_SIZE;
    KBDReady = 1;            // flag code available
  }

  // 2. sample the inputs clock and data at the same time
  d = PS2DAT;
  k = PS2CLK;

  // 3. Keyboard state machine
  if ( KState)
  { // previous time clock was high KState 1
    if ( !k)                // PS2CLK == 0
    { // falling edge detected,
      KState = 0;           // transition to State0
      KTimer = KMAX;        // restart the counter
```

```
switch( PS2State){
default:
case PS2START:
  if ( !d)// PS2DAT == 0
  {
    KCount = 8;          // init bit counter
    KParity = 0;         // init parity check
    PS2State = PS2BIT;
  }
  break;

case PS2BIT:
  KBDBuf >>= 1;          // shift in data bit
  if ( d)                // PS2DAT == 1
    KBDBuf += 0x80;
  KParity ^= KBDBuf;     // calculate parity
  if ( --KCount == 0)    // all bit read
    PS2State = PS2PARITY;
  break;

case PS2PARITY:
  if ( d)                // PS2DAT == 1
    KParity ^= 0x80;
  if ( KParity & 0x80)   // parity odd, continue
    PS2State = PS2STOP;
else
    PS2State = PS2START;
break;

case PS2STOP:
  if ( d)                // PS2DAT == 1
  {
    KCB[ KBW] = KBDBuf;  // write in the buffer
    // check if buffer full
    if ( (KBW+1)%KB_SIZE != KBR)
      KBW++;             // else increment ptr
    KBW %= KB_SIZE;      // wrap around
  }
  PS2State = PS2START;
  break;
```

```
      } // switch
    } // falling edge
    else
    { // clock still high, remain in State1
      KTimer--;
      if ( KTimer == 0)         // timeout
        PS2State = PS2START;  // reset data SM
    } // clock still high
  } // Kstate 1
  else
  { // Kstate 0
    if ( k)                     // PS2CLK == 1
    { // rising edge, transition to State1
      KState = 1;
    } // rising edge
    else
    { // clocl still low, remain in State0
      KTimer--;
      if ( KTimer == 0)         // timeout
        PS2State = PS2START;  // reset data SM
    } // clock still low
  } // Kstate 0

  // 4. clear the interrupt flag
  mT4ClearIntFlag();

//_RA2 = 0;
} // T4 Interrupt
```

Key Code Decoding

So far we have been talking exclusively about key codes, and you might have assumed
that they match the ASCII codes for each key—say, if you press the A key on the keyboard
you would expect the corresponding ASCII code (0x41) to be sent. But things are not that
simple. To maintain a level of layout neutrality, all PC keyboards use *scan codes*, where
each key is assigned a numerical value that is related to the original implementation of
the keyboard scanning firmware of the first IBM PC, circa 1980. The translation from
scan codes to actual ASCII characters happens at a higher level according to specific
(international) keyboard layouts and, nowadays, is performed by Windows drivers. Keep
in mind also that for historical reasons there are at least three different and partially

compatible "scan code sets." Fortunately, by default, all keyboards support the scan code set #2, which is the one we will focus on in the following discussion.

Each time a key is pressed (any key, including a Shift or Ctrl key), the scan code associated with it is sent to the host; this is called the *make* code. As soon as the same key is released, a new (sequence of) codes is sent to the host; this is called the *break* code. The break code is typically composed of the same make code but prefixed with 0xF0. Some keys have a 2-byte-long make code (typically the Ctrl, Alt, and arrow keys) and consequently the break code is 3 bytes long (see Table 12.6).

Table 12.6: Example of make and break codes used in Scan Code Set 2 (default).

Key	Make Code	Break Code
A	1C	F0, 1C
5	2E	F0, 2E
F10	09	F0, 09
Right Arrow	E0, 74	E0, F0, 74
Right Ctrl	E0, 14	E0, F0, 14

To process this information and translate the scan codes intro proper ASCII, we will need a table that will help us map the basic scan codes for a given keyboard layout. The following code will illustrate the translation table for a common U.S. English keyboard layout:

```
// PS2 keyboard codes (standard set #2)
const char keyCodes[128]={
      0,   F9,    0,  F5,   F3,  F1,   F2,  F12,   //00
      0,  F10,   F8,  F6,   F4, TAB,  '`',    0,   //08
      0,    0,L_SHFT,  0,L_CTRL, 'q', '1',    0,   //10
      0,    0,  'z', 's',  'a', 'w',  '2',    0,   //18
      0,  'c',  'x', 'd',  'e', '4',  '3',    0,   //20
      0,  ' ',  'v', 'f',  't', 'r',  '5',    0,   //28
      0,  'n',  'b', 'h',  'g', 'y',  '6',    0,   //30
      0,    0,  'm', 'j',  'u', '7',  '8',    0,   //38
      0,  ',',  'k', 'i',  'o', '0',  '9',    0,   //40
      0,  '.',  '/', 'l',  ';', 'p',  '-',    0,   //48
      0,    0, '\'',   0,  '[', '=',    0,    0,   //50
```

```
        CAPS, R_SHFT,ENTER,  ']',     0,0x5c,   0,    0,         //58
           0,     0,    0,     0,     0,   0, BKSP,    0,         //60
           0,   '1',    0,   '4',   '7',   0,    0,    0,         //68
           0,   '.',  '2',   '5',   '6', '8',  ESC,  NUM,        //70
         F11,   '+',  '3',   '-',   '*', '9',    0,    0         //78
      };
```

Notice that the array has been declared as const so that it will be allocated in program memory space to save precious RAM space.

It will also be convenient to have available a similar table for the Shift function of each key:

```
const char keySCodes[128] = {
           0,    F9,    0,    F5,   F3,   F1,    F2,  F12,        //00
           0,   F10,   F8,    F6,   F4,  TAB,   '~',    0,        //08
           0,     0,L_SHFT,    0,L_CTRL,'Q', '!',    0,          //10
           0,     0,  'Z',   'S',  'A',  'W',   '@',    0,        //18
           0,   'C',  'X',   'D',  'E',  '$',   '#',    0,        //20
           0,   ' ',  'V',   'F',  'T',  'R',   '%',    0,        //28
           0,   'N',  'B',   'H',  'G',  'Y',   '^',    0,        //30
           0,     0,  'M',   'J',  'U',  '&',   '*',    0,        //38
           0,   '<',  'K',   'I',  'O',  ')',   '(',    0,        //40
           0,   '>',  '?',   'L',  ':',  'P',   '_',    0,        //48
           0,     0, '\"',     0,  '{',  '+',     0,    0,        //50
        CAPS, R_SHFT,ENTER,  '}',     0,  '|',    0,    0,        //58
           0,     0,    0,     0,     0,    0, BKSP,    0,        //60
           0,   '1',    0,   '4',  '7',    0,    0,    0,         //68
           0,   '.',  '2',   '5',  '6',  '8',  ESC,  NUM,        //70
         F11,   '+',  '3',   '-',  '*',  '9',    0,    0         //78
      };
```

For all the ASCII characters, the translation is straightforward, but we will have to assign special values to the function, Shift, and Ctrl keys. Only a few of them will find a corresponding code in the ASCII set:

```
// special function characters
#define TAB     0x9
#define BKSP    0x8
#define ENTER   0xd
#define ESC     0x1b
```

For all the others we will have to create our own conventions or, until we have a use for them, we might just ignore them and assign them a common code (0):

```
#define L_SHFT    0x12
#define R_SHFT    0x12
#define CAPS      0x58
#define L_CTRL    0x0
#define NUM       0x0
#define F1        0x0
#define F2        0x0
#define F3        0x0
#define F4        0x0
#define F5        0x0
#define F6        0x0
#define F7        0x0
#define F8        0x0
#define F9        0x0
#define F10       0x0
#define F11       0x0
#define F12       0x0
```

The getC() function will perform the basic translations for the most common keys and it will keep track of the Shift keys status as well as the Caps key toggling:

```
int CapsFlag=0;
char getC( void)
{
  unsigned char c;

  while( 1)
  {

    while( !KBDReady);// wait for a key to be pressed
    // check if it is a break code
    while (KBDCode == 0xf0)
    { // consume the break code
      KBDReady = 0;
      // wait for a new key code
      while ( !KBDReady);
      // check if the shift button is released
      if ( KBDCode == L_SHFT)
        CapsFlag = 0;
```

```
      // and discard it
      KBDReady = 0;
      // wait for the next key
      while ( !KBDReady);
    }
    // check for special keys
    if ( KBDCode == L_SHFT)
    {
      CapsFlag = 1;
      KBDReady = 0;
    }
    else if ( KBDCode == CAPS)
    {
      CapsFlag = !CapsFlag;
      KBDReady = 0;
    }

    else // translate into an ASCII code
    {
      if ( CapsFlag)
        c = keySCodes[KBDCode%128];
      else
        c = keyCodes[KBDCode%128];
      break;
    }

  }
  // consume the current character
  KBDReady = 0;
  return ( c);
} // getC
```

Debriefing

Today we explored several popular mechanisms used in embedded control to obtain user input. Starting from basic buttons and mechanical switch debouncing, we explored rotary encoders and analyzed the challenges of interfacing to (PS/2) computer keyboards. This gave us the perfect opportunity to exercise two new peripheral modules: Input Capture and Change Notification. We discussed methods to implement a FIFO circular buffer, and we polished our interrupt management skills a little. We managed to learn something new about the MPLAB SIM simulator as well, using for the first time asynchronous input stimuli to test our code. Throughout the entire day our focus has been constantly on balancing the use of resources and the performance offered by each solution.

Notes for the PIC24 Experts

The IC module of the PIC32 is mostly identical to the PIC24 peripheral, yet some important enhancements have been included in its design. Here are the major differences that will affect your code while porting an application to the PIC32:

1. The ICxCON register now follows the standard peripheral module layout and offers an ON control bit that allows us to disable the module when not used, for reduced power consumption.

2. The ICxC32 control bit allows 32-bit capture resolution when the module is used in conjunction with a timer pair (forming a 32-bit timer).

3. The ICxFEDGE control bit allows the selection of the first edge (rising or falling) when the IC module operates in the new mode 6 (ICxM=110).

The CN module of the PIC32 is mostly identical to the PIC24 peripheral, yet some important enhancements have been included in its design. Here are the major differences that will affect your code while porting an application to the PIC32:

1. A new CNCON register has been added to offer a standard set of control bit, including ON, FRZ and IDL to better manage the module behavior in low-power consumption modes.

2. The CNEN (32-bit) control register now groups all the input pin enable bits previously contained in two separate (16-bit) registers of the PIC24 (CNEN1 and CNEN2).

3. Similarly, the CNPUE (32-bit) control register groups all the pull-up enable bits previously contained in two separate (16-bit) registers of the PIC24 (CNPUE1 and CNPUE2).

Tips & Tricks

Each PS/2 keyboard has an internal FIFO buffer 16 key codes deep. This allows the keyboard to accumulate the user input, even when the host is not ready to receive. The host, as we mentioned at the beginning of this chapter, has the option to stall the communication by pulling low the Clock line at any given point in time (for at least 100us) and can hold it low for the desired period of time. When the Clock line is released, the keyboard resumes transmissions. It will retransmit the last key code, if it had been interrupted, and will offload its FIFO buffer.

To exercise our right to stall the keyboard transmissions as a host, we have to control the Clock line with an output using an open drain driver. Fortunately, this is easy with the PIC32, thanks to its configurable I/O port modules. In fact, each I/O port has an associated control register (ODCx) that can individually configure each pin output driver to operate in open-drain mode.

Note that this feature is extremely useful in general to interface PIC32 outputs to any 5V device. In our example, turning the PS/2 Clock line into an open-drain output would require only a few lines of code:

```
_ODG13 = 1;    // cfg PORTG pin 13 output in open-drain mode
_LATG13 = 1;   // initially let the output in pull up
_TRISG13 = 0;  // enable the output driver
```

Note that, as usual for all PIC microcontrollers, even if a pin is configured as an output, its current status can still be read as an input. So there is no reason to switch continuously between input and output when we alternate sending commands and receiving characters from the keyboard.

Exercises

1. Add a function to send commands to the keyboard to control the status LEDs and set the key repeat rate.

2. Replace the stdio.h library input helper function _mon_getc() to redirect the keyboard input as the stdin stream input.

3. Add support for a PS/2 mouse interface.

Books

Nisley, Ed. *The Embedded PCs ISA Bus* (Annabooks/Rtc Books, 1997). Speaking of legacy interfaces, the ISA bus, the heart of every IBM PC for almost two decades, is today interestingly surviving in some industrial control "circles" (like the PC104 platform) and embedded applications.

Links

www.computer-engineering.org. This is an excellent Web site where you will find a lot of useful documentation on the PS/2 keyboard and mouse interface.

www.pc104.com/whatis.html. The PC104 platform, one of the first attempts at bringing the IBM PC architecture to single-board computers for embedded control.

UTube

The Plan

Thanks to the recent advancements in the so-called *chip-on-glass* (COG) technology and the mass adoption of LCD displays in cell phones and many consumer applications, small displays with integrated controllers are becoming more and more common and inexpensive. The integrated controller takes care of the image buffering and performs simple text and graphics commands for us, offloading our applications from the hard work of maintaining the display. But what about those cases when we want to have full control of the screen to produce animations and or simply bypass any limitation of the integrated controller?

In today's exploration we will consider techniques to interface directly to a TV screen or, for that matter, any display that can accept a standard composite video signal. It will be a good excuse to use new features of several peripheral modules of the PIC32 and review new programming techniques. Our first project objective will be to get a nice dark screen (a well-synchronized video frame), but we will soon see to fill it up with several useful and (why not?) entertaining graphical applications.

Preparation

In addition to the usual software tools, including the MPLAB® IDE, the MPLAB C32 compiler, and the MPLAB SIM simulator, this lesson will require the use of the Explorer 16 demonstration board and In-Circuit Debugger of your choice). You will also need a soldering iron and a few components at hand to expand the board capabilities using the prototyping area or a small expansion board. You can check on the companion Web site (www.exploringPIC32.com) for the availability of expansion boards that will help you with the experiments.

The Exploration

There are many different formats and standards in use in the world of video today, but perhaps the oldest and most common one is the so called "composite" video format. This is what was originally used by the very first TV sets to appear in the consumer market. Today it represents the minimum common denominator of every video display, whether a modern high-definition flat-screen TV of the latest generation, a DVD player, or a VHS tape recorder. All video devices are based on the same basic concept, that is, the image is "painted," one line at a time, starting from the top left corner of the screen and moving horizontally to the right edge, then quickly jumping back to the left edge at a lower position and painting a second line, and so on and on in a zigzag motion until the entire screen has been scanned. Then the process repeats and the entire image is refreshed fast enough for our eyes to be tricked into believing that the entire image is present at the same time, and if there is motion, that it is fluid and continuous (see Figure 13.1).

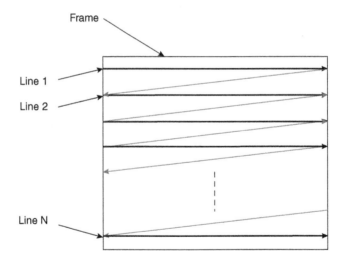

Figure 13.1: Video image scanning.

In different parts of the world, slightly incompatible systems have been developed over the years, but the basic mechanism remains the same. What changes is the number of lines composing the image, the refreshing frequency, and the way the color information is encoded.

Table 13.1 illustrates three of the most commonly used video standards adopted in the United States, Europe, and Asia. All those standards encode the "luminance" information (that is, the underlying black-and-white image) together with synchronization information in a similarly defined composite signal. Figure 13.2 shows the NTSC composite signal in detail.

Table 13.1: International video standard examples.

	United States	Europe and Asia	France and Others
Standard	NTSC	PAL	SECAM
Frames per second	29.97*	25	25
Number of lines	525	625	625

*NTSC used to be 30 frames per second, but the introduction of the new color standard changed it to 29.97, to accommodate a specific frequency used by the "color subcarrier" crystal oscillator.

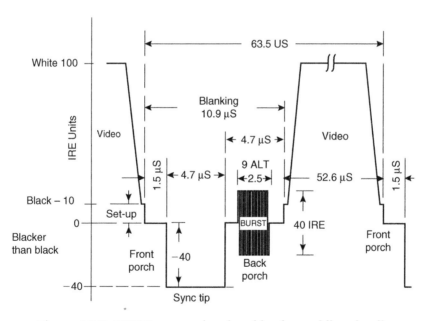

Figure 13.2: NTSC composite signal horizontal line detail.

The term *composite* is used to describe the fact that this video signal is used to combine and transmit in one three different pieces of information: the actual luminance signal and both horizontal and vertical synchronization information.

The horizontal line signal is in fact composed of:

1. The horizontal synchronization pulse, used by the display to identify the beginning of each line

2. The so-called "back porch," the left edge of the dark frame around the image

3. The actual line luminosity signal; the higher the voltage, the more luminous the point

4. The so-called "front porch," producing the right edge of the image

The color information is transmitted separately, modulated on a high-frequency subcarrier. A short burst of pulses in the middle of the back porch is used to help synchronize with the subcarrier. The three main standards differ significantly in the way they encode the color information but, if we focus on a black-and-white display, we can ignore most of the differences and remove the color subcarrier burst altogether.

All these standard systems utilize a technique called *interlacing* to provide a (relatively) high-resolution output while requiring a reduced bandwidth. In practice only half the number of lines is transmitted and painted on the screen in each frame. Alternate *frames* present only the odd or the even lines composing the picture so that the entire image content is effectively updated at the nominal rate (25 Hz and 30 Hz, respectively). The actual frame rates are effectively double. This is effective for typical TV broadcasting but can produce an annoying flicker when text and especially horizontal lines are displayed, as is often the case in computer monitor applications.

For this reason, all modern computer displays are not using interlaced but instead use *progressive* scanning. Most modern TV sets, especially those using LCD and plasma technologies, perform a *deinterlacing* of the received broadcast image. In our project we will avoid interlacing as well, but we'll sacrifice half the image resolution in favor of a more stable and readable display output. In other words, we will transmit frames of 262 lines (for NTSC) at the double rate of 60 frames per second. Readers who have easier access to PAL or SECAM TV sets/monitors will find it relatively easy to modify the project for a 312-line resolution with a refresh rate of 50 frames per second. A complete video frame signal is represented in Figure 13.3.

Notice that, of the total number of lines composing each frame, three line periods are filled by prolonged *synchronization pulses* to provide the vertical synchronization information, identifying the beginning of each new frame. They are preceded and

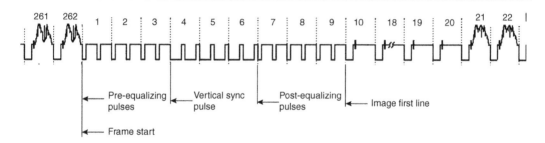

Figure 13.3: A complete video frame signal.

followed by groups of three additional lines, referred to as the pre- and post-equalization lines.

Generating the Composite Video Signal

If we limit the scope of the project to generating a simple black-and-white image (no gray shades, no color) and a noninterlaced image as well, we can considerably simplify our project's hardware and software requirements. In particular, the hardware interface can be reduced to just three resistors of appropriate value connected to two digital I/O pins. One of the I/O pins will generate the synchronization pulses and the other I/O pin will produce the actual luminance signal (see Figure 13.4).

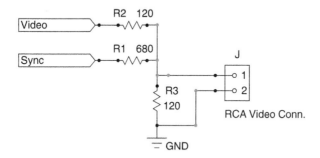

Figure 13.4: Simple hardware interface for composite video output.

The values of the three resistors must be selected so that the relative amplitudes of the luminance and synchronization signals are close to the standard specifications, the signal total amplitude is close to 1 V peak to peak, and the output impedance of the circuit is approximately 75 ohms. With the standard resistor values shown in the previous figure, we can satisfy such requirements and generate the three basic signal levels required to produce a black-and-white image (see Table 13.2 and Figure 13.5).

Table 13.2: Generating luminance and synchronization pulses.

Signal Feature	Sync	Video
Synch pulse	0	0
Black level	1	0
White level	1	1

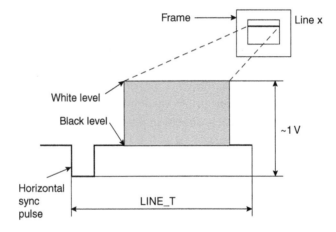

Figure 13.5: Simplified composite video signal.

Since we are not going to utilize the interlacing feature, we can also simplify the pre-equalization, vertical synchronization, and post-equalization pulses by producing a single horizontal synchronization pulse per each period, as illustrated in Figure 13.6

The problem of generating a complete video output signal can now be reduced to (once more) a simple state machine that can be driven by a fixed period time base produced by a single timer interrupt. The state machine will be quite trivial because each state will be associated with one type of line composing the frame, and it will repeat for a fixed amount of times before transitioning to the next state (see Figure 13.7).

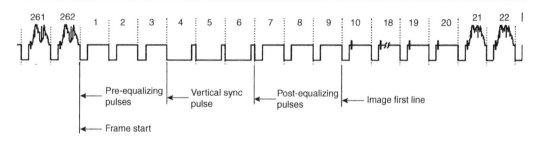

Figure 13.6: Simplified composite video frame (noninterlaced).

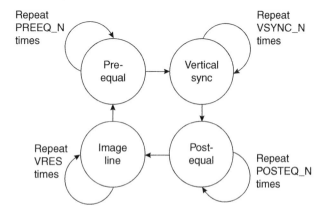

Figure 13.7: Vertical state machine graph.

A simple table will help describe the transitions from each state (see Table 13.3).

Table 13.3: Video state machine transitions table.

State	Repeat	Transition to
Pre-equal	PREEQ_N times	Vertical Sync
Vertical Sync	3 times	Post-equal
Post-equal	POSTEQ_N times	Image line
Image line	VRES times	Pre-equal

Although the number of vertical synchronization lines is fixed and prescribed by the video standard of choice (NTSC, PAL, and so on), the number of lines effectively composing the image inside each frame is up to us to define (within limits, of course). In fact, although in theory we could use all the lines available to display the largest possible amount of data on the screen, we will have to consider some practical limitations, in particular the amount of RAM we are willing to allocate to store the video image inside the PIC32 microcontroller (see Figure 13.8). These limitations will dictate a specific number of lines (VRES) to be used for the image, whereas all the remaining lines (up to the standard line count) will be left blank.

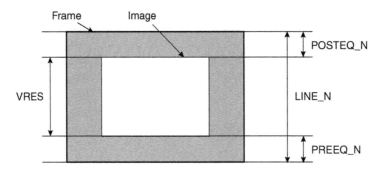

Figure 13.8: Defining frame and image resolution.

In practice, if LINE_N is the total number of lines composing a video frame and VRES is the desired vertical resolution, we will determine a value for PREEQ_N and POSTEQ_N as follows:

```
// timing for composite video vertical state machine
#ifdef NTSC
#define LINE_N .262       // number of lines in NTSC frame
#define LINE_T .2284      // Tpb clock in a line (63.5us)
#else
#define LINE_N. 312       // number of lines in PAL frame
#define LINE_T. 2304      // Tpb clock in a line (64us)
#endif

// count the number of remaining black lines top+bottom
#define VSYNC_N 3              // V sync lines
#define VBLANK_N (LINE_N -VRES -VSYNC_N)
#define PREEQ_N VBLANK_N/2         // preeq+bottom blank
#define POSTEQ_N VBLANK_N -PREEQ_N // posteq + top blank
```

If we choose Timer3 to generate our time base, set to match the horizontal synchronization pulse period (LINE_T) as shown in Figure 13.5, we can use the timer's associated interrupt

service routine to execute the vertical state machine. Here is a skeleton of the interrupt service
routine on top of which we can start flashing the complete composite video logic:

```
// next state table
int VS[4] = { SV_SYNC, SV_POSTEQ, SV_LINE, SV_PREEQ};
// next counter table
int VC[4] = { VSYNC_N,    POSTEQ_N,    VRES,    PREEQ_N};

void __ISR( _TIMER_3_VECTOR, ipl7) T3Interrupt( void)
{
  // advance the state machine
  if ( --VCount == 0)
  {
    VCount=VC[ VState&3];
    VState=VS[ VState&3];
  }
  // vertical state machine
  switch ( VState) {
    case SV_PREEQ:
      // horizontal sync pulse
      ...
      break;

    case SV_SYNC:
      // vertical sync pulse
      ...
      break;

    case SV_POSTEQ:
      // horizontal sync pulse
      ...
      break;

    default:
    case SV_LINE:
      ...
      break;
  } //switch

  // clear the interrupt flag
  mT3ClearIntFlag();

} // T3Interrupt
```

To generate the actual horizontal synch pulse output, there are several options we can explore:

1. Control an I/O pin directly and use various delay loops.

2. Control an I/O and use a second timer (interrupt) to produced the required timings.

3. Use the Output Compare modules and the associated interrupt service routines.

The first solution is probably the simplest to code but has the clear disadvantage of keeping the processor constantly tied in endless loops, preventing it from performing any useful work while the video signal is being generated.

The second solution is clearly more efficient, and by now we have ample experience in using timers and their interrupt service routines to execute small state machines.

The third solution involves the use of a new peripheral we have not yet explored and deserves a little more attention.

The Output Compare Modules

The PIC32MX family of microcontrollers offers a set of five Output Compare peripheral modules that can be used for a variety of applications, including single pulse generation, continuous pulse generation, and pulse width modulation (PWM). Each module can be associated to one of two 16-bit timers (Timer2 or Timer3) or a 32-bit timer (obtained by combining Timer2 and Timer3) and has one output pin that can be configured to toggle and produce rising or falling edges as necessary (see Figure 13.9). Most importantly, each module has an associated and independent interrupt vector.

The basic configuration of the Output Compare modules is performed by the OCxCON register where a small number of control bits, in a layout that we have grown familiar with, allow us to choose the desired mode of operation (see Figure 13.10).

When used in continuous pulse mode (OCM=101) in particular, the OCxR register is used to determine the instant (relative to the value of the associated timer) when the output pin will be set, while the OCxRS register determines when the output pin will be cleared (see Figure 13.11).

Choosing the OC3 module, we can now connect the associated output pin RD2 directly as our Synch output, shown in Figure 13.4.

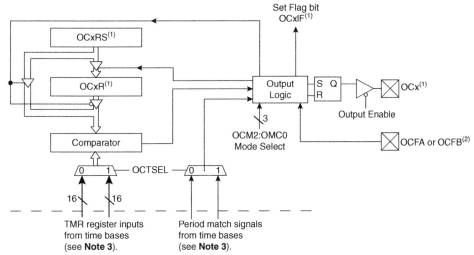

Note 1: Where 'x' is shown, reference is made to the registers associated with the respective output compare channels
2: OCFA pin controls OC1-OC4 channels.
3: Each output compare channel can use one of two selectable time bases. Refer to the device data sheet for the time bases associated with the module.

Figure 13.9: Output Compare module block diagram.

U-0	U-0	U-0	U-0	U-0	U-0	U-0	U-0
—	—	—	—	—	—	—	—
Bit 31							Bit 24

U-0	U-0	U-0	U-0	U-0	U-0	U-0	U-0
—	—	—	—	—	—	—	—
Bit 23							Bit 16

R/W-0	R/W-0	R/W-0	U-0	U-0	U-0	U-0	U-0
ON	FRZ	SIDL	—	—	—	—	—
Bit 15							Bit 8

U-0	U-0	R/W-0	R-0	R/W-0	R/W-0	R/W-0	R/W-0
—	—	OC32	OCFLT	OCTSEL	OCM<2:0>		
Bit 7							Bit 0

Figure 13.10: Output Compare Control register OCxCON.

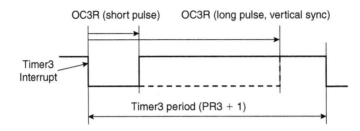

Figure 13.11: Output Compare module continuous pulse mode.

We can also start flashing the vertical state machine body to make sure that in each
state the OC3 produces pulses of the correct width. In fact, though during normal, pre-
equalization, and post-equalization lines, the horizontal synch pulse is short (approx.
5us), during the three lines devoted to the vertical synchronization the pulse must be
widened to cover most of the line period (see lines 4, 5 and 6 in Figure 13.6):

```
. . .
  // vertical state machine
  switch ( VState) {
    case SV_SYNC: // 1
      // vertical sync pulse
        OC3R=LINE_T - HSYNC_T - BPORCH_T;
        break;

    case SV_POSTEQ: // 2
      // horizontal sync pulse
        OC3R=HSYNC_T;
        break;

    case SV_PREEQ: // 0
      // prepare for the new frame
      VPtr=VA;
      break;

    default:
    case SV_LINE: // 3
      VPtr += HRES/32;
      break;
  } //switch
. . .
```

Image Buffers

So far we have been working on the generation of the synchronization signals `Synch` connected to our simple hardware interface (refer back to Figure 13.4). The actual image represented on the screen will be produced by mixing in a second digital signal. Toggling the `Video` pin, we can alternate segments of the line that will be painted in white (`1`) or black (`0`). Since the NTSC standard specifies a maximum luminance signal bandwidth of about 4.2 MHz (PAL has very similar constraints) and the space between front and back porch is 52us wide, it follows that the maximum number of alternate segments (cycles) of black and white we can display is 218, (52 × 4.2), or in other words, our maximum theoretical horizontal resolution is 436 pixels per line (assuming the screen is completely used from side to side). The maximum vertical resolution is given by the total number of lines composing each frame minus the minimum number of equalization (6) and vertical synchronization (3) lines. (This gives 253 lines for the NTSC standard.)

If we were to generate the largest possible image, it would be composed of an array of 253 × 436 pixels, or 110,308 pixels. If 1 bit is used to represent each pixel, a complete frame image would require us to allocate an array of 13.5 K bytes, using up almost 50 percent of the total amount of RAM available on the PIC32MX360. In practice, though it is nice to be able to generate a high-resolution output, we need to make sure that the image will fit in the available RAM, possibly leaving enough space for an application to run comfortably along and allowing for adequate room for stack and variables. There are an almost infinite number of possible combinations of the horizontal and vertical resolution values that will give an acceptable memory size, but there are two considerations that we will use to pick the perfect numbers:

- A horizontal resolution value multiple of 32 will make the math involved in determining the position of each pixel in the image buffer easier and will maximize the use of the microcontroller's 32-bit bus.

- A ratio between the horizontal and vertical resolution close to 4:3 will avoid geometrical distortions of the image—circles drawn on the screen will look like circles rather than ovals.

Choosing a horizontal resolution of 256 pixels (`HRES`) and a vertical resolution of 200 lines (`VRES`) we obtain an image memory requirement of 6,400 bytes (256 × 200/8), representing roughly 20 percent of the total amount of RAM available. Using the

MPLAB C32 compiler, we can easily allocate a single array of integers (grouping 32 pixels at a time in each word) to contain the entire image memory map:

```
int VMap[VRES * (HRES/32)];
```

Serialization, DMA, and Synchronization

If each image line is represented in memory in the VMap array by a row of (eight) integers, we will need to serially output each bit (pixel) in a timely fashion in the short amount of time (52us) between the back and the front porch part of the composite video waveform. In other words, we will need to set or clear the chosen Video output pin with a new pixel value every 200 ns or faster. This would translate into about 14 instruction cycles between pixels, way too fast for a simple shift loop, even if we plan on coding it directly in assembly. Worse, even assuming we managed to squeeze the loop so tight, we would end up using an enormous percentage of the processing power for the video generation, leaving very few processor cycles for the main application.

Fortunately, we already know one peripheral of the PIC32 that can help us efficiently serialize the image data: It's the SPI synchronous serial communication module. In a previous chapter we used the SPI2 module to communicate with a serial EEPROM device. In that chapter we noted how the SPI module is in fact composed of a simple shift register that can be clocked by an external clock signal (when in slave mode) or by an internal clock (when in master mode). Today we can use the SPI1 module as a master connecting the SDO (serial data output, RF8) pin directly to the Video pin of the video hardware interface, leaving the SDI (data input) and SCK (clock output) pins unused. Among the many new and advanced features of the PIC32 SPI module and the PIC32 in general there are two that fit our video application particularly well:

- The ability to operate in 32-bit mode

- The connection to another powerful peripheral, the Direct Memory Access (DMA) controller

Operating in 32-bit mode, we can practically quadruple the transfer speed of data between the image memory map and the SPI module. By leveraging the connection with the DMA controller, we can completely offload the microcontroller core from any activity involving the serialization of the video data.

The bad news is that the DMA controller of the PIC32 is an extremely powerful and complex module that requires as many as 20 separate control registers for its configuration. But the good news is that all this power can be easily managed by an equally powerful and well-documented library, dma.h, which is easily included as part of plib.h.

The DMA module shares the 32-bit-wide system bus of the PIC32 and operates at the full system clock frequency. It can perform data transfers of any size to and from any of the peripherals of the PIC32 and any of the memory blocks. It can generate its own set of specific interrupts and can multitask, so to speak, since each one of its four channels can operate at the same time (interleaving access to the bus) or sequentially (channels activity can be chained so that the completion of a transfer initiates another). See Figure 13.12.

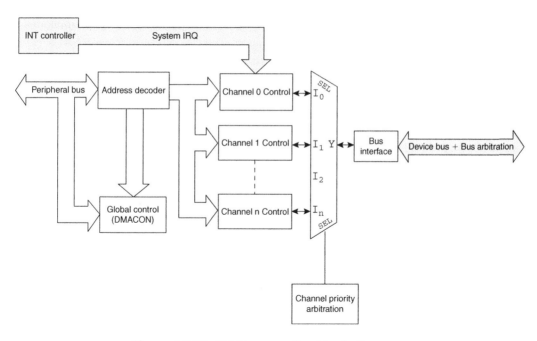

Figure 13.12: DMA controller block diagram.

The arbitration for the use of the system bus is provided by the BMX module (which we have encountered before) and happens seamlessly. In particular, when the microcontroller cache system is enabled and the pre-fetch cache is active, the effect on the performance of the microcontroller can hardly be noticed. In fact, when an application requires a fast data transfer, nothing beats the performance and efficiency of the DMA controller.

The DMA module initialization requires just a couple of function calls:

- DmaCHOpen(), enables the module and prepares it for "normal" data transfers, those to and from peripherals normally requiring up to a maximum of 256 bytes of data at a time, or extended ones, those from memory to memory that extend for up to 64 K bytes.

- DmaChnSetEventControl(), determines which peripheral event (interrupt) will be used to trigger the transfer of each block of data.

- DmaChnSetTxfer(), informs the controller of where the data will be coming from, where it will be transferred to, how many bytes at a time should be sent, and how many bytes in total will need to be transferred.

- DmaChnSetControl(), allows us to chain multiple channels for sequential execution.

So, for example, we can initialize channel 0 of the DMA controller to respond to the SPI1 module requests (interrupt on transmit buffer empty), transferring 32 bits (4 bytes) at a time for a total of 32 bytes per line, with the following three lines of code:

```
DmaChnOpen( 0, 0, DMA_OPEN_NORM);
DmaChnSetEventControl( 0, DMA_EV_START_IRQ_EN |
                       DMA_EV_START_IRQ(_SPI1_TX_IRQ));
DmaChnSetTxfer( 0, (void*)VPtr, (void *)&SPI1BUF,
               HRES/8, 4, 4);
```

All we need to do is have the PIC32 initiate the first SPI transfer, writing the first 32-bit word of data to the SPI1 module data buffer (SPI1BUF), and the rest will be taken care of automatically by the DMA module to complete the rest of the line.

Unfortunately, this creates a new efficiency problem. Between the Timer3 interrupt marking the beginning of a new line period, and the beginning of the SPI1 transfer, there is a difference of about 10us. Not only this is an incredibly long time to "wait out" for a microcontroller operating at 72 Mhz (up to 720 useful instructions could be executed in that time), but the timing of this delay must be extremely accurate. Even a discrepancy of a single clock cycle would be amplified by the video synchronization circuitry of the TV and would result in a visible "indentation" of the line. Worse, if the discrepancy were not absolutely deterministic, as it could/would be if the PIC32 cache were enabled (the cache behavior is by its very definition unpredictable), this would result in a noticeable oscillation of the left edge of the screen. Since we are not willing to sacrifice such a key element of the PIC32 performance, we need to find another way to close the gap while maintaining absolute synchronization between horizontal synch pulse and SPI data serialization transfer start (see Figure 13.13).

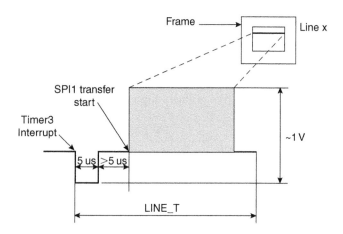

Figure 13.13: Synchronization of synch pulse and SPI transfer start.

By looking more carefully at the SPI module and comparing it with previous PIC® architectures, you will discover that there is one particular new feature that seems to have been added exactly for this purpose. It is called the Framed Slave mode and it is enabled by the FRMEN bit in the SPIxCON register. Not to be confused with the bus master and slave mode of operation of the SPI port, there are in fact two new *framed* modes of operation for the SPI. In framed mode, the SS pin, otherwise used to select a specific peripheral on the SPI bus, changes roles. It becomes a synchronization signal of sorts:

- When a framed master mode is selected, it acts as an output, flagging the first bit of a new transfer.

- When a framed slave mode is selected, it acts as an input, triggering the beginning of an impending data transfer.

Note that the SPI port can now be configured in a total of four modes:

- SPI bus master, framed master

- SPI bus master, framed slave

- SPI bus slave, framed master

- SPI bus slave, framed slave

In particular, we are interested in the second case, where the SPI port is a bus master and so does not require an external clock signal to appear on the SCK pin, but it is a framed slave, so it will wait for the SS pin to become active before starting a data transfer. As a final nice touch, you will discover that it is possible to select the polarity of the SS frame signal.

Our synchronization problem is now completely solved (see Figure 13.14). We can connect (directly or via a small value resistor) the OC3 output (RD2 pin) to the SPI1 module SS input (RB2 pin) with active polarity high.

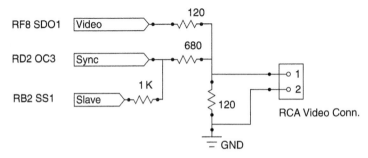

Figure 13.14: Composite video interface.

Note

One of the many functions assigned to the RB2 pin is the channel 2 input to the ADC. As with all such pins, it is by default configured as an analog input at power-up. When it's in such a configuration, its digital input value is always 1 (high). Before using it as an effective framed slave input, we will need to remember to reconfigure it as a digital input pin.

With this connection, the rising edge of the horizontal synchronization pulse produced by the OC3 module will trigger the beginning of the transmission by the SPI1 module, provided we preloaded its output buffer with data ready to be shifted out. But it is still too early to start sending out the line data (the image pixels from the video map). We have to respect the back-porch timing and then leave some additional time to center our image on the screen. One quick way to do this is to begin every line preloading the SPI1 module buffer with a data word containing all zeros. The SPI1 module will be shifting out data, but since the first 32 bits are all zeros, we will buy some precious time and we will let the DMA take care of the real data later.

But how much time does one word of 32 bits take to be serialized by the SPI1 module? If we are operating at 72 MHz with a peripheral clock divider by 2, and assuming an SPI baud rate divider by 4 (SPI1BRG = 1), we are talking of just 3.5us. That's definitely below the minimum specs for the NTSC back porch. There are two practical methods to extend the back-porch timing further:

- Add one more column to the image map, one more word that is always set to 0 and is never used to paint any actual image.

- Use another DMA channel always pointing to a string of words (as many as we desire) set to zero and queue the two DMA channels execution automatically.

Both methods add cost to our application, since both are using precious resources. Adding one column implies using more RAM, 800 bytes more for the precision. Using a second channel of DMA (out of the four total) seems also a high price to pay. My choice goes to the DMA, though, because to me it seems there's never enough RAM, and this way we get to experiment with yet another cool feature of the PIC32 DMA controller: DMA channel chaining.

It turns out that there is another friendly function call, `DmaChnSetControl()`, that can quickly perform just what we need, triggering the execution of a specific channel DMA transfer to the completion of a previous channel DMA transfer. Here is how we link the execution of channel 0 (the one drawing a line of pixels) to the previous execution of channel 1:

```
// chain DMA0 to completion of DMA1 transfer
DmaChnSetControl( 0, DMA_CTL_CHAIN_EN | DMA_CTL_CHAIN_DIR);
```

Notice that only contiguous channels can be chained. Channel 0 can be chained only to channel 1; channel 1 can be chained to channel 0 or channel 2 (you decide the "direction" up or down), and so on.

We can now configure the DMA channel 1 to feed the SPI1 module with some more bytes of zero; four more will take our total back-porch time to 7us:

```
// DMA 1 configuration  back porch extension
DmaChnOpen( 1, 1, DMA_OPEN_NORM);
DmaChnSetEventControl( 1, DMA_EV_START_IRQ_EN |
                    DMA_EV_START_IRQ(_SPI1_TX_IRQ));
DmaChnSetTxfer( 1, (void*)zero, (void *)&SPI1BUF,
              8, 4, 4);
```

The symbol `zero`, used here could be a reference to a 32-bit integer variable that needs to be initialized to zero or an array of such integers to allow us to extend the back porch and further center the image on the screen.

Now that we have identified all the pieces of the puzzle, we can write the complete initialization routine for all the modules required by the video generator:

```
void initVideo( void)
{

  // 1. init the SPI1
  // select framed slave mode to synch SPI with OC3
  SpiChnOpen( 1, SPICON_ON | SPICON_MSTEN | SPICON_MODE32
              | SPICON_FRMEN | SPICON_FRMSYNC | SPICON_FRMPOL
              , PIX_T);

  // 2. make SS1(RB2) a digital input
  AD1PCFGSET = 0x0004;

  // 3. init OC3 in single pulse, continuous mode
  OpenOC3( OC_ON | OC_TIMER3_SRC | OC_CONTINUE_PULSE,
           0, HSYNC_T);

  // 4. Timer3 on, prescaler 1:1, internal clock, period
  OpenTimer3( T3_ON | T3_PS_1_1 | T3_SOURCE_INT, LINE_T-1);

  // 5. init the vertical sync state machine
  VState = SV_LINE;
  VCount = 1;

  // 6. init the active and hidden screens pointers
  VA = VMap1;

  // 7. DMA 1 configuration back porch extension
  DmaChnOpen( 1, 1, DMA_OPEN_NORM);
  DmaChnSetEventControl( 1, DMA_EV_START_IRQ_EN |
                    DMA_EV_START_IRQ(_SPI1_TX_IRQ));
  DmaChnSetTxfer( 1, (void*)zero, (void *)&SPI1BUF,
                  8, 4, 4);
```

```
// 8. DMA 0 configuration image serialization
DmaChnOpen( 0, 0, DMA_OPEN_NORM);
DmaChnSetEventControl( 0, DMA_EV_START_IRQ_EN |
                          DMA_EV_START_IRQ(_SPI1_TX_IRQ));
DmaChnSetTxfer( 0, (void*)VPtr, (void *)&SPI1BUF,
                HRES/8, 4, 4);
// chain DMA0 to completion of DMA1 transfer
DmaChnSetControl( 0, DMA_CTL_CHAIN_EN | DMA_CTL_CHAIN_DIR);

// 9. Enable Timer3 Interrupts
// set the priority level 7 to use shadow register set
mT3SetIntPriority( 7);
mT3IntEnable( 1);
} // initVideo
```

Completing a Video Library

We can now complete the coding of the entire video state machine, adding all the definitions and pin assignments necessary:

```
/*
**    graphic.c
**    Composite Video using:
**    T3        time based
**    OC3       Horizontal Synchronization pulse
**    DMA0      image data
**    DMA1      back porch extension
**    SPI1 in Frame Slave Mode
*/
#include <p32xxxx.h>
#include <plib.h>
#include <string.h>
#include <graphic.h>

// timing for composite video vertical state machine
#ifdef NTSC
#define LINE_N  262    // number of lines in NTSC frame
#define LINE_T  2284   // Tpb clock in a line (63.5us)
```

```
#else
#define LINE_N   312            // number of lines in PAL frame
#define LINE_T   2304           // Tpb clock in a line (64us)
#endif

// count the number of remaining black lines top+bottom
#define VSYNC_N 3                      // V sync lines
#define VBLANK_N (LINE_N -VRES -VSYNC_N)
#define PREEQ_N VBLANK_N/2             // preeq + bottom blank
#define POSTEQ_N VBLANK_N -PREEQ_N     // posteq + top blank

// definition of the vertical sync state machine
#define SV_PREEQ   0
#define SV_SYNC    1
#define SV_POSTEQ  2
#define SV_LINE    3

// timing for composite video horizontal state machine
#define PIX_T     4             // Tpb clock per pixel
#define HSYNC_T   180           // Tpb clock width horizontal pulse
#define BPORCH_T  340           // Tpb clock width back porch

int VMap1[ VRES*(HRES/32)];     // image buffer
int *VA = VMap1;                // pointer to the Active VMap

volatile int *VPtr;
volatile short VCount;
volatile short VState;

// next state table
short VS[4] = { SV_SYNC, SV_POSTEQ, SV_LINE, SV_PREEQ};
// next counter table
short int VC[4]={ VSYNC_N, POSTEQ_N, VRES, PREEQ_N};

int zero[2]= {0x0, 0x0};

void __ISR( _TIMER_3_VECTOR, ipl7) T3Interrupt( void)
{
```

```
// advance the state machine
if ( --VCount == 0)
{
  VCount = VC[ VState&3];
  VState = VS[ VState&3];
}

// vertical state machine
switch ( VState) {
  case SV_SYNC:      // 1
    // vertical sync pulse
    OC3R = LINE_T - HSYNC_T - BPORCH_T;
    break;

  case SV_POSTEQ:    // 2
    // horizontal sync pulse
    OC3R = HSYNC_T;
    break;

  case SV_PREEQ:     // 0
    // prepare for the new frame
    VPtr = VA;
    break;

  default:
  case SV_LINE: // 3
    // preload of the SPI waiting for SS (Synch high)
    SPI1BUF = 0;
    // update the DMA0 source address and enable it
    DCH0SSA = KVA_TO_PA((void*) VPtr);
    VPtr += HRES/32;
    DmaChnEnable( 1);
    break;
} //switch

// clear the interrupt flag
mT3ClearIntFlag();

} // T3Interrupt
```

Notice how at the beginning of each line containing actual image data (SV_LINE) we update the DMA source pointer (DCH0SSA) to point to the next line of pixels, but in doing so, we take care to translate the address to a physical address using the KVA_TO_PA() inline function, which involves a simple bit-masking exercise. As you can understand, the DMA controller does not need to be concerned with the way we have remapped the memory and the peripheral space; it demands a physical address. Normally it is the DMA library that takes care of such a low-level detail, and we could have once more used the DmaChnSetTxfer() function to get the job done, but I could not help it—I just needed an excuse to show you how to directly manipulate the DMA controller registers and in the process save a few instruction cycles.

To make it a complete graphic library module, we need to add a couple of accessory functions, such as:

```
void clearScreen( void)
{ // fill with zeros the Video array
  memset( VA, 0, VRES*( HRES/8));
} //clearScreen

void haltVideo( void)
{
  T3CONbits.TON = 0;  // turn off the vertical state machine
} //haltVideo
```

In particular, clearScreen() will be useful to initialize the image memory map, the VMap array. However, haltVideo() will be useful to suspend the video generation, should an important task/application require 100 percent of the PIC32 processing power.

Save all the preceding functions in a file called **graphic.c** and place it in our *lib* directory, I can foresee an extensive use of its functions in this and the next few chapters. Also add this file to a new project called **Video**.

Then create a new file and add the following definitions:

```
/*
** graphic.h
**
** Composite video and graphic library
**
*/
#define NTSC    // comment if PAL required
```

```
#define VRES    200    // desired vertical resolution
#define HRES    256    // desired horizontal resolution pixel

void initVideo( void);

void haltVideo( void);

void clearScreen( void);
```

Notice how the horizontal resolution and vertical resolution values are the only two parameters exposed. Within reasonable limits (due to timing constraints and the many considerations exposed in the previous sections), they can be changed to adapt to specific application needs, and the state machine and all other mechanisms of the video generator module will adapt their timing as a consequence.

Save this file as **graphic.h** and add it to the common *include* directory.

Testing the Composite Video

To test the composite video module we have just completed, we need only the MPLAB SIM simulator tool and possibly a few more lines of code for a new main module, to be called **GraphicTest.c**:

```
/*
** GraphicTest.c
**
** A dark screen
**
*/
// configuration bit settings, Fcy=72 MHz, Fpb=36 MHz
#pragma config POSCMOD=XT, FNOSC=PRIPLL
#pragma config FPLLIDIV=DIV_2, FPLLMUL=MUL_18, FPLLODIV=DIV_1
#pragma config FPBDIV=DIV_2, FWDTEN=OFF, CP=OFF, BWP=OFF

#include <p32xxxx.h>
#include <plib.h>
#include <explore.h>
#include <graphic.h>
```

```
main()
{
  // initializations
  initEX16();    // init and enable vectored interrupts
  clearScreen(); // init the video map
  initVideo();   // start the video state machine

  // main loop
    while( 1)
    {

    } // main loop
} // main
```

Remember to add the **explore.c** module from the *lib* directory, then save the project and use the Build Project checklist to build and link all the modules.

Open the **Logic Analyzer** window and use the Logic Analyzer checklist to add the OC3 signal (sync) and the SDO1 (video) to the analyzer channels.

At this point you could run the simulator for a few seconds and, after pressing the **halt** button, switch to the **Logic Analyzer** output window to observe the results (see Figure 13.15). The trace memory of the simulator is of limited capacity (unless you have configured it to use the extended buffers) and can visualize only a small subset of an entire video frame. In other words, it is very likely that you will be confronted with a relatively uninteresting display containing a regular series of sync pulses. Unfortunately, the MPLAB SIM simulator does not yet simulate the output of the SPI port, so for that, we'll have to wait until we run the application on real hardware.

Regarding the sync line, there is one interesting time we would like to observe: that is when we generate the vertical synchronization signal with a sequence of three long horizontal synch pulses at the beginning of each frame. By setting a breakpoint on the first line of the SV_POSTEQ state inside the Timer3 interrupt service routine, you can make sure that the simulation will stop close to the beginning of a new frame.

You can now zoom in the central portion to verify the proper timing of the sync pulses in the pre/post and vertical sync lines (see Figure 13.16).

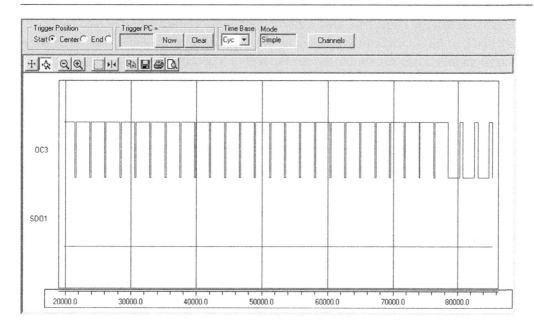

Figure 13.15: Screen capture of the Logic Analyzer window, Vertical Sync pulses.

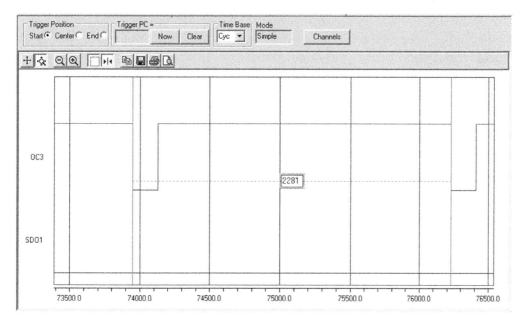

Figure 13.16: Zoomed view of a single pre-equalization line.

Keep in mind that the Logic Analyzer window approximates the reading to the nearest screen pixel, so the accuracy of your reading will depend on the magnification (improving as you zoom in) and the resolution of your PC screen. Naturally, if what you need is to determine with absolute precision a time interval, the most direct method is to use the Stopwatch function of the MPLAB SIM software simulator together with the appropriate breakpoint settings.

Measuring Performance

It might be interesting to get an idea of the actual processor overhead caused by the video module. Using the Logic Analyzer we can visualize and attempt to estimate the percentage of time the processor spends inside the interrupt service routine.

As we did before, we will use a pin of PORTA (RA2) as a flag that will be set to indicate when we are inside the interrupt service routine and cleared when we are executing the main loop.

```
void __ISR() T3Interrupt( void)
{
_RA2=1;
. . .
_RA2=0;
} // T3Interrupt
```

After recompiling and adding RA2 to the channels captured by the Logic Analyzer tool (see Figure 13.17), we can zoom in a single horizontal line period. Using the cursors, we can measure the approximate duration of an interrupt service routine. We obtain a value of 35 cycles out of a line period of 2284 cycles, representing an overhead of less than 1.5 percent of the processor time—a remarkable result due in great part to the support of the DMA controller!

Seeing the Dark Screen

Playing with the simulator and the Logic Analyzer tool can be entertaining for a little while, but I am sure at this point you will feel an itch for the real thing! You'll want to test the video interface on a real TV screen or any other device capable of receiving an composite video signal, connected with the simple (resistors only) interface to an actual PIC32. If you have an Explorer 16 board, this is the time to take out the soldering iron

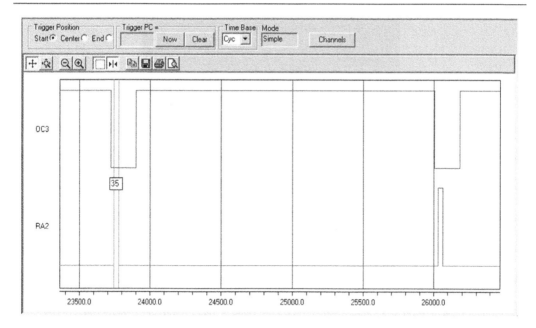

Figure 13.17: Screen capture of the Logic Analyzer output, measuring performance.

and connect the three resistors to a standard RCA video jack using the small prototyping area in the top-right corner of the demo board. Alternatively, if you feel your electronic hobbyist skills are up to the task, you could even develop a small PCB for a daughter board (a PICTail™) that would fit in the expansion connectors of the Explorer 16.

Check the companion Web site (www.pic32explorer.com) for the availability of expansion boards that will allow you to follow all the advanced projects presented in the third part of the book.

Whatever your choice, the experience will be breathtaking.

Or . . . not (see Figure 13.18)! In fact, if you wire all the connections just right, when you power up the Explorer 16 board you are going to be staring at just a blank or, I should say, "black" screen. Sure, this is an achievement; in fact, this already means that a lot of things are working right, since both the horizontal and vertical synchronization signals are being decoded correctly by the TV set and a nice and uniform black background is being displayed.

Figure 13.18: The dark screen.

Test Pattern

To spice things up, let's start filling that video array with something worth looking at, possibly something simple that can give us immediate feedback on the proper functioning of the video generator. Let's create a new test program as follows:

```
/*
** GraphicTest2.c
**
** A test pattern
**
*/
// configuration bit settings, Fcy=72MHz, Fpb=36MHz
#pragma config POSCMOD=XT, FNOSC=PRIPLL
#pragma config FPLLIDIV=DIV_2, FPLLMUL=MUL_18, FPLLODIV=DIV_1
#pragma config FPBDIV=DIV_2, FWDTEN=OFF, CP=OFF, BWP=OFF

#include <p32xxxx.h>
#include <plib.h>
#include <explore.h>
#include <graphic.h>

extern int * VA;    // pointer to the image buffer

main()
{
  int x, y;
```

```
// initializations
initEX16();      // init and enable vectored interrupts
clearScreen();   // init the video map
initVideo();     // start the video state machine

// fill the video memory map with a pattern
for( y=0; y<VRES; y++)
   for (x=0; x<HRES/32; x++)
     VA[y*HRES/32+x]= y;
// main loop
   while( 1)
   {

   } // main loop
} // main
```

Instead of calling the `clearScreen()` function, this time we used two nested `for` loops to initialize the `VMap` array. The external (`y`) loop counts the vertical lines, and the internal (`x`) loop moves horizontally, filling the eight words (each containing 32 bits) with the same value: the line count. In other words, on the first line, each 32-bit word will be assigned the value 0; on the second line, each word will be assigned the value 1, and so on until the last line (200th), where each word will be assigned the value 199 (`0x000000C7` in hexadecimal).

If you build the new project and test the video output you should be able to see the pattern shown in Figure 13.19.

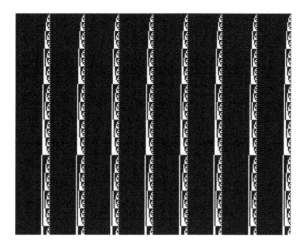

Figure 13.19: A screen capture of the test pattern.

In its simplicity, there is a lot we can learn from observing the test pattern. First, we notice that each word is visually represented on the screen in binary, with the most significant bit presented on the left. This is a consequence of the order used by the SPI module to shift out bits: that is, MSb first. Second, we can verify that the last row contains the expected pattern, 0x000000c7, so we know that all rows of the memory map are being displayed. Finally, we can appreciate the detail of the image. Different output devices (TV sets, projectors, LCD panels, and so on) will be able to lock the image more or less effectively and/or will be able to present a sharper image, depending on the actual display resolution and their input stages bandwidth. In general, you should be able to appreciate how the PIC32 can generate effectively straight vertical lines. This is not a trivial achievement.

This does not mean that on the largest screens you will not be able to notice small imperfections here and there as small echoes and possibly minor visual artifacts in the output image. Realistically, the simple three-resistor interface can only take us so far.

Ultimately the entire composite video signal interface could be blamed for a lower-quality output. As you might know, S-Video, VGA, and most other video interfaces keep luminance and synchronization signals separate to provide a more stable and clean picture.

Plotting

Now that we are reassured about the proper functioning of the graphic display module, we can start focusing on putting it to good use. The first natural step is to develop a function that allows us to light up one pixel at a precise coordinate pair (*x, y*) on the screen. The first thing to do is derive the line number from the *y* coordinate. If the *x* and *y* coordinates are based on the traditional Cartesian plane representation, with the origin located in the bottom-left corner of the screen, we need to invert the address before accessing the memory map so that the first row in the memory map corresponds to the *y* maximum coordinate VRES-1 or 199 while the last row in the memory map corresponds to the *y* coordinate 0. Also, since our memory map is organized in rows of eight words, we need to multiply the resulting line number by 32 to obtain the address of the first word on the given line. This can be obtained with the following expression:

```
VH[ (VRES-1 -y) *8]
```

where VH is a pointer to the image buffer.

Pixels are grouped in 32-bit words, so to resolve the *x* coordinate we first need to identify the word that will contain the desired pixel. A simple division by 32 will give us the word offset on the line. Adding the offset to the line address as we calculated will provide us with the complete word address inside the memory map:

```
VH[ (VRES-1 -y)*8 + (x/32)]
```

To optimize the address calculation, we can use shift operations to perform the multiplication and divisions as follows:

```
VH[ ((VRES-1 -y)<<3)+(x>>5)]
```

To identify the bit position inside the word corresponding to the required pixel, we can use the reminder of the division of x by 32, or more efficiently, we can mask out the lower 5 bits of the x coordinate. Since we want to turn the pixel on, we will need to perform a binary OR operation with an appropriate mask that has a single bit set in the corresponding pixel position. Remembering that the display puts the MSb of each word to the left (the SPI module shifts bits MSb first), we can build the mask with the following expression:

```
(0x80000000 >> ( x & 0x1f))
```

Putting it all together, we obtain the core of the plot function:

```
VH[ ((VRES-1-y)<<3)+(x>>5)] |= ( 0x80000000>>(x&0x1f));
```

As a final touch we can add "clipping"—that is, a simple safety check, just to make sure that the coordinates we are given are in fact valid and within the current screen map limits.

Add the following few lines of code to the **graphic.c** module we saved in the *lib* directory:

```
void plot( unsigned x, unsigned y)
{
  if ((x<HRES) && (y<VRES) )
    VH[ ((VRES-1-y)<<3)+(x>>5)] |= ( 0x80000000>>(x&0x1f));
} // plot
```

By defining the x and y parameters as unsigned integers, we guarantee that, should negative values be passed along, they will be discarded too because they will be considered large integers outside the screen resolution.

Now let's remember to add the function prototype to the **graphic.h** file in the *include* directory:

```
void plot( unsigned x, unsigned y);
```

Watch Out

The `plot()` function as defined is efficient, but it is *not* scalable. In other words, if you change the HRES or VRES parameters in the graphic.h file, you will have to rethink the way you compute the address and bit position of a pixel for a given *x, y* pair of coordinates.

A Starry Night

To test the newly developed `plot()` function, let's once more modify the Video project. We will include the graphic.c and graphic.h files, but we will also use the pseudo-random number-generator functions available in the standard C library stdlib.h. By using the pseudo-random number generator to produce random *x* and *y* coordinates for 1,000 points, we will test both the `plot()` function and, in a way, the random generator itself with the following simple code:

```
/*
** GraphicTest3.c
**
** A starry night
*/
// configuration bit settings, Fcy=72MHz, Fpb=36MHz
#pragma config POSCMOD=XT, FNOSC=PRIPLL
#pragma config FPLLIDIV=DIV_2, FPLLMUL=MUL_18, FPLLODIV=DIV_1
#pragma config FPBDIV=DIV_2, FWDTEN=OFF, CP=OFF, BWP=OFF
#include <p32xxxx.h>
#include <plib.h>
#include <explore.h>
#include <graphic.h>

main()
{

   int i;
```

```
// initializations
initEX16();      // init and enable vectored interrupts
clearScreen();   // init the video map
initVideo();     // start the video state machine

for( i=0; i<1000; i++)
{
  plot( rand()%HRES, rand()%VRES);
}
// main loop
  while( 1)
  {

  } // main loop

} // main
```

Save the file as **GraphicTest3.c** and add it to the Video project to replace the previous demo. Once you build the project and program the Explorer 16 board with your in circuit emulator of choice, the output on your video display should look like a nice starry night, as in the screen shot captured in Figure 13.20.

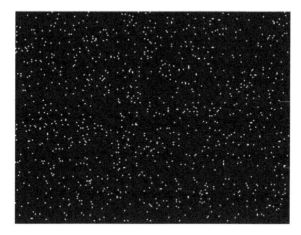

Figure 13.20: Screen capture: plotting a starry night.

A starry night it is, but not a realistic one, you'll notice, since there is no recognizable trace of any increased density of stars around a belt—in other words, there is no Milky Way!

This is a good thing! This is a simple proof that our pseudo-random number generator is in fact doing the job it is supposed to do.

Line Drawing

The next obvious step is drawing lines, or I should say line *segments*. Granted, horizontal and vertical line segments are not a problem; a simple `for` loop can take care of them. But drawing oblique lines is a completely different thing. We could start with the basic formula for the line between two points that you will remember from school days:

```
y=y0 + (y1-y0)/(x1-x0) * ( x-x0)
```

where `(x0,y0)` and `(x1,y1)` are, respectively, the coordinates of two generic points that belong to the line.

This formula gives us, for any given value of *x*, a corresponding *y* coordinate. So we might be tempted to use it in a loop for each discreet value of *x* between the starting and ending point of the line, as in the following example:

```c
/*
** LineTest1.c
**
** testing the basic line drawing function
*/
// configuration bit settings, Fcy=72 MHz, Fpb=36 MHz
#pragma config POSCMOD=XT, FNOSC=PRIPLL
#pragma config FPLLIDIV=DIV_2, FPLLMUL=MUL_18, FPLLODIV=DIV_1
#pragma config FPBDIV=DIV_2, FWDTEN=OFF, CP=OFF, BWP=OFF
#include <p32xxxx.h>
#include <plib.h>
#include <explore.h>
#include <graphic.h>

main()
{
  int x;
  float x0 = 10.0,   y0 = 20.0;
  float x1 = 200.0,  y1 = 150.0;
  float x2 = 20.0,   y2 = 150.0;
```

```
// initializations
initEX16();      // init and enable vectored interrupts
clearScreen();   // clear the image buffer
initVideo();     // start the video state machine

// draw an oblique line (x0,y0) - (x1,y1)
for( x = x0; x<x1; x++)
  plot( x, y0 + (y1-y0)/(x1-x0)* (x-x0));

// draw a second (steeper) line (x0,y0) - ( x2,y2)
for( x = x0; x<x2; x++)
  plot( x, y0+(y2-y0)/(x2-x0)* (x-x0));

// main loop
  while( 1)
  {
  } // main loop

} // main // main
```

The output produced (Figure 13.21) is an acceptably continuous segment only for the first (shallower) line, where the horizontal distance (x1-x0) is greater than the vertical distance (y1-y0). In the second, much steeper, line, the dots appear disconnected and we are clearly unhappy with the result. Also, we had to perform floating-point arithmetic,

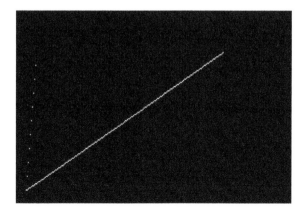

Figure 13.21: Screen capture: drawing oblique lines.

a computationally expensive proposition compared to integer arithmetic, as we have seen in previous chapters.

Bresenham Algorithm

Back in 1962, when working at IBM in the San José development lab, Jack E. Bresenham developed a line-drawing algorithm that uses exclusively integer arithmetic and is today considered the foundation of any computer graphic program. Its approach is based on three optimization "tricks":

1. Reduction of the drawing direction to a single case (left to right)

2. Reduction of the line steepness to the single case where the horizontal distance is the greatest

3. Multiply both sides of the equation by the horizontal distance (`deltax`) to obtain only integer quantities

The resulting line-drawing code is compact and extremely efficient; here is an adaptation for our video module:

```
#define abs( a) (((a)> 0) ? (a) : -(a))

void line( short x0, short y0, short x1, short y1)
{

    short steep, t ;
    short deltax, deltay, error;
    short x, y;
    short ystep;

  // simple clipping
    if (( x0 < 0) || (x0 > HRES))
      return;
    if (( x1 < 0) || (x1 > HRES))
      return;
    if (( y0 < 0) || (y0 > VRES))
      return;
    if (( y1 < 0) || (y1 > HRES))
      return;
```

```
   steep = ( abs(y1 - y0) > abs(x1 - x0));

   if ( steep )
   { // swap x and y
     t = x0; x0 = y0; y0 = t;
     t = x1; x1 = y1; y1 = t;
   }
   if (x0 > x1)
   { // swap ends
     t = x0; x0 = x1; x1 = t;
     t = y0; y0 = y1; y1 = t;
   }

   deltax = x1 - x0;
   deltay = abs(y1 - y0);
   error = 0;
   y = y0;

   if (y0 < y1) ystep = 1; else ystep = -1;
   for (x = x0; x < x1; x++)
   {
     if ( steep) plot(y,x); else plot(x,y);
     error += deltay;
     if ( (error<<1) >= deltax)
     {
       y += ystep;
       error -= deltax;

     } // if
   } // for

} // line
```

We can add this function to the video module **graphic.c** and add its prototype to the *include* file **graphic.h**:

```
void line( short x0, short y0, short x1, short y1);
```

To test the efficiency of the Bresenham algorithm, we can create a new small project and, once more, use the pseudo-random number-generator function. The following example code will first draw a frame around the screen and then exercise the line-drawing routine, producing 100 lines at randomly generated coordinates:

```
/*
** Bresenham.c
**
** Fast line drawing algorithm example
*/
// configuration bit settings, Fcy=72 MHz, Fpb=36 MHz
#pragma config POSCMOD=XT, FNOSC=PRIPLL
#pragma config FPLLIDIV=DIV_2, FPLLMUL=MUL_18, FPLLODIV=DIV_1
#pragma config FPBDIV=DIV_2, FWDTEN=OFF, CP=OFF, BWP=OFF

#include <p32xxxx.h>
#include <plib.h>
#include <explore.h>
#include <graphic.h>

main()
{
  int i;

  // initializations
  initEX16();      // init and enable vectore interrupts
  initVideo();     // start the state machines

  // main loop
    while( 1)
    {
    clearScreen();
    line( 0, 0, 0, VRES-1);
    line( 0, VRES-1, HRES-1, VRES-1);
    line( HRES-1, VRES-1, HRES-1, 0);
    line( 0, 0, HRES-1, 0);

    for( i=0; i<100; i++)
      line( rand()%HRES, rand()%VRES,
        rand()%HRES, rand()%VRES);
```

```
    // wait for a button to be pressed
    getKEY();
    } // main loop

} // main
```

The main loop also uses the getKey() function, developed in the previous chapters and added to the explore.h module, to wait until a button is pressed before the screen is cleared and a new set of 100 random lines is drawn on the screen (see Figure 13.22).

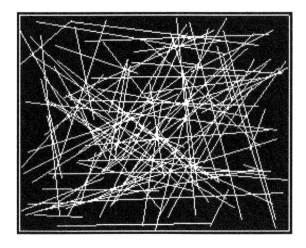

Figure 13.22: Screen capture: Bresenham line-drawing test.

You will be impressed by the speed of the line-drawing algorithm. Even when increasing the number of lines drawn to batches of 1,000, the PIC32 performance will be apparent.

Plotting Math Functions

With the completed graphic module we can now start exploring some interesting applications that can take full advantage of its visualization capabilities. One classical application could be plotting a graph based on data logged from a sensor, or more simply for our demonstration purposes, calculated on the fly from a given math function.

For example, let's assume that the function is a sinusoid (with a twist), as in the following:

```
y(x) = x * sin( x)
```

Let's also assume that we want to plot its graph for values of *x* between 0 and 8*PI.

With minor manipulations, we can scale the function to fit our screen, remapping the input range from 0 to 200 and the output range to the +75/-75 value range.

The following program example will plot the function after tracing the *x* and *y* axes:

```
/*
** graph1d.c
**
** Plotting a function graph
*/
// configuration bit settings, Fcy=72MHz, Fpb=36MHz
#pragma config POSCMOD=XT, FNOSC=PRIPLL
#pragma config FPLLIDIV=DIV_2, FPLLMUL=MUL_18, FPLLODIV=DIV_1
#pragma config FPBDIV=DIV_2, FWDTEN=OFF, CP=OFF, BWP=OFF
#include <p32xxxx.h>
#include <plib.h>
#include <explore.h>
#include <graphic.h>
#include <math.h>

#define X0 10
#define Y0 (VRES/2)

main( void)
{
    int x, y;
    float xf, yf;

    // initializations
    initEX16();     // init and enable vectored interrupts
    clearScreen();
    initVideo();    // init video state machine
```

```
   // draw the x and y axes crossin in (X0,Y0)
   line( X0, 10, X0, VRES-10);     // y axes
   line( X0-5, Y0, HRES-10, Y0);  // x axes

   // plot the graph of the function for
   for( x=0; x<200; x++)
   {
     xf = (2 * M_PI / 50) * (float) x;
     yf = 75.0 / ( 8 * M_PI) * xf * sin( xf);
     plot( x+X0, yf+Y0);
   }

   // main loop
   while( 1);

} // main
```

Notice the inclusion of the math.h library to obtain the prototypes of the `sin()` function and some useful definitions, among which is the value of pi, or I should say `M_PI`.

Save the file as **graph1d.c** and replace it as the main module of the Video project. Build the project and program the Explorer 16 board with your in-circuit debugger of choice. Quick, the new function graph will appear on the screen (see Figure 13.23)!

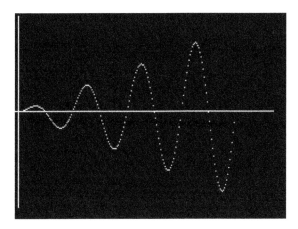

Figure 13.23: Screen capture: a sinusoidal function graph.

Should the points on the graph become too sparse, we have the option now to use the line-drawing algorithm to connect each point to the previous one.

Two-Dimensional Function Visualization

More interesting and perhaps entertaining could be plotting two-dimensional function graphs. This adds the thrill of managing the perspective distortion and the challenge of connecting the calculated points to form a visually pleasant grid.

The simplest method to squeeze the third axis in a two-dimensional image is to utilize what is commonly known as an *isometric projection*, a method that requires minimal computational resources while providing a small visual distortion. The following formulas applied to the *x*, *y*, and *z* coordinates of a point in a three-dimensional space produce the px and py coordinates of the projection on a two-dimensional space (our video screen; see Figure 13.24).

```
px = x + y/2;
py = z + y/2;
```

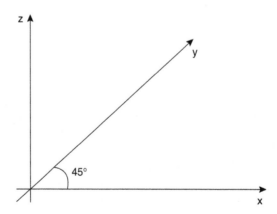

Figure 13.24: Isometric projection.

To plot the three-dimensional graph of a given function: z = f(x,y) we proceed on a grid of points equally spaced in the *x* and *y* plane using two nested for loops. For each point we compute the function to obtain the *z* coordinate, and we apply the isometric projection to obtain a (px,py) coordinate pair. Then we connect the newly calculated point with a segment to the previous point on the same row (previous column). A second segment needs to be drawn to connect the point to a previously computed point in the same column and the previous row (see Figure 13.25).

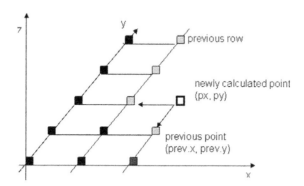

Figure 13.25: Drawing a grid to enhance a two-dimensional graph visualization.

Although it is a trivial task to keep track of the coordinates of the previously computed point on the same row, recording the coordinates of the points on "each" previous row might require significant memory space. If, for example, we are using a grid of 20×20 points, we would need to store the coordinates of up to 400 points. Requiring two integers each, that would add up to 800 words, or 3,200 bytes of precious RAM. In reality, as should be evident from the preceding picture, all we really need is the coordinates of the points on the "edge" of the grid as painted so far. Therefore, with a little care, we can reduce the memory requirement to just 20 coordinate pairs by maintaining a small (rolling) buffer.

The following example code visualizes the graph of the function:

```
z(x,y) = 1/ sqrt( x2 + y2) * cos ( sqrt( x2 + y2)
```

for values of *x* and *y* in the range -3*PI to +3*PI:

```
/*
** graph2d.c
**
** 07/02/06 v1.0 LDJ
** 11/21/07 v2.0 LDJ PIC32 porting
*/
```

```c
// configuration bit settings
#pragma config POSCMOD=XT, FNOSC=PRIPLL
#pragma config FPLLIDIV=DIV_2, FPLLMUL=MUL_18, FPLLODIV=DIV_1
#pragma config FWDTEN=OFF, CP=OFF, BWP=OFF

#include <p32xxxx.h>
#include <explore.h>
#include <graphic.h>
#include <math.h>

#define X0       10     // graph offset
#define Y0       10
#define NODES    20     // define grid
#define SIDE     10
#define STEP     1      // movement increment

typedef struct {
    int x;
    int y;
  } point;

point edge[NODES], prev;

main( void)
{
  int i, j, x, y, z;
  float xf, yf, zf, sf;
  int px, py;
  int xoff, scale;

  // initializations
  initEX16();
  clearScreen();
  initVideo();

  xoff = 100;

  scale = 75;

  while (1)
  {
    // clear hidden screen
    clearScreen();
```

```
// draw the x, y and z axes crossing in (X0,Y0)
line( X0, 10, X0, 10);               // z axis
line( X0-5, Y0, HRES-10, Y0);        // x axis
line( X0-2, Y0-2, X0+120, Y0+120);   // y axis

// init the array of previous egde points
for( j=0; j<NODES; j++)
{
  edge[j].x = X0+ j*SIDE/2;
  edge[j].y = Y0+ j*SIDE/2;
}

// plot the graph of the function for
for( i=0; i<NODES; i++)
{
  // transform the x range to 0..200 offset 100
  x = i * SIDE;
  xf = (6 * M_PI/200) * (float)(x-xoff);
  prev.y = Y0;
  prev.x = X0 + x;

  for ( j=0; j<NODES; j++)
  {
    // transform the y range to 0..200 offset 100
    y = j * SIDE;
    yf = (6 * M_PI / 200) * (float)(y-100);

    // compute the function
    sf = sqrt( xf * xf + yf * yf);
    zf = 1/(1+ sf) * cos( sf );

    // scale the output
    z = zf * scale;

    // apply isometric perspective and offset
    px = X0 + x+ y/2;
    py = Y0 + z + y/2;

    // plot the point
    plot( px, py);
```

```
            // draw connecting lines to visualize the grid
            line( px, py, prev.x, prev.y);
            line( px, py, edge[j].x, edge[j].y);

            // update the previous points
            prev.x = px;
            prev.y = py;
            edge[j].x = px;
            edge[j].y = py;
        } // for j
    } // for i

    // wait for a button
    getKEY();

    } // main loop
} // main
```

Save the file as **graph2d.c** and replace it in the Video project as the main source. After building the project and programming the Explorer 16 demo board, you will notice how quickly the PIC32 can produce the output graph, although significant floating-point math is required because the function is applied sequentially to 400 points and as many as 800 line segments are drawn on the video (see Figure 13.26).

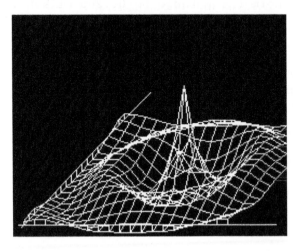

Figure 13.26: Screen capture: graph of a two-dimensional function.

Fractals

Fractals is a term coined by Benoit Mandelbrot, a mathematician and fellow researcher at the IBM Pacific Northwest Labs, back in 1975 to denote a large set of mathematical objects that presented an interesting property: that of appearing *self-similar* at all scales of magnification, as though constructed recursively with an infinite level of detail. There are many examples of fractals in nature, although their self-similarity property is typically extended over a finite scale. Examples include clouds, snowflakes, mountains, river networks, and even the blood vessels in our bodies.

Since it lends itself to impressive computer visualizations, the most popular example of mathematical fractal object is perhaps the Mandelbrot set. It's defined as a subset of the complex plane where the quadratic function $z^2 + c$ is iterated. By exclusion, points c of the complex plane for which the iteration does not "diverge" are considered to be part of the set. Since it is easy to prove that once the modulus of z is greater than 2, the iteration is bound to diverge; hence the given point is not part of the set, we can proceed by elimination. The problem is that as long as the modulus of z remains smaller than 2, we have no way of telling when to stop the iteration and declare the point part of the set. So, typically, computer algorithms that depict the Mandelbrot set use an approximation by setting an arbitrary maximum number of iterations past which a point is simply assumed to be part of the set.

Here is an example of how the inner iteration can be coded in C language:

```
// initialization
  x = x0;
  y = y0;
  k = 0;

// core iteration
do {
    x2 = x*x;
    y2 = y*y;
    y = 2*x*y+y0;
    x = x2-y2+x0;
    k++;
  } while ( (x2 + y2 < 4) && ( k < MAXIT));
```

```
            // check if the point belongs to the Mandelbrot set
            if ( k == MAXIT) plot( j, i);
```

where x0 and y0 are the coordinates in the complex space of the point c.

We can repeat this iteration for each point of a squared subset of the complex plane to obtain an image of the entire Mandelbrot set. The considerations we made on the modulus of c imply that the entire set must be contained in the disc of radius 2 centered on the origin, so, as we develop a first program, we will scan the complex plan in a grid of HRES xVRES points (to use the fullscreen resolution of our video module), making sure to include the entire disc:

```
/*
** Mandelbrot.c
**
** Mandelbrot Set graphic demo
*/
// configuration bit settings, Fcy=72 MHz, Fpb=36 MHz
#pragma config POSCMOD=XT, FNOSC=PRIPLL
#pragma config FPLLIDIV=DIV_2, FPLLMUL=MUL_18, FPLLODIV=DIV_1
#pragma config FPBDIV=DIV_2, FWDTEN=OFF, CP=OFF, BWP=OFF


#include <p32xxxx.h>
#include <plib.h>
#include <explore.h>
#include <graphic.h>

#define SIZE VRES
#define MAXIT 64

void mandelbrot( float xx0, float yy0, float w)
{
   float x, y, d, x0, y0, x2, y2;
   int i, j, k;
   // calculate increments
   d = w/SIZE;
```

```
  // repeat on each screen pixel
  y0 = yy0;
  for (i=0; i<SIZE; i++)
  {
    x0 = xx0;
    for (j=0; j<SIZE; j++)
    {
      // initialization
        x = x0;
        y = y0;
        k = 0;

      // core iteration
        do {
          x2 = x*x;
          y2 = y*y;
          y = 2*x*y + y0;
          x = x2-y2 + x0;
          k++;
        } while ( (x2 + y2 < 4) && ( k < MAXIT));

      // check if the point belongs to the Mandelbrot set
      if ( k == MAXIT) plot( j, i);

      // compute next point x0
      x0 += d;
    } // for j
    // compute next y0
    y0 += d;
  } // for i
} // mandelbrot

int main( void)
{
  float x, y, w;
  int c;
  // initializations
  initEX16();     // init and enable vectored interrupts
  initVideo();    // init the video state machine
```

```
// intial coordinates lower left corner of the grid
x = -2.0;
y = -2.0;
// initial grid size
w = 4.0;

clearScreen();          // clear the screen
mandelbrot( x, y, w);// draw new image

while( 1);
} // main
```

Save this file as **Mandelbrot.c** and add it to a new project that we will call **Mandelbrot**. Make sure that all the other required modules are added to the project too, including **graphic.c**, **graphic.h**, and **explore.c**. Build the project, program the Explorer 16 board using your in-circuit debugger of choice, and if all is well, when you let the program run you will see the so-called Mandelbrot "cardiod" appear on your screen (see Figure 13.27).

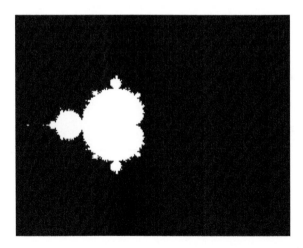

Figure 13.27: Screen capture: Mandelbrot cardiod.

I will confess that since when, as a kid, I bought my first personal computer—actually, *home computer* was the term used back then for the Sinclair ZX Spectrum—I have been playing with fractal programs. So I have a vivid memory of the long hours I used to spend staring at the computer screen, waiting for the old trusty ZX80 processor (running at the whopping speed of 3.5 MHz) to paint this same images. A few years later, my first IBM PC, an XT clone running on a 8088 processor at a not much higher clock speed of 4 MHz, was not faring much better and, although the screen resolution of my monochrome Hercules graphic card was higher, I would still launch programs in the evening to watch the results the following morning after what amounted sometimes to up to eight hours of processing.

WOW

Clearly, the amount of computation required to paint a fractal image varies enormously with the chosen area and the number of maximum iterations allowed (MAXIT), but, although I have seen this program run by several other processors, including the PIC24 (at 32 MHz), the first time I saw the PIC32 paint the cardiod in less than 5 seconds, I got really excited again!

The real fun has just begun. The most interesting parts of the Mandelbrot set are at the fringes, where we can increase the magnification and zoom in to discover an infinitely complex world of details. By visualizing not just the points that belong to the set but also the ones that diverge at its edges, and by assigning each point a "color" that depends on how fast they do diverge, we can further improve "aesthetically" the resulting image. Since we have only a monochrome display, we will simply use alternate bands of black and white assigned to each point according to the number of iterations it took before it either reached the maximum modulus or the maximum number of iterations. Simply enough, this means we will have to modify just one line of code from our previous example:

```
// check if the point belongs to the Mandelbrot set
if ( k & 2) plot( j, i);
```

Also, since the best way to play with Mandelbrot sets is to explore them by selecting new areas and zooming in the details, we can modify the main program loop to let us select a portion of the image by pressing one of the four buttons on the Explorer 16 board. We can imagine splitting the image into four corresponding quadrants, numbered clockwise starting from the top left, and doubling the resolution by halving the grid dimension (w) (see Figure 13.28).

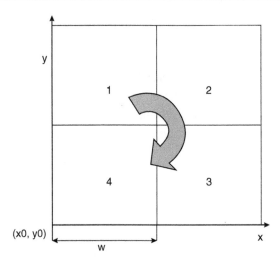

Figure 13.28: Splitting the screen into four quadrants.

```c
int main( void)
{
  float x, y, w;
  int c;

  // initializations
  initEX16();
  initVideo();    // start the state machines

  // intial coordinates lower left corner of the grid
  x = -2.0;
  y = -2.0;
  // initial grid size
  w = 4.0;

  while( 1)
  {
    clearScreen();        // clear the screen
    mandelbrot( x, y, w);// draw new image
    // wait for a button to be pressed
    c = getKEY();
    switch ( c){
    case 8:    // first quadrant
      w/= 2;
      y += w;
      break;
```

```
   case 4:    // second quadrant
     w/= 2;
     y += w;
     x += w;
     break;

   case 2:    // third quadrant
     w/= 2;
     x += w;
     break;

   default:
   case 1:    // fourth quadrant
     w/= 2;
     break;
   } // switch
  } // main loop
} // main
```

Figure 13.29 shows a selection of interesting areas you will be able to explore with a little patience.

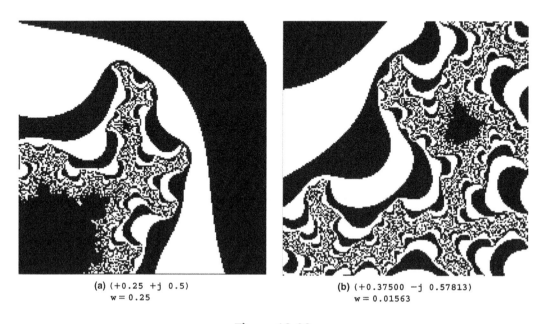

(a) (+0.25 +j 0.5)
w = 0.25

(b) (+0.37500 −j 0.57813)
w = 0.01563

Figure 13.29.

(c) $(-1.28125 +j\ 0.3125)$
$w = 0.3125$

(d) $(+0.34375 +j\ 0.56250)$
$w = 0.3125$

(e) $(-1.28125 +j\ 0.4688)$
$w = 0.01563$

Figure 13.29: (Contiuned).

Text

So far we have been focusing on simple graphical visualizations, but on more than one occasion you might feel the desire to actually augment the information presented on the screen with some text. Writing text on the video memory is no different than plotting points or drawing lines; in fact, it can be achieved using a variety of methods, including the plotting and line-drawing functions we have already developed. But for greater performance and to require the smallest possible amount of code, the easiest way to text on our graphic display is to develop a fixed spacing font. Each character can be drawn in an 8 × 8 pixel box; this way 1 byte will encode each row and 8 bytes will encode the entire character. We can then assemble a basic set of alphabetical, numerical, and punctuation characters, using the order in which they are appear in the ASCII character set, as a single array of `char` integers that will constitute our simple *font* (see Figure 13.30).

0	0	0	1	1	1	0	0
0	0	1	0	0	0	1	0
0	0	1	0	0	0	1	0
0	0	1	1	1	1	1	0
0	0	1	0	0	0	1	0
0	0	1	0	0	0	1	0
0	0	1	0	0	0	1	0
0	0	0	0	0	0	0	0

Figure 13.30: The letter *A* as represented in a simple 8 × 8 font.

To save space, we don't need to create the first 32 codes of the ASCII set that correspond mostly to commands and special synchronization codes used by teletypewriters and modems of the old days.

```
/*
** 8 x 8 Simple Character Font
**
*/
```

```
#define F_OFFS   0x20 // initial offset
#define F_SIZE   96 // define only the first 96 characters

const char Font8x8[]={
// 20 - SPACE
   0x00,          // 0b 0000000,
   0x00,          // 0b 0000000,
   0x00,          // 0b 0000000,
   0x00,          // 0b 0000000,
   0x00,          // 0b 0000000,
   0x00,          // 0b 0000000,
   0x00,          // 0b 0000000,
   0x00,          // 0b 0000000,
// 1 - !
   0x18,          // 0b 0011000,
   0x18,          // 0b 0011000,
   0x18,          // 0b 0011000,
   0x18,          // 0b 0011000,
   0x18,          // 0b 0011000,
   0x00,          // 0b 0000000,
   0x18,          // 0b 0011000,
   0x00,          // 0b 0000000,
   ...
} // Font 8x8[]
```

Notice that the Font8x8[] array is defined with the attribute const because its contents are supposed to remain (mostly) unchanged during the execution of the program, and it is best allocated in the Flash memory of the PIC32 to save precious RAM memory space.

Of course, the definition of the shape of each character can be a matter of personal taste. You are welcome to modify the Font8x8[] array contents to suit your preferences.

Note

Defining a new font is a long and detailed work, but it is one that gives a lot of space to creativity, and I know that some of you will find it pretty entertaining. A complete listing of the font.h file would waste several pages of this book, so I decided to omit it here. You can find it on the companion CD-ROM.

Printing a character on the screen is now a matter of copying 8 bytes from the font array to the desired position on the screen. In the simplest case, characters can be aligned to the words that compose the image buffer of the graphics module. In this way the character positions are limited to 32 characters per line (256/8, assuming HRES = 256) and a maximum of 25 rows of text could be displayed (200/8, assuming VRES = 200).

A more advanced solution would call for absolute freedom in positioning each character at any given pixel coordinate. This would require a type of manipulation, often referred to as *BitBLT* (an acronym that stands for *bit block transfer*) that is common in computer graphics, particularly in video game design. In the following we will stick to the simpler approach. looking for the solution that requires the smallest amount of resources to get the job done.

Printing Text on Video

When printing text on video we need the assistance of a cursor, a virtual placeholder to keep track of where on the screen we are going to place the next character. As we print, it is easy to advance the cursor to mimic somewhat the behavior of a typewriter as it zigzags across the sheet and as it scrolls the paper.

OOPS

As I am writing this, it occurs to me that many of you might have never used a typewriter in real life and that the beauty of this parallel is going to be totally lost on you. Maybe it feels like I might be talking of reed pens and kalamoi or parchment . . .

Our cursor will be made of two integers, holding the *x* and *y* of a new coordinate system that is now upside down with respect to the traditional Cartesian orientation and much more coarse as it counts rows and columns rather than individual pixels:

- cx, will indicate the current column, counting from left to right, from 0 to 31.

- cy, will indicate the row, counting from top to bottom, from 0 to 24.

To print one ASCII character on the screen at the current cursor position, we will create the `putcV()` function that will perform the following simple steps:

1. Check whether the character requested is within the range of characters for which we have font definition (from ASCII code `0x20` all the way up to `0x7F`):

```
void putcV( char a)
{
    int i, j, *p;
    const char *pf;

    // 1. check if char in range
    if ( a < F_OFFS)
        return;
    if ( a >= F_OFFS+F_SIZE)
        return;
```

2. Check whether the cursor position is within the screen boundaries, wrapping around and scrolling as necessary:

```
    // 2. check page boundaries and wrap or scroll as
    necessary
    if ( cx >= HRES/8)        // wrap around x
    {
      cx = 0;
      cy++;
    }
    if ( cy >= VRES/8)        // scroll up y
    {
    int *pd = VH;
    int *ps = pd+(HRES/32)*8;
    for( i=0; i<(HRES/32)*(VRES-8); i++)
      *pd++ = *ps++;
    for( i=0; i<(HRES/32)*8; i++)
      *pd++ = 0;
    // keep cursor within boundary
    cy=VRES/8-1;
    }
```

3. Find the address inside the image buffer corresponding to the cursor location (p), and find the character definition inside the `Font8x8[]` array (pf):

```
// 3. set pointer to word in the video map
p = &VH[ cy * 8 * HRES/32 + cx/4];
// set pointer to first row of the character in font
array
pf = &Font8x8[ (a-F_OFFS) << 3];
```

4. Copy the character byte after byte, taking care to clear the background image before overimposing each character row:

```
// 4. copy one by one each line of the character on
screen
for ( i=0; i<8; i++)
{
   j = (3-(cx & 3))<<3;     // consider MSB first
   *p &= ~(0xff << j);      // clear background
   *p |= ((*pf++) << j);    // overimposed character

   // point to next row
   p += HRES/32;
} // for
```

5. Finally, advance the cursor position:

```
// 5. advance cursor position
cx++;
} // putcV
```

Add this function to the bottom of the **graphic.c** module and its prototype to the bottom of the **graphic.h** include file:

```
void putcV( char a);
```

For our convenience we can now create a small function that will print an entire (zero-terminated) ASCII string on the screen:

```
void putsV( char *s)
{
  while (*s)
    putcV( *s++);
```

```
  // advance to next line
  cx=0; cy++;} // putsV
} // putsV
```

Add this function to the **graphic.c** library module and its prototype to the **graphic.h**:

```
void putsV( char *s);
```

Since we're at it, let's add a couple more useful macros to the **graphic.h** file:

```
#define Home()          { cx=0; cy=0;}
#define Clrscr()        { clearScreen(); Home();}
#define AT( x, y)       { cx = (x); cy =(y);}
```

- Home() will simply position the cursor on the upper-left corner of the screen.

- Clrscr() will clear the screen first and then reposition the cursor to the top.

- AT(x, y) will position the cursor at the desired column (x) and row (y).

Text Test

To quickly test the effectiveness of the new text functions, we can now create a short program that, after printing a small banner on the first line of the screen, will print out each character defined in the 8 × 8 font:

```
//*
** TextTest.c
**
*/
// configuration bit settings, Fcy=72MHz, Fpb=36MHz
#pragma config POSCMOD=XT, FNOSC=PRIPLL
#pragma config FPLLIDIV=DIV_2, FPLLMUL=MUL_18, FPLLODIV=DIV_1
#pragma config FPBDIV=DIV_2, FWDTEN=OFF, CP=OFF, BWP=OFF
#include <p32xxxx.h>
#include <explore.h>
#include <graphic.h>

main( void)
{
  int i;
```

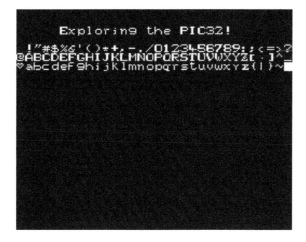

Figure 13.31: Screen capture: text test.

```
// initializations
initEX16();  // init and enable vectored interrupts
initVideo(); // start the state machines

Clrscr();

AT( 5, 2);
putsV( "Exploring the PIC32!");

AT( 0, 4);
for( i=0; i<128; i++)
  putcV( i);
while (1);

} // main
```

Save this file as **TextTest.c** and add it to a new project that we will call **TextTest**. Make sure that all the other required modules are added to the project too, including **graphic. c**, **graphic.h**, and **explore.c**. Build the project, program the Explorer 16 board using your in-circuit debugger of choice, and if all is well, when you run you will see the screen come alive with a nice welcome message (see Figure 13.31).

The Matrix Reloaded

To further test the new text page video module, we will modify an example we saw in a previous chapter: the Matrix. Back then, we were using the asynchronous serial communication module (UART1) to communicate with a VT100 computer terminal or, more likely, a PC running the HyperTerminal program configured for emulation of the historical DEC VT100 terminal's protocol. Now we can replace the `putcU()` function calls used to send a character to the serial port, with `putcV()` function calls directed to the graphic interface.

Let's modify the **TextTest** project by replacing the **TextTest.c** main module with the new **Matrix2.c** module, modified as follows:

```
/*
** Matrix2.c
*/
// configuration bit settings, Fcy=72MHz, Fpb=36MHz
#pragma config POSCMOD=XT, FNOSC=PRIPLL
#pragma config FPLLIDIV=DIV_2, FPLLMUL=MUL_18, FPLLODIV=DIV_1
#pragma config FPBDIV=DIV_2, FWDTEN=OFF, CP=OFF, BWP=OFF
#include <p32xxxx.h>
#include <graphic.h>

#define COL     HRES/8
#define ROW     VRES/8

main()
{
  int v[ COL];   // vector containing length of each string
  int i,j,k;

  // 1. initializations
  initEX16();
  initVideo();
  Clrscr();       // clear the screen

  // 2. init each column length
  for( j =0; j < COL; j++)
        v[j] = rand()%ROW;

  // 3. main loop
  while( 1)
  {

    // 3.1 refresh the screen with random columns
```

```
    for( i=0; i<ROW; i++)
  {
    AT( 0, i);
    // refresh one row at a time
    for( j=0; j<COL; j++)
    {
      // fill random char down to each column length
      if ( i < v[j])
         putcV( '!'+(rand()%15));
      else
        putcV(' ');
      } // for j
    } // for i

    // 3.2 randomly increase or reduce each column length
    for( j=0; j<COL; j++)
    {
       switch ( rand()%3){
       case 0: // increase length
           v[j]++;
           if (v[j]>ROW)
              v[j]=ROW;
          break;

    case 1:   // decrease length
          v[j]--;
          if (v[j]<1)
             v[j]=1;
          break;

    default:// unchanged
        break;
      } // switch

    } // for j
  } // main loop
} // main
```

After saving and building the project, program the Explorer 16 board using your in-circuit debugger of choice and run the program (see Figure 13.32). You will notice how much faster the screen updates can be compared because the program now has direct

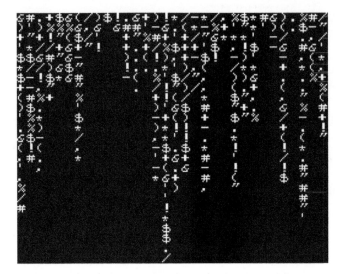

Figure 13.32: Screen capture, the Matrix . . . Reloaded.

access to the video memory and no serial connection limits the information transfers (as fast as the 115,200 baud connection was in our previous demo project; that was our bottleneck). The demo will run so fast that you will need to add a delay of a few extra milliseconds to give your eyes time to focus.

```
// 3.3 delay to slow down the screen update
Delayms( 5);
```

Debriefing

Today we have explored the possibility of producing a video output using a minimal hardware interface composed in practice of only three resistors. We learned to use four peripheral modules together to build the complex mechanism required to produce a properly formatted NTSC composite video signal. Combining a 16-bit-timer, an output compare module, one SPI port, and a couple of channel of the DMA module we have obtained video capabilities at the cost of just 1.5% processor overhead. After developing basic graphic functions to plot individual pixels first and efficiently draw lines, we explored some of the possibilities offered by the availability of a graphic video output, including unidimensional and two-dimensional function graphing. We completed our explanations with a brief foray in the world of fractals and learning to display text on top of graphics.

Notes for the PIC24 Experts

The OC module of the PIC32 is mostly identical to the PIC24 peripheral, yet some important enhancements have been included in its design. Here are the major differences that will affect your code while porting an application to the PIC32:

1. The OCxCON control register layout has been updated to resemble more closely the layout of most other peripherals so that the module ON, FRZ, and IDL bits are now available to better control operation in the low-power modes.

2. The OC32 control bit has been added to enable a 32-bit mode of operation when the module is associated with a 32-bit timer pair.

Tips & Tricks

The final touch, to complete our brief excursion into the world of graphics, would be to add some animation capabilities to our graphic libraries. To make the motion fluid and avoid an annoying flicker of the image on the screen, a technique known as *double buffering* is often used. This requires the allocation of two image buffers of identical size. One, the "active" buffer, is shown on the screen; the other, the "hidden" buffer, is where the drawing takes place. When the drawing on the hidden buffer is completed, the two are swapped. What used to be the active buffer is not visible anymore. The (now) hidden buffer can be cleared without fear of producing any flicker, and the drawing process can restart.

With the current image resolution settings (256 × 200), the RAM usage grows to a total of 12,800 bytes (256*200*2/8), which represents only approximately 40 percent of the total RAM available on the PIC32MX360.

To extend our graphic libraries and support double buffering, we can implement the following simple modifications:

- At the top of the **graphic.c** module, add the declaration of a second image buffer;

```
#ifdef DOUBLE_BUFFER
int VMap2[ VRES*(HRES/32)]; // second image buffer
#endif
```

- Inside the `initVideo()` function, where the `VA` and `VH` pointers were assigned initial values, add a new conditional assignment. (Now you can understand why I had chosen to use two separate pointers to the same image buffer.)

```
// 6. init the active and hidden screens pointers
VA = VMap1;
#ifdef DOUBLE_BUFFER
  VH = VMap2;
#else
  VH = VA;
#endif
```

- Add the new function `clearHScreen()`, to clear the hidden buffer in double-buffering mode:

```
void clearHScreen( void)
{ // fill with zeros the Hidden Video array
  memset( VH, 0, VRES*( HRES/8));
  // reset text cursor position
  cx = cy = 0;
} //clearHScreen
```

- Add the `swapV()` function to swap the two buffers (it's just the pointers that get swapped):

```
void swapV( void)
{
  int * V;

  if ( VState == SV_LINE)    // wait end of the frame
    while ( VCount != 1);

  V = VA; VA = VH; VH = V;   // swap the pointers
  VPtr = VA;
} // swapV
```

Notice that care must be taken not to perform the swap in the middle of a frame but synchronized with the end of one frame and the beginning of the next.

One last utility function can be added for all those cases when the animation needs to be suspended and the display has to return to a simple buffering mode:

```
void singleV( void)
{ // make all functions work on a single image buffer
  VA = VMap1;
  VH = VA;
}
```

Remember to add all the corresponding function prototypes to the **graphic.h** include file and additionally, at the top, declare the new symbol DOUBLE_BUFFER:

```
#define DOUBLE_BUFFER   // comment if single buffering required
void clearHScreen( void);
void swapV(void);
void singleV( void);
```

Note

All the examples developed in this chapter and the previous one can now be recompiled using the newly extended graphic modules with the condition that the DOUBLE_BUFFER declaration is either commented out or the singleV() function is called immediately after the initVideo() call!

Exercises

1. Modify the **Mandelbrot.c** demo to use a 32-bit timer to self-time the PIC32 performance and display the time and coordinates of the image on the screen.

2. Create a combined demo project that uses the PS/2 keyboard input and the graphic libraries to provide a terminal console.

3. Modify the **graph2D.c** demo and allow the user to "manipulate" the function using the four buttons on the Explorer 16 demo board to increase and decrease the scaling factor and change the position of the "peak" while refreshing the screen using the double-buffering animation technique.

4. Experiment with 3D geometry, drawing objects in perspective and rotating them in space.

Books

Mandelbrot, Benoit, B., *The Fractal Geometry of Nature* (W. H. Freeman, 1982). This is "the" book on fractals, written by the man who contributed most to the rediscovery of fractal theory.

Hofstadter, Douglas, *Godel, Escher, Bach: An Eternal Golden Braid*, 20th Anniversary Edition (Basic Books, 1999). One of the most inspiring books in my library. At 777 pages, it's not easy reading, but it will take you on a journey through graphics, math, and music and the surprising connections among the three.

Links

http://en.wikipedia.org/wiki/Fractals. A starting point for you to begin the online exploration of the world of fractals.

http://en.wikipedia.org/wiki/Zx_spectrum. The Sinclair ZX Spectrum was one of the first personal computers (home computers, as they used to be called) launched in the early 1980s. Its graphic capabilities were very similar to those of the graphic libraries we developed in this project. Although it used several custom logic devices to produce a video output, its processing power was less than a tenth that of the PIC32. Still, the limited ability to produce color (only 16 colors with a resolution of a block of 8 × 8 pixels) enticed many programmers to create thousands of challenging and creative video games.

Mass Storage

The Plan

In many embedded-control applications, you might find a need for a larger nonvolatile data storage space well beyond the capabilities of the common Serial EEPROM devices and the Flash program memory available inside the microcontroller itself. You might be looking for orders of magnitude more—hundreds of megabytes and possibly gigabytes. If you own a digital camera, an MP3 player, or even just a cell phone, you have probably become familiar with the storage requirements of consumer multimedia applications and with the available mass storage technologies. Hard disk drives have become smaller and less power thirsty, but also a multitude of solid state solutions (based once more on Flash technologies such as Compact Flash, Smart Media, Secure Digital, Memory Stick, and others) have flooded the market, and because of the volumes absorbed by the consumer applications, the price range has been reduced to a point where it is possible, if not convenient, to integrate these devices into embedded-control applications.

In this lesson we will learn how to interface one of the most common and inexpensive mass storage device types to a PIC32 microcontroller using the smallest amount of processor resources.

Preparation

In addition to the usual software tools, including the MPLAB® IDE, the MPLAB C32 compiler, and the MPLAB SIM simulator, this lesson will require the use of the Explorer 16 demonstration board, an In-Circuit Debugger of your choice), and a soldering iron and a few components you'll need ready at hand to extend the board capabilities using the prototyping area or a small expansion board. You can check on the book's companion

Web site (www.exploringPIC32.com) for the availability of expansion boards that will help you with the experiments.

The Exploration

Each one of the many competing mass storage technologies has its strengths and weaknesses, since each was designed for a somewhat different target application. We will choose the ideal mass storage media for our applications according to the following criteria:

- Availability of the memory and required connectors

- Pin count required by the physical interface (possibly serial)

- Memory capacity

- Open specifications available

- Ease of implementation

- Cost of the memory and the required connectors

The Secure Digital (SD) card standard compares favorably in all those aspects; today it is one of the most commonly adopted mass storage media for digital cameras and many other multimedia consumer applications. The SD card specifications represent an evolution of a previous technology known as Multi Media Card, or MMC, with which they are still partially (backward) compatible both electrically and mechanically. The Secure Digital Card Association (SDCA) owns and controls the technical specification standards for SD memory cards, and they require all companies that plan to actively engage in the design, development, manufacture, or sale of products that utilize the SD specifications to become members of the association. As of this writing, a general SDCA membership will cost you $2,000 in annual fees. The Multi Media Card Association (MMCA), on the other side, does not require implementers to become members and makes copies of the MMC specifications available for sale starting at $500. So both technologies are far from free or "open" by any means.

Fortunately there is a "subset" of the SD specifications that has been released to the public by the SDCA in the form of a "simplified physical specification". This information is truly all we need to develop a basic understanding of the SD/MMC memory technology and get started designing a PIC32 mass storage interface.

The Physical Interface

SD cards require only nine electrical contacts and an SD/MMC compatible connector, which can be purchased through most online catalogs for less than a couple of dollars. The connector requires only a couple of pins more to account for insertion detection and write protection switch sensing. Two main modes of communication are available: the first one (known as the SD bus) is original to the SD/MMC standard and it requires a nibble (4-bit) wide bus interface; the second mode is serial and is based on the popular SPI bus standard. It is this second mode that makes the SD/MMC mass storage devices particularly appealing for all embedded-control applications, since most microcontrollers will either have a hardware SPI interface available or will be able to easily emulate one (bit-banging) with a reduced number of I/Os. Finally, the physical specifications of the SD/MMC cards indicate an operating voltage range of 2.0 V to 3.6 V that is ideally suited for all application with modern microcontrollers implemented in advanced CMOS geometries, as is the case of the PIC32MX family (see Figure 14.1).

Note

Logically and electrically, miniSD cards, microSD cards, and SD cards are identical. Only the form factor, size, and number of pins are different from the original standard. Both the miniSD cards and microSD cards were designed to meet special size requirements. With an adapter or an appropriate connector, they can be used in the following application.

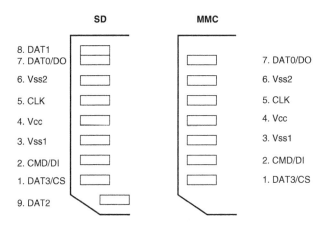

Figure 14.1: SD card and MMC card connectors pin-out.

Interfacing to the Explorer 16 Board

Unfortunately, although the number of electrical connections required for the SPI interface is very small, all SD/MMC card connectors available on the market are designed for surface-mount applications only, which makes it almost impossible to breadboard a card interface or use the prototyping area of the Explorer 16 demonstration board.

Since in the previous chapters we used the first SPI peripheral module (SPI1) to produce a video output and the application does not allow for sharing the resource, we will share instead the second SPI module (SPI2) between the SD card interface and the EEPROM interface using separate Chip Select (cs) signals for the two. In addition to the usual SCK, SDI, and SDO pins, we will provide pull-ups for the unused pins (reserved for the 4-bit-wide SD bus interface) of the SD/MMC connector and for two more pins that will be dedicated to the Card Detect and Write Protect signals (see Figure 14.2).

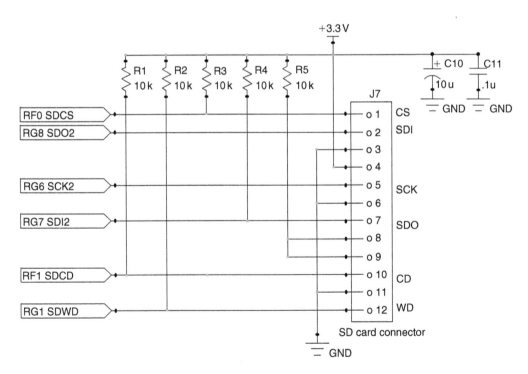

Figure 14.2: SD/MMC card interface to Explorer 16 demo board.

Note

Microchip has recently made available an expansion board known as the PICTail® Daughter Board for SD and MMC Cards (AC164122) that can be effectively used to complete all the projects presented in this chapter. An alternative set of pin assignments to support the new PICTail board will be offered on the companion web site: www.ExploringPIC32.com.

Starting a New Project

After creating a new project that we will obviously call **SDMMC**, let's start writing the basic initialization routines for all the necessary I/Os and the configuration of the SPI2 module:

```
/*
** SDMMC.c SD card interface
*/
#include <p32xxxx.h>
#include <sdmmc.h>

// I/O definitions
#define SDWP    _RG1    // Write Protect input
#define SDCD    _RF1    // Card Detect input
#define SDCS    _RF0    // Card Select output

void initSD( void)
{
  SDCS = 1;             // initially keep the SD card disabled
  _TRISF0 = 0;          // make Card select an output pin

  // init the SPI2 module for a slow (safe) clock speed first
  SPI2CON = 0x8120;     // ON, CKE=1; CKP=0, sample middle
  SPI2BRG = 71;// clock = Fpb/144 = 250kHz
} // initSD
```

In particular, in the SPI2CON register we need to configure the SPI module to operate in master mode with the proper clock polarity, clock edge, input sampling point, and an initial clock frequency. The clock output (SCK) must be enabled and set low when idle.

The sampling point for the SDI input must be centered. The frequency is controlled by means of the SPI baud rate generator (SPI2BRG) that divides the peripheral clock (Tpb). After power-up and until the SD card is properly initialized, we will have to keep the SPI clock speed to a safe setting, below 400 kHz; therefore we will use a setting of Tpb/144 to obtain a 250 kHz clock signal. This is just a temporary arrangement, though; after sending only the first few commands, we will be able to speed up the communication considerably.

Notice how only the SDCS signal (RF0 pin) needs to be manually configured as an output pin, whereas SCK2 and SDO2 (corresponding to the RG6 and RG8 pins) are automatically configured as outputs as soon as we enable the SPI 2 peripheral.

Selecting the SPI Mode of Operation

When an SD/MMC card is inserted in the connector and powered up, it starts in the default mode of communication: the SD bus mode. To inform the card that we intend to communicate using the alternative SPI mode, all we need to do is to select the card (SDCS pin low) and start sending the first reset command. We can be assured that once it's entered the SPI mode, the card will not be able to change back to the SD bus mode unless the power supply is cycled. However, this means that if the card is removed from the slot without our knowledge and then reinserted, we will have to make sure that the initialization routine or at least the reset command are repeated, to get back to the SPI mode. We can detect the card presence at any time by checking the status of the SDCD line (RF1 input pin).

Sending Commands in SPI Mode

In SPI mode, commands are sent to an SD/MMC card as packets of 6 bytes, and all responses from the SD card are provided with multiple byte data blocks of variable length. So all we need to communicate with the memory card is the usual basic SPI routine to send and receive (the two operations are really the same, as we have seen in the previous chapters) a byte at a time:

```
// send one byte of data and receive one back at the same time
unsigned char writeSPI( unsigned char b)
{
  SPI2BUF=b;                       // write to buffer for TX
  while( !SPI2STATbits.SPIRBF);    // wait transfer complete
  return SPI2BUF;                  // read the received value
}// writeSPI
```

For improved code readability and convenience, we will also define two more macros that will mask the same writeSPI() function as a pure readSPI(), or just as a clock output function clockSPI(). Both macros will send a dummy byte of data (0xFF):

```
#define readSPI()  writeSPI( 0xFF)
#define clockSPI() writeSPI( 0xFF)
```

To send a command, we will start selecting the card (SDCS low) and send through the SPI port a packet composed of three parts:

- The first part is a single byte containing a command index. The following definitions cover all the commands we will be using for this project:

```
// SD card commands
#define RESET          0    // a.k.a. GO_IDLE (CMD0)
#define INIT           1    // a.k.a. SEND_OP_COND (CMD1)
#define READ_SINGLE    17
#define WRITE_SINGLE   24
```

- The command index is followed by a 32-bit memory address. It is an unsigned integer (32-bit) value that must be sent MSB first.

- Finally, the command packet is completed by a single byte CRC.

The Cyclic Redundancy Check (CRC) feature is always used in SD bus mode to make sure that every command and every block of data transmitted on the bus is free from error. But, as soon as we switch to the SPI mode after sending the reset command, the CRC protection is automatically disabled and the CRC value is ignored. In fact, from that moment on, the card assumes that a direct and reliable connection to the host, the PIC32 in our case, is available. By taking advantage of this default behavior, we can simplify our code by using a single precomputed value. This will be the CRC code of the RESET command. For all the subsequent commands, the CRC field will be a "don't care." Here is the first part of the sendSDCmd() function that we will use to send all commands to the SD card:

```
int sendSDCmd( unsigned char c, unsigned a)
// c command code
// a byte address of data block
{
  int i, r;

  // enable SD card
  enableSD();
```

```
// send a comand packet  (6 bytes)
writeSPI( c | 0x40);      // send command
writeSPI( a>>24);         // msb of the address
writeSPI( a>>16);
writeSPI( a>>8);
writeSPI( a);             // lsb

writeSPI( 0x95);          // send CMD0 CRC
```

After sending all 6 bytes to the card, we are supposed to wait for a response byte. In fact, it is important that we keep sending "dummy" data continuously clocking the SPI port. The response will be 0xFF; basically, the SDI line will be kept high until the card is ready to provide a proper response code. The specifications indicate that up to 64 clock pulses, or 8 bytes, might be necessary before a proper response is received. Should we exceed this limit, we would have to assume a major malfunctioning of the card and abort communication:

```
// now wait for a response, allow for up to 8 bytes delay
for( i=0; i<8; i++)
{
  r=readSPI();
  if ( r != 0xFF)
  break;
}
return ( r);

// NOTE CSCD is still low!
} // sendSDCmd
```

If we receive a response code, each bit, if set, will provide us with an indication of a possible problem (see Table 14.1).

Notice that, on return, the sendSDCmd() function leaves the SD card still selected (SDCS low) so that commands such as Block Write and Block Read, which require additional data to be sent to or received from the card, will be able to proceed. In all other commands that do not require additional data transfers, we will have to remember to deselect the card (set SDCS high) immediately after the function call. Furthermore, since we want to share the SPI2 port with other peripherals such as the Serial EEPROM mounted on the Explorer 16 board, we need to make sure that the SD/MMC card receives a few more clock cycles (eight will suffice) immediately after the rising edge of the chip select line (SDCS). According to the SD/MMC specifications, this will allow the card to

Table 14.1: SD card
Command Response codes.

Bit	Description
0	Idle state
1	Erase Reset
2	Illegal command
3	Communication CRC error
4	Erase sequence error
5	Address error
6	Parameter error
7	0 (always)

complete a few important housekeeping chores, including the proper release of the SDO line, essential to allow other devices on the same bus to communicate properly.

Here is another pair of macros that will help us perform this consistently:

```
#define disableSD() SDCS = 1; clockSPI()
#define enableSD() SDCS = 0
```

Completing the SD Card Initialization

Before the card can be effectively used for mass storage applications, a well-defined sequence of commands needs to be completed. This sequence is defined in the original MMC card specifications and has been modified only slightly by the SD card specifications. Since we are not planning on using any of the advanced features specific to the SD card standard, we will use the basic sequence as defined for MMC cards for maximum compatibility. There are five steps in a sequence that starts as soon as the card is inserted in the connector and powered up:

1. The CS line is initially kept high (the card is not selected).

2. More than 74 clock pulses must be provided before the card becomes capable of receiving commands.

3. The card must then be selected.

4. The RESET (CMD0) command is sent; the card should respond by entering the Idle state and (activating the SPI mode).

5. An INIT (CMD1) command is provided and repeated until the card exits the Idle state.

The following segment of the function initMedia() will perform exactly those initial five steps:

```
int initMedia( void)
// returns 0 if successful
//          E_COMMAND_ACK   failed to acknowledge reset command
//          E_INIT_TIMEOUT  failed to initialize
{
  int i, r;

  // 1. with the card NOT selected
  disableSD();

  // 2. send 80 clock cycles start up
  for ( i=0; i<10; i++)
    clockSPI();

  // 3. now select the card
  enableSD();

  // 4. send a single RESET command
  r = sendSDCmd( RESET, 0); disableSD();
  if ( r != 1)              // must return Idle
  return E_COMMAND_ACK;     // comand rejected

  // 5. send repeatedly INIT until Idle terminates
  for (i=0; i<I_TIMEOUT; i++)
  {
    r = sendSDCmd( INIT, 0); disableSD();
    if ( !r)
      break;
  }
  if ( i == RI_TIMEOUT)
    return E_INIT_TIMEOUT; // init timed out
```

The initialization command can require quite some time, depending on the size and type of memory card, normally measured in several tenths of a second. Since we are operating at 250 kb/s, each byte sent will require 32 us. If we consider 6 bytes for every command retry, using a count up to 10,000 will provide us with a generous timeout limit (I_TIMEOUT) of approximately three tenths of a second as per SD card specifications.

It is only upon successful completion of the preceding sequence that we will be allowed to finally switch gear and dramatically increase the clock speed to the highest possible value supported by our hardware. With minimal experimentation you will find that an Explorer 16 board, with a properly designed daughter board providing the SD/MMC connector, can easily sustain a clock rate as high as 18 MHz. This value can be obtained by reconfiguring the SPI baud rate generator for a 1:2 ratio. We can now complete the initMedia() function with the last segment:

```
    // 6. increase speed
    SPI2CON = 0;        // disable the SPI2 module
    SPI2BRG = 0;        // Fpb/(2*(0+1))= 36/2 = 18MHz
    SPI2CON = 0x8120;   // re-enable the SPI2 module
    return 0;
} // init media
```

Reading Data from an SD/MMC Card

SD/MMC cards are solid-state devices typically containing large arrays of Flash memory, so we would expect to be able read and write any amount of data (within the card capacity limits) at any desired address. In reality, compatibility considerations with many previous (legacy) mass storage technologies have imposed a number of constraints on how we can access the memory. In fact, all operations are defined in blocks of a fixed size that by default is 512 bytes. It is not a coincidence that 512 bytes is the exact standard size of a data "sector" of a typical personal computer hard disk. Although this can be changed with an appropriate command, we will maintain the default setting so that later we will be able to take advantage of this compatibility. In the next chapter we will develop a set of routines that will allow us to implement a complete file system compatible with the most common PC operating systems. This way we will be able to access files written on the SD card by a personal computer, and vice versa, a personal computer will be able to access files written by our applications onto an SD card.

The READ_SINGLE (CMD17) is all we need to initiate a transfer of a single sector from a given address in memory. The command takes as an argument a 32-bit "byte address," but when accessing sectors of data, we will be constantly referring to *logical block addresses*, or LBAs, borrowing from a term used in other mass storage applications.

```
typedef unsigned LBA;    // logic block address, 32 bit wide
```

To avoid confusion, in the following we will uniformly use only LBAs or block addresses, and we will obtain an actual byte address by multiplying the LBA value by 512 just before passing the parameter to the READ_SINGLE command.

Writing a sector of data to an SD card requires the following five steps:

1. Send a READ_SINGLE command.

2. Wait for the SD card to respond with a specific token: DATA_START. This will be the card's way to tell us it is ready to send the block of data.

    Since the card might need a little time to locate the block of data, just like during the initialization phase, it is important to impose a timeout. Since only the readSPI() function is called repeatedly, sending/receiving only 1 byte at a time (@18 MHz) while waiting for the data token, a timeout counter of 25,000 (R_TIMEOUT) will provide an effective time limit of less than one millisecond.

3. Once the DATA_START token is received, we can confidently read in a rapid sequence all 512 bytes composing the requested block of data.

4. They will be followed by a 16-bit CRC value that we should read, but otherwise we can discard. It is only at this point that we will deselect the memory card and terminate the entire read command sequence.

The following routine readSECTOR() performs the entire sequence in a few lines of code:

```
#define DATA_START    0xFE
int readSECTOR( LBA a, char *p)
// a       LBA of sector requested
// p       pointer to sector buffer
// returns TRUE if successful
{
  int r, i;
  // 1. send READ command
  r = sendSDCmd( READ_SINGLE, ( a << 9));
  if ( r == 0)    // check if command was accepted
  {
```

```
    // 2. wait for a response
    for( i=0; i<R_TIMEOUT; i++)
    {
      r = readSPI();
      if ( r == DATA_START)
        break;
    }

    // 3. if it did not timeout, read 512 byte of data
    if ( i != R_TIMEOUT)
    {
      i = 512;
      do{
        *p++ = readSPI();
      } while (--i>0);

      // 4. ignore CRC
      readSPI();
      readSPI();

    } // data arrived
  } // command accepted

  // 5. remember to disable the card
  disableSD();

  return ( r == DATA_START);    // return TRUE if successful
} // readSECTOR
```

Note

To provide a visual indication of activity on the memory card similarly to hard drives and diskette drives, we could assign one of the LEDs available on the Explorer 16 board as the "read" LED, hoping this will help prevent a user from removing the card while in use. The LED can be turned on before each read command and turned off at the end.

Other strategies are possible, though. For example, similarly to common practice on USB Flash drives, an LED could be turned on as soon as the card is initialized, regardless of whether an actual command is performed on it at any given point in time. Only calling a deinitialization routine would turn the LED off and indicate to the user that the card can be removed.

Writing Data to an SD/MMC Card

Based on the same considerations we made for the readSECTOR() function, we will develop a writeSECTOR() function that will be similarly constrained to operate on 512-byte-wide blocks of data. The write sequence we will use, as you would expect, is the WRITE_SINGLE command and will be composed of five steps. However, this time the data transfer will be in the opposite direction:

1. Send a WRITE_SINGLE command and check the SD card response to make sure that the command is accepted.

2. Send the DATA_START token and immediately after it, in a short loop, all 512 bytes of data.

3. Send 2 bytes for the 16-bit CRC (any dummy value will do) since the CRC check is not enabled in SPI mode.

4. Check the SD card response. The token DATA_ACCEPT will confirm that the entire block of data has been received and the write operation has started.

5. Wait for the completion of the write command. While the card is busy writing, it will keep the SDO line low. So we will wait for the SDO line to return high. Once more a timeout must be imposed to limit the amount of time allowed to the card to complete the operation. Since all SD/MMC memories are based on Flash memory technology, we can expect the time typically required for a write operation to be considerably longer than that required for a read operation. A timeout value of 250,000 (W_TIMEOUT) will provide us with a 100ms limit that is more than sufficient to accommodate even the slowest memory card on the market.

It is only at this point that we will deselect the memory card and terminate the entire write command sequence:

```
#define DATA_ACCEPT             0x05

int writeSECTOR( LBA a, char *p)
// a       LBA of sector requested
// p       pointer to sector buffer
// returns TRUE if successful
{
```

```c
  unsigned r, i;

  // 1. send WRITE command
  r = sendSDCmd( WRITE_SINGLE, ( a << 9));
  if ( r == 0)      // check if command was accepted
  {

    // 2. send data
    writeSPI( DATA_START);

    // send 512 bytes of data
    for( i=0; i<512; i++)
      writeSPI( *p++);

    // 3. send dummy CRC
    clockSPI();
    clockSPI();

    // 4. check if data accepted
    r = readSPI();
    if ( (r & 0xf) == DATA_ACCEPT)
    {

      // 5. wait for write completion
      for( i=0; i<W_TIMEOUT; i++)
      {
        r = readSPI();
        if ( r != 0 )
          break;
      }
    } // accepted
    else
      r = FAIL;

  } // command accepted

  // 6. remember to disable the card
  disableSD();

  return ( r);      // return TRUE if successful

} // writeSECTOR
```

Note

Similarly to the readSECTOR() function, a second LED can be assigned to indicate when a write operation is being performed and alert the user. Should the card be removed during the write sequence, data will most likely be lost or corrupted.

Save the source we developed so far in a file called **SDMMC.c** to be placed inside the *lib* directory. We will have ample use for it in the next few chapters.

As a final nice touch, we should add the following two functions to manage the SD/MMC connector switches:

```
// SD card connector presence detection switch
int getCD( void)
// returns  TRUE card present
//          FALSE card not present
{
  return !SDCD;
}
```

When a card is inserted in the connector, the Card Detect switch is closed and the SDCD input pin is pulled low. The getCD() function will allow us to detect the card's presence by returning TRUE when the card is inserted and ready for use.

Similarly, when the Write Protection tab on the card is *not* in the "lock" position and the card is inserted, the Write Protect switch will close and the corresponding SDWP input pin will be pulled low.

```
// card Write Protect tab detection switch
int getWP( void)
// returns  TRUE write protect tab on LOCK
//          FALSE write protection tab OPEN
{
  return SDWP;
}
```

The getWP() function, called when the card is properly inserted, will return TRUE if the card is locked.

Notice that the Write Protect tab on the SD/MMC card is similar to cassette and VHS tape protection tabs. It is merely suggesting that the device should not be written to.

So it is our responsibility to respect the user's desire and implement a check for the WP switch at the beginning of our `writeSECTOR()` function and abort immediately if the lock is set.

```
// 0. check Write Protect
  if ( getWP())
    return FAIL;
```

Finally, let's create a new include file called **SDMMC.h** that we will save in a common *include* directory to provide the prototypes and basic definitions used in the SD/MMC interface module:

```
/*
** SDMMC.h SD card interface
*/
#define FAIL    FALSE
// Init ERROR code definitions
#define E_COMMAND_ACK    0x80
#define E_INIT_TIMEOUT    0x81

typedef unsigned LBA;      // logic block address, 32 bit wide

void initSD( void);        // initializes I/O pins and SPI
int initMedia( void);      // initializes the SD/MMC memory device
int getCD();               // chech card presence
int getWP();               // check write protection tab
int readSECTOR ( LBA, char *); // reads a block of data
int writeSECTOR( LBA, char *); // writes a block of data
```

Testing the SD/MMC Interface

Whether you believe it or not, the four minuscule routines we just developed are all we need to gain access to the seemingly unlimited amount of "storage space" offered by the SD/MMC memory cards. For example, a 1GB SD card would provide us with approximately 2,000,000 (yes, that is 2 million) individually addressable memory blocks (sectors), each 512 bytes large. Note that as of this writing, SD/MMC cards of this capacity are normally offered for retail in the United States for less than $20!

Let's develop a small test program to demonstrate the use of the SD/MMC module. The idea is to simulate a somewhat typical application that is required to save some large amount of data on the SD/MMC memory card. A fixed number of blocks of data will be written in a predetermined range of addresses and then read back to verify the successful completion of the process. We will use the LCD to report diagnostic information and track the progress.

Let's create a new source file that we will call **RWTest.c**, and let's start by adding the usual header and processor specific include files, followed by the new **sdmmc.h** library:

```
/*
**      RWTest.c
**
*/
// configuration bit settings, Fcy=72MHz, Fpb=36MHz
#pragma config POSCMOD=XT, FNOSC=PRIPLL
#pragma config FPLLIDIV=DIV_2, FPLLMUL=MUL_18, FPLLODIV=DIV_1
#pragma config FPBDIV=DIV_2, FWDTEN=OFF, CP=OFF, BWP=OFF

#include <p32xxxx.h>
#include <explore.h>
#include <LCD.h>
#include <SDMMC.h>
```

Then let's define two byte arrays, each the size of a default SD/MMC memory block that is 512 bytes:

```
#define B_SIZE      512    // data block size
char data[ B_SIZE];
char buffer[ B_SIZE];
```

The test program will fill the first array with a specific and easy to recognize pattern and will repeatedly write its contents onto the memory card. The chosen address range will be defined by two constants:

```
#define START_ADDRESS  10000   // start block address
#define N_BLOCKS       10      // number of blocks
```

The LED2 connected on the PORTA RA2 pin on the Explorer 16 demonstration board will provide us with visual feedback about the SD card usage status. Notice that this I/O is available even if you are using the PIC32 Starter Kit and therefore the JTAG port is enabled:

```
#define LED    _RA2
```

The first few lines of the main program can now be written to initialize the I/Os required by the SD/MMC module on the LCD:

```
main( void)
{
  LBA addr;
  int i, j, r;
  // 1. initializations
  initEX16();
  initLCD();             // init LCD module
  initSD();              // init SD/MMC module

  // 2. fill the buffer with pattern
  for( i=0; i<B_SIZE; i++)
    data[i]= i;
```

The next code segment will prompt the user to insert the card in the slot and will check for the presence of the SD card in a loop. After a short debouncing delay, the initialization routine is performed to prepare the card to receive SPI commands:

```
  // 3. wait for the card to be inserted
  putsLCD( "Insert card..");
  while( !getCD());      // check CD switch
  Delayms( 100);         // wait contacts de-bounce
  if ( initMedia())      // init card
  { // if error code returned
    clrLCD();
    putsLCD( "Failed Init");
    goto End;
  }
```

When ready, we proceed with the actual data writing phase. The LED is turned on to indicate that the SD card is in use, and a status message is printed on the first line of the LCD display. Two nested loops repeatedly call the writeSECTOR() function to write 16 groups of 10 sectors starting at the absolute LBA=10,000. Every 10 sectors (approx. 5 KBytes) a brick character (black box) is added on the second line of the LCD display to form a progress bar. Should any write command fail, the procedure is immediately aborted and an error message is reported on the LCD:

```
  // 4. fill 16 groups of N_BLOCK sectors with data
  LED = 1;               // SD card in use
```

```
clrLCD();
putsLCD( "Writing\n");
addr = START_ADDRESS;
for( j=0; j<16; j++)
{
  for( i=0; i<N_BLOCKS; i++)
  {
    if (!writeSECTOR( addr+i*j, data))
    { // writing failed
      putsLCD( "Failed to Write");
      goto End;
    }
  } // i
  putLCD( 0xff);
} // j
```

Then it is time to read back each sector of data and verify its content. After the LCD is updated to reflect the new phase, the same two nested loops perform the reading and verification in groups of 10 sectors. After each group of sectors is read and verified, a new brick (black bar) is added to the display to indicate the progress. Should any of these steps fail, the procedure is immediately aborted and an error message is displayed on the LCD:

```
// 5. verify the contents of each sector written
clrLCD();
putsLCD( "Verifying\n");
addr = START_ADDRESS;
for( j=0; j<16; j++)
{
  for( i=0; i<N_BLOCKS; i++)
  { // read back one block at a time
    if (!readSECTOR( addr+i*j, buffer))
    { // reading failed
      putsLCD( "Failed to Read");
      goto End;
    }

    // verify each block content
    if ( memcmp( data, buffer, B_SIZE))
    { // mismatch
      putsLCD( "Failed to Match");
```

```
     goto End;
   }
 } // i
 putLCD(0xff);
} // j
```

Notice how the memcmp() function, part of the standard C string.h library, is used to efficiently perform the data comparison. It returns a zero value when the two buffers' content is identical, a nonzero value otherwise.

If all went well, a success message is printed on the LCD and the LED is turned off, since the SD card is no more in use and can now be removed:

```
// 7. indicate successful execution
clrLCD();
putsLCD( " Success!");

End:
 LED = 0;    // SD card not in use
 // main loop
 while( 1);

} // main
```

Make sure to add all the required source files—**SDMMC.h**, **SDMMC.c**, **LCDlib.c**, **explore.c**, and **RWTest.c**—to the project, then build all and program the Explorer 16 board with your in-circuit debugger of choice. You will need a daughter board with the SD/MMC connections as described at the beginning of the lesson and an empty SD card to perform the test.

Warning

This is the real thing! When you run the RWTest program, the contents of the SD card will be modified, overwriting any data on the card and potentially corrupting any files. Make sure you have saved all the family photos and your favorite MP3 files somewhere else! Only in the next chapter we will develop a library compatible with common PC "file systems." It will allow us to share the SD card without risk of damaging existing files, reading and writing data using a common format.

As you run the code, the efforts of building the SD/MMC interface (or the expense of purchasing one) will be more than compensated by the joy of seeing the PIC32 perform the test flawlessly in a few seconds.

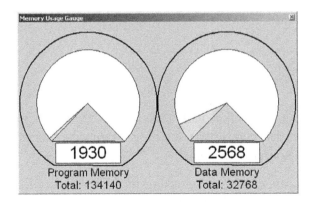

Figure 14.3: MPLAB memory usage gauges.

Also, admire how small the overall amount of code and resources we used was (see Figure 14.3)!

All together, the test program and the SD/MMC access library module have used only 1.930 words of the processor Flash program memory—that is, less than 2 percent of the total available memory. Not to mention that, as in all previous lessons, this result was obtained with all compiler optimization options turned off.

Debriefing

In my personal opinion, it does not get cheaper or easier than this with any other mass storage technology. After all, we can use only a handful of pull-up resistors, a cheap connector, and just a few I/O pins to enormously expand the storage capabilities of our applications. In terms of PIC32 resources required, only the SPI peripheral module has been used, and even that could be shared with other applications.

The simplicity of the approach has his obvious limitations, though. Data can be written only in blocks of fixed size, and its position inside the memory array will be completely application specific. In other words, there will be no way to share data with a personal computer or other device capable of accessing SD/MMC memory cards unless a "custom" application is developed. Worse, if an attempt is made to use a card already used by a PC, PC data would likely be corrupted and the entire card might require complete reformatting. In the next lesson, we will address these issues by developing a complete file system library.

Tips & Tricks

The choice of operating on the default block size of 512 bytes was dictated mostly by historical reasons. By making the low-level access routines in this lesson conform with the standard size adopted by most other mass storage media devices (including hard drives), we made developing the next layer (the file system) easier. But if we were looking for maximum performance, this could have been the wrong choice. In fact, if we were looking for faster write performance, typically the bottleneck of every Flash memory media, we would be better off looking at much larger data blocks.

Flash memory typically offers very fast access to data (reading) but is relatively slow when it comes to writing. Writing requires two steps: First, a large block of data (often referred to as a *page*) must be erased; then the actual writing can be performed on smaller blocks. The larger the memory array, the larger, proportionally, the erase page size will be. For example, on a 512 Mbyte memory card, the erase page can easily exceed 2k bytes. Although these details are typically hidden from the user as the main controller inside the card takes care of the erase/write sequencing and buffering, they can have an impact on the overall performance of the application. In fact, if we assume a specific SD card has a 2k byte page, writing any amount of data (<2k) would require the internal card controller to perform the following steps:

- Read the contents of an entire 2k byte block in an internal buffer.

- Erase it, and wait for the erase time.

- Replace a portion of the buffer content with the new data.

- Write back the entire 2k byte block, and wait for the write time.

By performing write operations only on blocks of 512 bytes each, to write 2k bytes of data our library would have to ask the SD card controller to perform the entire sequence four times, whereas it could be done in just one sequence by changing the data block length or using a multiple-block write command. Although this approach could theoretically increase the writing speed by 400 percent in our example, consider the option carefully because the price to pay could be quite high. In fact, consider the following drawbacks:

- The actual memory page size might not be known or guaranteed by the manufacturer, although betting on increasing densities of Flash media (and therefore increasing page size) is pretty safe.

- The size of the RAM buffer to be allocated inside the PIC32 application is increased, and this is a precious resource in any embedded application.

- The higher software layers (which we will explore in the next lesson) might be more difficult to integrate if the data block size varies.

- The larger the buffer, the larger the data loss if the card is removed before the buffer is flushed.

Exercises

1. Experiment with various data block sizes to identify where your SD card provides the best write performance. This will give you an indirect indication of the actual page size of the Flash memory device used by the card manufacturer.

2. Experiment with multiple-block write commands or by changing the block length to verify how the internal buffering is performed by the SD card controller and if the two methods are equivalent.

Books

Schmidt, F., *The SCSI Bus and IDE Interface: Protocols, Applications, and Programming,* second edition (Addison-Wesley Professional, New york, 1999). If the SD card interface has intrigued you for its simplicity, you might now be curious about the interfaces used on most of the older (nonsolid-state) mass storage devices used in the world of personal computers. You will see they were not that much more complex.

Axelson, J., *USB Mass Storage: Designing and Programming Devices and Embedded Hosts* (Lakeview Research, WI, 2006). This book continues the excellent series on USB by Jan Axelson. Low-level interfacing directly to an SD/MMC card was easy, as you have seen in this chapter, but creating a proper USB interface to a mass storage device is a project of a much higher order of complexity.

Links

www.mmca.org/home. The official Web site of the MultiMedia Card Association (MMCA).

www.sdcard.org. The official Web site of the Secure Digital Card Association (SDCA).

www.sdcard.org/Sdio/Simplified%20SDIO%20Card%20Specification.pdf. The simplified SDIO card specifications. With SDIO, the SD interface is no longer used only for mass storage but is also a viable interface for a number of advanced peripherals and gizmos, such as GPS receivers, digital cameras, and more.

File I/O

The Plan

Just yesterday, we developed a basic interface module (both software and hardware) to gain access to an SD/MMC card and support applications that require large amounts of data storage. A similar interface could be built for several other types of mass storage media, but in this lesson we will instead focus on the algorithms and data structures required to properly share information on the mass storage device with the most common PC operating systems (DOS, Windows, and some Linux distributions). In other words, we will develop a module for access to a standard file system known commonly as FAT16.

The first FAT file system was created by Bill Gates and Marc McDonald in 1977 for managing disks in Microsoft Disk BASIC. It used techniques that had been available in file systems many years before that, and it has continued to evolve in numerous versions over the last few decades to accommodate ever larger-capacity mass storage devices and new features. Among the many versions still in use today, the FAT12, FAT16, and FAT32 are the most common ones. FAT16 and FAT32, in particular, are recognized by practically every PC operating system currently in use; the choice between the two is mostly dictated by efficiency considerations and the capacity of the media. Ultimately, for most Flash mass storage devices of common use in consumer multimedia applications, FAT16 is the file system of choice.

Preparation

Today's exploration continues using the hardware platform used in the previous chapter. You will need an Explorer 16 or equivalent demo board with an additional expansion board or prototyped circuit to connect an SD card connector and a few pull-up resistors. Check the companion Web site at www.exploringPIC32.com for a list of expansion options available to facilitate the experiments presented in this chapter.

The Exploration

The term *FAT* is an acronym for *File Allocation Table*, which is also the name of one of the most important data structures used in this file system. After all, a file system is just a method for storing and organizing computer files and the data they contain to make it easy to find and access them. Unfortunately, as often is the case in the history of personal computing, standards and technologies are the fruit of constant evolutionary progress rather than original creation. For this reason many of the details of the FAT file system we will reveal in the following discussion can only be explained in the context of a struggle to continue and maintain compatibility with an enormous mass of legacy technologies and software over many years.

Sectors and Clusters

Still, the basic ideas at the root of a FAT file system are quite simple. As we saw in the previous lesson, most mass storage devices follow a "tradition" derived from the hard disk technology of managing memory space in blocks of a fixed size, 512 bytes, commonly referred to as *sectors*. In a FAT file system, a small number of these sectors are reserved and used as a sort of general index: the File Allocation Table. The remaining sectors (the majority) are available for proper data storage, but instead of being handled individually, small groups of contiguous sectors are handled jointly to form new, larger entities known as *clusters*. Clusters can be as small as one single sector or can be formed by as many as 64 sectors. It is the use of each cluster and its position that is tracked inside the File Allocation Table. Therefore, clusters are the true smallest unit of memory allocation in a FAT file system (see Figure 15.1).

The simplified diagram illustrates a hypothetical example of a FAT file system formatted for 1,022 clusters, each composed of 16 sectors. (Notice that the data area always starts with cluster number 2.) In this example, each cluster would contain 8 KB of data and the total storage capacity would be about 8 MB.

Note that the larger clusters are, the fewer will be required to manage the entire memory space and the smaller the allocation table required, hence the higher efficiency of the file system. On the contrary, if many small files are to be written, the larger the cluster size,

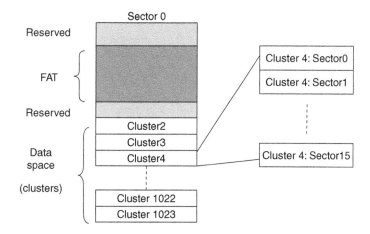

Figure 15.1: Simplified example of a FAT file system layout.

the more space will be wasted. It is typically the responsibility of the operating system, when formatting a storage device for use with a FAT file system, to decide the ideal cluster size to be used for an optimal balance.

The File Allocation Table

In a FAT16 file system, the File Allocation Table is essentially an array of 16-bit integers. Each element of the table represents one cluster. If a cluster is considered empty and available, the corresponding entry in the table will contain the value 0x0000. If a cluster is in use and it contains an entire file of data, its corresponding entry in the table will contain the value 0xFFFF. If a file is larger than the size of a single cluster, a chain of clusters is formed. In the FAT each element will contain the index of the next cluster in the chain. The last cluster in the chain will have the corresponding entry set to 0xFFFF.

Additionally, certain unique values are used to mark reserved clusters (0x0001) and bad clusters (0xFFF7). Since 0x0000 and 0x0001 have been assigned special meanings (free and reserved, respectively), this explains why the convention wants the cluster counting to start in the data area with cluster number 2. Inside the FAT, the corresponding first two entries are similarly reserved.

In Figure 15.2, you can see an example of a FAT for the system presented in our previous example in Figure 15.1. Clusters 0 and 1 are reserved. Cluster 2 appears to contain some data, meaning that some or all of the (16) sectors forming the cluster have been filled with data from a file whose size must have been less than 8 KB.

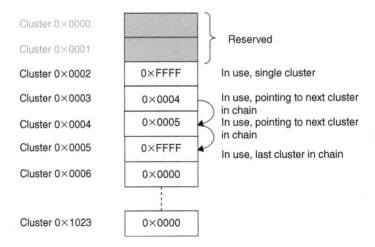

Figure 15.2: Cluster chains in a File Allocation Table.

Cluster 3 appears to be the first cluster in a chain of three that also includes Clusters 4 and 5. All of Cluster 3 and 4 sectors and some or all of Cluster 5 sectors must have been filled with data from a file whose size (we can only assume so far) was more than 16 KB but less than 24 KB. All following clusters appear to be empty and available.

Notice that the size of a FAT itself is dictated by the total number of clusters multiplied by 2 (2 bytes per cluster) and that it can spread over multiple sectors. In our previous example, a FAT of 1,024 clusters would have required 2,048 bytes, or four sectors of 512 bytes each. Also, since the file allocation table is perhaps the most critical structure in the entire FAT file system, multiple copies (typically two) are maintained and allocated one after the other before the beginning of the data space.

The Root Directory

The role of the FAT is to keep track of how and where data is allocated. It does not contain any information about the nature of the file to which the data belonged. For that purpose there is another structure, called the *root directory*, whose sole purpose is that of storing filenames, sizes, dates, times, and a number of other attributes. In a FAT16 file system, the root directory, or simply the *root* from now on, is allocated in a fixed amount of space and a fixed position right between the FAT (second copy) and the first data cluster (cluster #2), as shown in Figure 15.3.

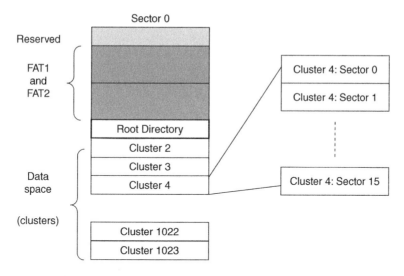

Figure 15.3: Example of a FAT file system layout.

Since both position and size (number of sectors) are fixed, the maximum number of files (or directory entries) in the root directory is limited and determined when formatting the media. Each sector allocated to the root will allow for 16 file entries to be documented where each entry will require a block of 32 bytes, as represented in Figure 15.4.

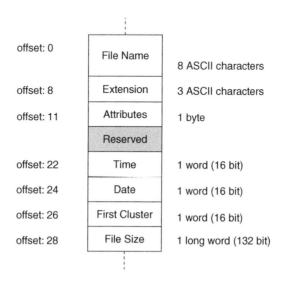

Figure 15.4: Basic Root Directory Entry structure.

The *Name* and *Extension* fields are the most obvious if you are familiar with the older Microsoft operating systems using the 8:3 conventions. The two fields need only to be padded with spaces and the dot can be discarded.

The *Attributes* field is composed of a group of flags with the meanings shown in Table 15.1.

Table 15.1: File attributes in a directory entry.

Bit	Mask	Description
0	0×01	Read only
1	0×02	Hidden
2	0×04	System
3	0×08	Volume label
4	0×10	Subdirectory
5	0×20	Archive

The *Time* and *Date* fields refer to the last time the file was modified and must be encoded in a special format to compress all the information in just two 16-bit words (see Tables 15.2 and 15.3).

Table 15.2: Time encoding in a directory entry field.

Bits	Description
15–11	Hours (0–23)
10–5	Minutes (0–59)
4–0	Seconds/2 (0–29)

Table 15.3: Date encoding in a directory entry field.

Bits	Description
15–9	Year (0 = 1980, 127 = 2107)
8–5	Month (1 = January, 12 = December)
4–0	Day (1–31)

Notice how the *Date* field encoding does not allow for the code 0x0000 to be interpreted as a valid date. This can provide clues to the file system when the field is not used or could be corrupted.

The *First Cluster* field provides the fundamental link with the FAT table. This 16-bit word contains the number of the first, and possibly only, cluster containing the file data.

Finally, the *Size* field, a 32-bit integer, contains the size (in bytes) of the file data.

Looking at the first character of the filename in a directory entry, we can also tell if and how the entry is currently in use:

- If it contains an ASCII printable character, the entry is valid and in use.

- If it is zero, the entry is empty. When browsing through a directory, we can also deduce that the list of files is terminated here as the file system proceeds sequentially using all entries in the directory table in strict sequential order.

There is a third possibility when a file is removed from the directory. In this case the first character of the filename is simply replaced by a special code (0xE5). This indicates that the contents of the entry are no longer valid and the entry can be reused for a new file at the next opportunity. However, when browsing through the list searching for a file, we should continue because more active entries might follow it.

There would be much more to say to fully document the structure of a FAT16 file system, but if you have followed the introduction so far, you should have a reasonable understanding of its core mechanisms and you will be ready to dive in for more detail since we will start soon writing some code.

The Treasure Hunt

So far we have maintained a certain level of simplification by ignoring some fundamental questions, such as:

- Where do we learn about the storage device capacity?

- How can we tell where the FAT is located?

- How can we tell how many sectors (1–64) compose each cluster?

- How can we tell where the data space starts?

The answers to all those questions will be found soon, but the process will resemble a treasure hunt more than a logical sequence of steps. In fact, you will find the first set of clues in Figure 15.5. By interpreting these clues we will gradually build a new function that will allow us to *mount* the file system and unlock its contents—the treasure.

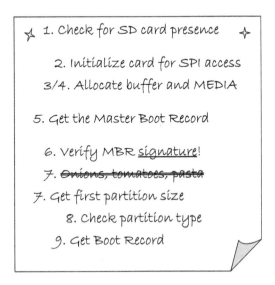

Figure 15.5: The first set of clues.

Using the **SDMMC.c** module functions developed in our previous explorations, we will start by initializing the I/Os with the initSD() function and checking for the presence of the card in the slot.

```
// 0. init the I/Os
initSD();

// 1. check if the card is in the slot
if (!detectSD())
{
   FError = FE_NOT_PRESENT;
   return NULL;
}
```

We will proceed by initializing the SD card for operation in SPI mode with the initMedia() function.

```
// 2. initialize the card
if ( initMedia())
{
  FError = FE_CANNOT_INIT;
  return NULL;
}
```

We will also use the standard C libraries `malloc()` function to dynamically allocate two data structures:

```
// 3. allocate space for a MEDIA structure
D = (MEDIA *) malloc( sizeof( MEDIA));
if ( D == NULL)            // report an error
{
  FError = FE_MALLOC_FAILED;
  return NULL;
}

// 4. allocate space for a temp sector buffer
buffer = (unsigned char *) malloc( 512);
if ( buffer == NULL)       // report an error
{
  FError = FE_MALLOC_FAILED;
  free( D);
  return NULL;
}
```

The first one is called MEDIA. It will be fully revealed to you later on but, for now, it will suffice to say that it will act as the repository for the many "answers" we are seeking. Perhaps a more appropriate name would've been CHEST?

The second structure, called `buffer`, is simply a 512-byte large array that will be used to retrieve sectors of data during the hunt.

Notice that to allow the `malloc()` function to successfully allocate memory, you must remember to inform the MPLAB® C32 linker to reserve some RAM space for the heap.

Hint

Follow the Project Build checklist to learn how to reach and modify the linker settings of your project.

Mostly historical reasons dictate that the first sector (LBA 0) of a mass storage device will contain what is commonly known as a *master boot record* (MBR).

Here is how we invoke the readSECTOR() function for the first time to access the MBR.

```
// 5. get the Master Boot Record
if ( !readSECTOR( 0, buffer))
{
  FError = FE_CANNOT_READ_MBR;
  free( D); free( buffer);
  return NULL;
}
```

A signature, consisting of a specific word value (0x55AA) present in the last word of the MBR sector, will confirm that we have indeed read the correct data.

```
#define FO_SIGN    0x1FE  // MBR signature location (55,AA)

// 6. check if the MBR sector is valid
//    verify the signature word
if (( buffer[ FO_SIGN] != 0x55) ||
    ( buffer[ FO_SIGN +1] != 0xAA))
{
  FError = FE_INVALID_MBR;
  free( D); free( buffer);
  return NULL;
}
```

Once upon a time, this record used to contain actual code to be executed by a PC upon power-up. No personal computer does this anymore, though, and certainly there is no use for that 8086 code in our PIC32 applications. Most of the time you will find the MBR (see Figure 15.6) to be completely filled with zeros except for a few locations where critical information used to be stored. For example, starting at offset 0x01BE, you will find what is called a *partition table*. This table is composed of only four entries of 16 bytes each. The role of a partition table is that of allowing for a single media device to host multiple operating systems and/or split the storage space in safe areas, where each one acts as a completely separate device.

Offset	0	1	2	3	4	5	6	7	8	9	A	B	C	D	E	F	Access ▼
00000000	00	00	00	00	00	00	00	00	00	00	00	00	00	00	00	00	. . . .
00000010	00	00	00	00	00	00	00	00	00	00	00	00	00	00	00	00	. . . .
00000020	00	00	00	00	00	00	00	00	00	00	00	00	00	00	00	00	. . . .
00000030	00	00	00	00	00	00	00	00	00	00	00	00	00	00	00	00	. . . .
00000040	00	00	00	00	00	00	00	00	00	00	00	00	00	00	00	00	. . . .
00000050	00	00	00	00	00	00	00	00	00	00	00	00	00	00	00	00	. . . .
00000060	00	00	00	00	00	00	00	00	00	00	00	00	00	00	00	00	. . . .
00000070	00	00	00	00	00	00	00	00	00	00	00	00	00	00	00	00	. . . .
00000080	00	00	00	00	00	00	00	00	00	00	00	00	00	00	00	00	. . . .
00000090	00	00	00	00	00	00	00	00	00	00	00	00	00	00	00	00	. . . .
000000A0	00	00	00	00	00	00	00	00	00	00	00	00	00	00	00	00	. . . .
000000B0	00	00	00	00	00	00	00	00	00	00	00	00	00	00	00	00	. . . .
000000C0	00	00	00	00	00	00	00	00	00	00	00	00	00	00	00	00	. . . .
000000D0	00	00	00	00	00	00	00	00	00	00	00	00	00	00	00	00	. . . .
000000E0	00	00	00	00	00	00	00	00	00	00	00	00	00	00	00	00	. . . .
000000F0	00	00	00	00	00	00	00	00	00	00	00	00	00	00	00	00	. . . .
00000100	00	00	00	00	00	00	00	00	00	00	00	00	00	00	00	00	. . . .
00000110	00	00	00	00	00	00	00	00	00	00	00	00	00	00	00	00	. . . .
00000120	00	00	00	00	00	00	00	00	00	00	00	00	00	00	00	00	. . . .
00000130	00	00	00	00	00	00	00	00	00	00	00	00	00	00	00	00	. . . .
00000140	00	00	00	00	00	00	00	00	00	00	00	00	00	00	00	00	. . . .
00000150	00	00	00	00	00	00	00	00	00	00	00	00	00	00	00	00	. . . .
00000160	00	00	00	00	00	00	00	00	00	00	00	00	00	00	00	00	. . . .
00000170	00	00	00	00	00	00	00	00	00	00	00	00	00	00	00	00	. . . .
00000180	00	00	00	00	00	00	00	00	00	00	00	00	00	00	00	00	. . . .
00000190	00	00	00	00	00	00	00	00	00	00	00	00	00	00	00	00	. . . .
000001A0	00	00	00	00	00	00	00	00	00	00	00	00	00	00	00	00	. . . .
000001B0	00	00	00	00	00	00	00	00	00	00	00	00	00	00	00	03	. . . .
000001C0	35	00	06	08	D8	C1	F1	00	00	00	0F	C9	0E	00	00	00	5 . . ØÁñ . . . . É . . . .
000001D0	00	00	00	00	00	00	00	00	00	00	00	00	00	00	00	00	. . . .
000001E0	00	00	00	00	00	00	00	00	00	00	00	00	00	00	00	00	. . . .
000001F0	00	00	00	00	00	00	00	00	00	00	00	00	00	00	55	AA	. . . . Uª

Figure 15.6: Hex dump of an MBR.

For our purposes it is safe to assume (demand) that the entire SD/MMC card is formatted with a single partition. Therefore, we need to focus only on the first entry (16-byte block) in the partition table. Of those 16 bytes, we need to access only a few to obtain:

- The partition size (should include the entire card)

- The starting sector

- Most importantly, the type of file system contained

A couple of macros will help us read the data from the partition table and assemble it into 16-bit and 32-bit words:

```
#define ReadW( a, f)  *(unsigned short*)(a+f)
#define ReadL( a, f)  *(unsigned short*)(a+f)+\
                     (( *(unsigned short*)(a+f+2))<<16)
```

Also, the following definitions will point us to the right offset in the MBR.

```
//-------------------------------------------------------------
// Master Boot Record key fields offsets
#define FO_MBR         0L     // master boot record sector LBA
#define FO_FIRST_P     0x1BE  // offset of first partition table
#define FO_FIRST_TYPE  0x1C2  // offset of first partition type
#define FO_FIRST_SECT  0x1C6  // first sector of first partition
#define FO_FIRST_SIZE  0x1CA  // number of sectors in partition
#define FO_SIGN        0x1FE  // MBR signature location (55,AA)

  // 7. read the number of sectors in partition
  psize = ReadL( buffer, FO_FIRST_SIZE);

  // 8. check if the partition type is acceptable
  i = buffer[ FO_FIRST_TYPE];
  switch ( i)
  {
    case 0x04:
    case 0x06:
    case 0x0E:
      // valid FAT16 options
      break;
    default:
      FError = FE_PARTITION_TYPE;
      free( D); free( buffer);
      return NULL;
  } // switch
```

For historical reasons, several codes correspond to different types of partitions. We will be able to correctly decode at least three types of FAT16 partitions, including `0x04`, `0x06`, and `0x0E`.

Getting access to the MBR and finding the partition table is a bit like getting a map with a new set of symbols and clues that need to be interpreted (see Figure 15.7).

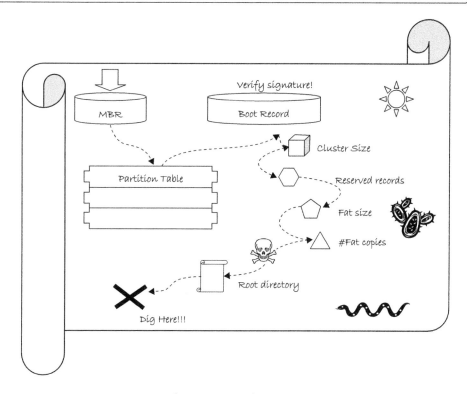

Figure 15.7: The map.

Extracting a 32-bit word found at offset FO_FIRST_SECT (0x1C6) as part of the first (and the only, in our assumptions) partition table entry, we obtain the address (LBA) of the very first sector of the partition.

```
// 9. get the first partition first sector -> Boot Record
firsts = ReadL( buffer, FO_FIRST_SECT);

// 10. get the sector loaded (boot record)
if ( !readSECTOR( firsts, buffer))
{
  free( D); free( buffer);
  return NULL;
}
```

It has a signature, similarly to the MBR, located in the last word of the sector, and we need to verify it before proceeding.

```
// 11. check if the boot record is valid
//      verify the signature word
if (( buffer[ FO_SIGN] != 0x55) ||
   ( buffer[ FO_SIGN +1] != 0xAA))
{
   FError = FE_INVALID_BR;
   free( D); free( buffer);
   return NULL;
}
```

It is called the (first partition) *boot record*, and once more it is supposed to contain actual executable code that is of no value to us (see Figure 15.8).

```
Offset    | 0  1  2  3  4  5  6  7   8  9  A  B  C  D  E  F  | Access ▼
0001E200  | EB 00 90 20 20 20 20 20  20 20 20 00 02 20 01 00 | ë.▮              . . .
0001E210  | 02 00 02 00 00 F8 77 00  3F 00 10 00 F1 00 00 00 | .....øw.?....ñ...
0001E220  | 0F C9 0E 00 80 00 29 13  18 FD E0 20 20 20 20 20 | .É. ▮.)..ýà
0001E230  | 20 20 20 20 20 20 46 41  54 31 36 20 20 20 00 00 |       FAT16    . .
0001E240  | 00 00 00 00 00 00 00 00  00 00 00 00 00 00 00 00 | . . . . . . . . . . . . . . . .
0001E250  | 00 00 00 00 00 00 00 00  00 00 00 00 00 00 00 00 | . . . . . . . . . . . . . . . .
0001E260  | 00 00 00 00 00 00 00 00  00 00 00 00 00 00 00 00 | . . . . . . . . . . . . . . . .
0001E270  | 00 00 00 00 00 00 00 00  00 00 00 00 00 00 00 00 | . . . . . . . . . . . . . . . .
0001E280  | 00 00 00 00 00 00 00 00  00 00 00 00 00 00 00 00 | . . . . . . . . . . . . . . . .
0001E290  | 00 00 00 00 00 00 00 00  00 00 00 00 00 00 00 00 | . . . . . . . . . . . . . . . .
0001E2A0  | 00 00 00 00 00 00 00 00  00 00 00 00 00 00 00 00 | . . . . . . . . . . . . . . . .
0001E2B0  | 00 00 00 00 00 00 00 00  00 00 00 00 00 00 00 00 | . . . . . . . . . . . . . . . .
0001E2C0  | 00 00 00 00 00 00 00 00  00 00 00 00 00 00 00 00 | . . . . . . . . . . . . . . . .
0001E2D0  | 00 00 00 00 00 00 00 00  00 00 00 00 00 00 00 00 | . . . . . . . . . . . . . . . .
0001E2E0  | 00 00 00 00 00 00 00 00  00 00 00 00 00 00 00 00 | . . . . . . . . . . . . . . . .
0001E2F0  | 00 00 00 00 00 00 00 00  00 00 00 00 00 00 00 00 | . . . . . . . . . . . . . . . .
0001E300  | 00 00 00 00 00 00 00 00  00 00 00 00 00 00 00 00 | . . . . . . . . . . . . . . . .
0001E310  | 00 00 00 00 00 00 00 00  00 00 00 00 00 00 00 00 | . . . . . . . . . . . . . . . .
0001E320  | 00 00 00 00 00 00 00 00  00 00 00 00 00 00 00 00 | . . . . . . . . . . . . . . . .
0001E330  | 00 00 00 00 00 00 00 00  00 00 00 00 00 00 00 00 | . . . . . . . . . . . . . . . .
0001E340  | 00 00 00 00 00 00 00 00  00 00 00 00 00 00 00 00 | . . . . . . . . . . . . . . . .
0001E350  | 00 00 00 00 00 00 00 00  00 00 00 00 00 00 00 00 | . . . . . . . . . . . . . . . .
0001E360  | 00 00 00 00 00 00 00 00  00 00 00 00 00 00 00 00 | . . . . . . . . . . . . . . . .
0001E370  | 00 00 00 00 00 00 00 00  00 00 00 00 00 00 00 00 | . . . . . . . . . . . . . . . .
0001E380  | 00 00 00 00 00 00 00 00  00 00 00 00 00 00 00 00 | . . . . . . . . . . . . . . . .
0001E390  | 00 00 00 00 00 00 00 00  00 00 00 00 00 00 00 00 | . . . . . . . . . . . . . . . .
0001E3A0  | 00 00 00 00 00 00 00 00  00 00 00 00 00 00 00 00 | . . . . . . . . . . . . . . . .
0001E3B0  | 00 00 00 00 00 00 00 00  00 00 00 00 00 00 00 00 | . . . . . . . . . . . . . . . .
0001E3C0  | 00 00 00 00 00 00 00 00  00 00 00 00 00 00 00 00 | . . . . . . . . . . . . . . . .
0001E3D0  | 00 00 00 00 00 00 00 00  00 00 00 00 00 00 00 00 | . . . . . . . . . . . . . . . .
0001E3E0  | 00 00 00 00 00 00 00 00  00 00 00 00 00 00 00 00 | . . . . . . . . . . . . . . . .
0001E3F0  | 00 00 00 00 00 00 00 00  00 00 00 00 00 00 55 AA | . . . . . . . . . . . . . . Uª
```

Figure 15.8: Hex dump of a boot record.

Fortunately, in the same record at fixed and known positions there are more of the answers we were looking for and new clues that will help us complete the map of the entire FAT16 file system. These are the key offsets in the boot record buffer:

```
// Partition Boot Record key fields offsets
#define BR_SXC      0xd    // (byte) sectors per cluster
#define BR_RES      0xe    // (word) reserved sectors
#define BR_FAT_SIZE 0x16   // (word) FAT size in sectors
#define BR_FAT_CPY  0x10   // (byte) number of FAT copies
#define BR_MAX_ROOT 0x11   // (odd word) max entries in root
```

With the following code we can calculate the size of a cluster:

```
// 12. determine the size of a cluster
D->sxc = buffer[ BR_SXC];
// this will also act as flag that the media is mounted
```

Determine the position of the FAT, its size, and the number of copies:

```
// 13. determine fat, root and data LBAs
// FAT = first sector in partition (boot record)
//      +reserved records
D->fat = firsts + ReadW( buffer, BR_RES);
D->fatsize = ReadW( buffer, BR_FAT_SIZE);
D->fatcopy = buffer[ BR_FAT_CPY];
```

Find the position of the root directory, too:

```
// 14. ROOT = FAT + (sectors per FAT * copies of FAT)
D->root = D->fat + ( D->fatsize * D->fatcopy);
```

But be careful now! As we get ready to make the last few steps, watch out for a trap!

```
// 15. MAX ROOT is the maximum number of entries
//in the root directory
D->maxroot = ReadW( buffer, BR_MAX_ROOT) ;
```

Can you see it? No? Okay, here's a hint: Look at the value of the BR_MAX_ROOT offset as defined a few lines before. You will notice that this is an odd address (0x11). This is all it takes for the ReadW() macro, which attempts to use it as a word address, to throw a processor exception (misaligned word access) and trap the PIC32 in the general exception handler!

We need a special macro (perhaps less efficient) that can assemble a word 1 byte at a time without falling into the trap!

```
// this is the safe versions of ReadW to be used on odd
address fields
#define ReadOddW( a, f) (*(a+f) + ( *(a+f+1) << 8))

  // 15. MAX ROOT is the maximum number of entries
  //     in the root directory
  D->maxroot = ReadOddW( buffer, BR_MAX_ROOT) ;
```

The last two pieces of information are easy to grab now. With them we learn where the data area (divided into clusters) begins and how many clusters are available:

```
  // 16. DATA = ROOT + (MAXIMUM ROOT *32/512)
  D->data = D->root + ( D->maxroot >> 4);
  // assuming maxroot % 16 == 0!!!

  // 17. max clusters in this partition
  //     = (tot sectors - sys sectors )/sxc
  D->maxcls = (psize - (D->data-firsts))/D->sxc;
```

It took us as many as 17 careful steps to get to the treasure, but now we have all the information we need to fully figure out the layout of the FAT16 file system present on the SD/MMC memory card or, practically, any other mass storage media formatted according to the FAT16 standard. The treasure, after all, is nothing more than another map—a map we will use from now on to find files on a mass storage device (see Figure 15.9).

It's time to organize the information we spent so much effort to retrieve. We will use the MEDIA structure, allocated on the heap at the very beginning.

```
typedef struct {
  LBA     fat;            // lba of FAT
  LBA     root;           // lba of root directory
  LBA     data;           // lba of the data area
  unsigned maxroot;       // max entries in root
  unsigned maxcls;        // max clusters in partition
  unsigned fatsize;       // number of sectors
  unsigned char fatcopy;  // number of FAT copies
  unsigned char sxc;      // number of sectors per cluster
  } MEDIA;
```

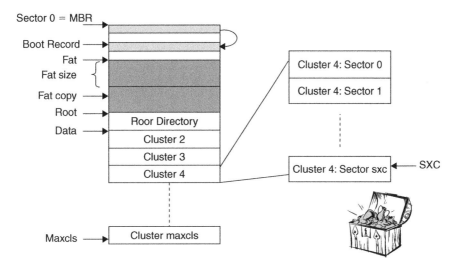

Figure 15.9: The FAT16 complete layout.

All the code we have developed can now be assembled in the mount() function. This is a name that will sound familiar to those of you who have experience in programming for the Linux family of operating systems.

For a mass storage device to be used in Linux, it must be first "mounted" on the file system or, in other words, attached as a new branch of the main (system) file system. Windows users might not be familiar with the concept because they don't have the option to choose if, when, or where a new device file system is mounted. All new mass storage devices are automatically and unconditionally "mounted" by Windows at power-up, or after insertion of any removable media, at the very root of the Windows file system by assigning them a unique, single-letter identifier (C:, D:, E:, and so on).

```
MEDIA * mount( void)
{
  LBA psize;       // number of sectors in partition
  LBA firsts;      // first sector inside the first partition
  int i;
  unsigned char *buffer;

  ... insert here all 17 steps of our treasure hunt

  // 18. free up the temporary buffer
  free( buffer);
  return D;

} // mount
```

Let's also define a global pointer D to a MEDIA structure. It will serve as the root for the entire file system in the assumption, for now, that only one storage device will be available at any given point in time (one connector/slot, one card).

```
// global definitions
MEDIA *D;
```

We will also define an unmount() function that will have the sole duty of releasing the space allocated for the MEDIA structure.

```
void unmount( void)
{
  free( D);
} // unmount
```

Opening a File

Now that we have unlocked the secret of the FAT16 file system, we can return to our original objective: accessing individual files and sharing them with a PC. In this section we will develop a set of high-level functions similar to those used for file manipulation in most operating systems. We will need a function to find a file location on the storage device, one for reading the data sequentially from the file, and possibly one more to write data and create new files.

In a logical order we will start developing what we will call the fopenM() function. Its role will be that of finding all possible information regarding a file (if present) and gathering it in a new structure that we will call MFILE.

Note

The name of this structure was chosen so to avoid conflicts with similar structures and functions defined inside the standard C library stdio.h.

```
typedef struct {
  MEDIA * mda;               // media structure pointer
  unsigned char * buffer;    // sector buffer
  unsigned short cluster;    // first cluster
  unsigned short ccls;       // current cluster in file
  unsigned short sec;        // sector in current cluster
```

```
unsigned short pos;            // position in current sector
unsigned short top;            // bytes in the buffer
int    seek;                   // position in the file
int    size;                   // file size
unsigned short time;           // last update time
unsigned short date;           // last update date
char    name[11];              // file name
char    mode;                  // mode 'r', 'w'
unsigned short fpage;          // FAT page currently loaded
unsigned short entry;          // entry position in cur dir
} MFILE;
```

I know, at first sight it looks like a lot—it is more than 40 bytes large—but as you will see in the discussion, we will end up needing all of them. You will have to trust me for now.

Mimicking standard C library implementations (common to many operating systems), the fopenM() function will receive two (ASCII) string parameters: the filename and a "mode" string, containing r or w, that will indicate whether the file is supposed to be opened for reading or writing.

```
MFILE *fopenM( const char *filename, const char *mode)
{
  char c;
  int i, r, e;
  unsigned char *b;
  MFILE *fp;
```

To optimize memory usage, an MFILE structure is allocated only when necessary, and it is in fact one of the first tasks of the fopenM() function. A pointer to the data structure is its return value. Should fopenM() fail, a NULL pointer will be returned.

Of course a prerequisite for opening a file is to have the storage device file system mapped out, and that is the responsibility of the mount() function. A pointer to a MEDIA structure must have already been deposited in the global D pointer.

```
  // 1. check if a storage device is mounted
  if ( D == NULL)    // unmounted
  {
  FError = FE_MEDIA_NOT_MNTD;
  return NULL;
  }
```

Since all activity with the storage device must be performed in blocks of 512 bytes, we will need that much space to be allocated for us to act as a read/write buffer.

```
// 2. allocate a buffer for the file
b = (unsigned char*)malloc( 512);
if ( b == NULL)
{
  FError = FE_MALLOC_FAILED;
  return NULL;
}
```

Only if that amount of memory is available can we proceed and allocate some more memory for the MFILE structure proper.

```
// 3. allocate a MFILE structure on the heap
fp = (MFILE *) malloc( sizeof( MFILE));
if ( fp == NULL)        // report an error
{
  FError = FE_MALLOC_FAILED;
  free( b);
  return NULL;
}
```

The buffer pointer and the MEDIA pointers can now be recorded inside the MFILE data structure.

```
// 4. set pointers to the MEDIA structure and buffer
fp->mda = D;
fp->buffer = b;
```

The filename parameter must be extracted and each character must be translated to uppercase (using the standard C library functions defined in ctype.h) and padded, if necessary, with spaces to an eight-character length.

```
// 5. format the filename into name
for( i=0; i<8; i++)
{
  // read a char and convert to upper case
  c = toupper( *filename++);
  // extension or short name noextension
  if (( c == '.') || ( c == '\0'))
    break;
```

```
    else
      fp->name[i] = c;
  } // for
  // if short fill the rest up to 8 with spaces
  while ( i<8) fp->name[i++] =' ';
```

Similarly, after discarding the dot, an extension of up to three characters must be
formatted and padded.

```
  // 6. if there is an extension
  if ( c != '\0')
  {
    for( i=8; i<11; i++)
    {
      // read char, convert to upper case
      c = toupper( *filename++);
      if ( c == '.')
        c = toupper( *filename++);
      if ( c == '\0')          // short extension
        break;
      else
        fp->name[i] = c;
    } // for
    // if short fill the rest up to 3 with spaces
    while ( i<11) fp->name[i++] = ' ';
  } // if
```

Though most C libraries provide extensive support for multiple "modes" of access to
files, such as distinguishing between text and binary files and offering an "append"
option, we will accept, at least initially, a subset consisting of just the two basic options:
r and w.

```
  // 7. copy the file mode character (r, w)
  if ((*mode == 'r')||(*mode == 'w'))
    fp->mode=*mode;
  else
  {
    FError = FE_INVALID_MODE;
    goto ExitOpen;
  }
```

With the filename properly formatted, we can now start searching the root directory of the storage device for an entry of the same name.

```
// 8. Search for the file in current directory
if ( ( r=findDIR( fp)) == FAIL)
{
  FError=FE_FIND_ERROR;
  goto ExitOpen;
}
```

Let's leave the details of the search out for now and trust the findDIR() function to return to us one of three possible values: FAIL, NOT_FOUND, and eventually FOUND. A possible failure must always be taken into account. After all, before we consider the possibility of major fatal failures of the storage device, there is always the possibility that the user simply removed the card from its slot without our knowledge. If that is the case, as in all prior error cases, we have no business continuing in the process. We'd better immediately release the memory allocated thus far and return with a NULL pointer after leaving an error code in the dedicated "mail box" FError, just as we did during the mount process.

However, if the search for the file is completed without error (whether it was found or not), we can continue initializing the MFILE structure.

```
// 9. init all counters to the beginning of the file
fp->seek = 0;          // first byte in file
fp->sec = 0;           // first sector in the cluster
fp->pos = 0;           // first byte in sector/cluster
```

The counter seek will be used to keep track of our position inside the file as we sequentially access its contents. Its value will be a 32-bit integer (unsigned) between 0 and the size of the entire file expressed in bytes.

The sec field will keep track of which sector inside the current cluster we are currently operating on. Its value will be an integer between 0 and sxc-1, the number of sectors composing each data cluster. Finally, pos will keep track of which byte inside the current buffer we are going to access next. Its value will be an integer between 0 and 511.

```
// 10. depending on the mode (read or write)
if ( fp->mode == 'r')
{
```

At this point, different things need to be done depending on whether an existing file needs to be opened for reading or a new file needs to be created for writing. Initially we will complete all the necessary steps for the fopenM() function when invoked in the read (r) mode, in which case the file had better be found.

```
// 10.1 'r' open for reading
if ( r == NOT_FOUND)
{
  FError = FE_FILE_NOT_FOUND;
  goto ExitOpen;
}
```

If it was indeed found, we trust the findDIR() function will have filled a couple more fields of the MFILE structure for us, including:

- Entry, indicating the position in the root directory where the file was found

- Cluster, indicating the number of the first data cluster used to store the file data as retrieved from the directory entry

- Size, indicating the number of bytes composing the entire file

- Time and date of creation

- The file attributes

The first cluster number will become our current cluster: ccls.

```
else
{ // found
// 10.2 set current cluster pointer on first cluster
  fp->ccls=fp->cluster;
```

Now we have all the information required to identify the first sector of data into the buffer. The function readDATA(), which we will describe in detail shortly, will perform the simple calculation required to convert the ccls and sec values into an absolute sector number inside the data area and will use the low-level readSECTOR() function to retrieve the data from the storage device.

```
// 10.3 read a sector of data from the file
  if ( !readDATA( fp))
  {
      goto ExitOpen;
  }
```

Notice that the file length is not constrained to be a multiple of a sector size, so it is perfectly possible that only a part of the data retrieved in the buffer belongs to the actual file. The MFILE structure field top will help us keep track of where the actual file data ends and padding possibly begins.

```
    // 10.4 determine how much data is really inside buffer
      if ( fp->size-fp->seek<512)
          fp->top=fp->size-fp->seek;
      else
          fp->top=512;
  } // found
} // 'r'
```

This is all we really need to complete the fopenM() function, so when opening a file for reading, we can return with the precious pointer to the MFILE structure.

```
    // 12. Exit with success
    return fp;
```

In case any of the previous steps failed, we will exit the function returning a NULL pointer after having released both the memory allocated for the sector buffer and the MFILE structure.

```
    // 12. Exit with error
ExitOpen:
    free( fp->buffer);
    free( fp);
    return NULL;
} // fopenM
```

In a top-down fashion, we can now complete the two accessory functions used during the development of fopenM(), starting with readDATA():

```
unsigned readDATA( MFILE *fp)
{
  LBA l;

  // calculate lba of cluster/sector
  l = fp->mda->data+(LBA)(fp->ccls-2) * fp->mda->sxc+fp->sec;
  fp->fpage = -1;           // invalidate FAT cache

  return( readSECTOR( l, fp->buffer));
} // readDATA
```

Ignoring for a moment the `fpage` field, notice how we use `data` and `sxc` from the `MEDIA` structure to compute the correct absolute address (LBA) of the desired data sector. Very simple!

Similarly, we create a function to read from the root directory a sector of data containing a given entry.

```
unsigned readDIR( MFILE *fp, unsigned e)
// loads current entry sector in file buffer
// returns    FAIL/TRUE
{
  LBA l;

  // load the root sector containing the DIR entry "e"
  l = fp->mda->root + (e >> 4);
  fp->fpage = -1;              // invalidate FAT cache

  return ( readSECTOR( l, fp->buffer));
} // readDIR
```

We know that each directory entry is 32 bytes large; therefore each sector will contain 16 entries.

The `findDIR()` function can now be quickly coded as a short sequence of steps enclosed in a search loop through all the available entries in the root directory.

```
unsigned findDIR( MFILE *fp)
// fp      file structure
// return  found/not_found/fail
{
  unsigned eCount;          // current entry counter
  unsigned e;               // current entry offset
  int i, a;
  MEDIA *mda = fp->mda;

  // 1. start from the first entry
  eCount = 0;

  // load the first sector of root
  if ( !readDIR( fp, eCount))
      return FAIL;
```

We start by loading the first root sector, containing the first 16 entries, in the buffer. For each entry we compute its offset inside the buffer.

```
    // 2. loop until you reach the end or find the file
    while ( 1)
    {
      // 2.0 determine the offset in current buffer
        e = (eCount&0xf) * DIR_ESIZE;
```

And we inspect the first character of the entry filename.

```
      // 2.1 read the first char of the file name
        a = fp->buffer[ e + DIR_NAME];
```

If its value is 0, indicating an empty entry and the end of the list, we can immediately exit, reporting that the filename was not found.

```
      // 2.2 terminate if it is empty (end of the list)
        if ( a == DIR_EMPTY)
        {
          return NOT_FOUND;
        } // empty entry
```

The other possibility is that the entry was marked as deleted, in which case we will skip it but we will continue searching.

```
      // 2.3 skip erased entries if looking for a match
        if ( a != DIR_DEL)
        {
```

Otherwise, it's a valid and healthy entry, and we should check the attributes to determine if it corresponds to a proper file or any other type of object. The possibilities include:

- Subdirectories
- Volume labels
- Long filenames

None of them is of our concern, since we will choose to keep things simple and we will steer clear of the most advanced and sometimes patented features of the more recent versions of the FAT file system standard.

```
        // 2.3.1 if not VOLume or DIR compare the names
          a = fp->buffer[ e + DIR_ATTRIB];
          if ( !(a & (ATT_DIR | ATT_HIDE)) )
          {
```

We will then compare the filenames character by character, looking for a complete match.

```
// compare file name and extension
for (i=DIR_NAME; i<DIR_ATTRIB; i++)
{
  if ( fp->buffer[ e+i] != fp-> name[i])
    break; // difference found
}
```

Only if every character matches will we extract the essential pieces of information from the entry and copy them into the MFILE structure, returning a FOUND code.

```
if ( i == DIR_ATTRIB)
{
  // entry found, fill the file structure
  fp->entry = eCount;     // store index
  fp->time = ReadW( fp->buffer, e+DIR_TIME);
  fp->date = ReadW( fp->buffer, e+DIR_DATE);
  fp->size = ReadL( fp->buffer, e+DIR_SIZE);
  fp->cluster = ReadL( fp->buffer, e+DIR_CLST);
  return FOUND;
}
} // not a dir nor a vol
} // not deleted
```

Should the filename and extension differ, we will simply continue our search with the next entry, remembering to load the next sector from the root directory after each group of 16 entries.

```
// 2.4 get the next entry
eCount++;
if ( (eCount & 0xf) == 0)
{ // load a new sector from the Dir
  if ( !readDIR( fp, eCount))
    return FAIL;
}
```

We know the maximum number of entries in the root directory (maxroot) and we need to terminate our search if we reach the end of the directory without a match indicating NOT_FOUND.

```
    // 2.5. exit the loop if reached the end or error
      if ( eCount >= mda->maxroot)
        return NOT_FOUND;     // last entry reached
  }// while
} // findDIR
```

Reading Data from a File

Finally, this is the moment we have been waiting for so long. The file system is mounted, a file is found and opened for reading. It is time to develop the freadM() function to freely read blocks of data from it.

```
unsigned freadM( void * dest, unsigned size, MFILE *fp)
// fp        pointer to MFILE structure
// dest      pointer to destination buffer
// count     number of bytes to transfer
// returns   number of bytes actually transferred
{
  MEDIA * mda = fp->mda;
  unsigned count=size;             // counts bytes to be transfer
  unsigned len;
```

The name, number, and sequence of parameters passed to this function are again supposed to mimic closely that of similarly named functions available in the standard C libraries. A destination buffer is supplied where the data read from the file will be copied, and a number of bytes is requested while passing the usual pointer to an open MFILE structure.

The freadM() function will do its best to read as many of the bytes requested as possible from the file and will return an unsigned integer value to report how many it effectively managed to get. In our simple implementation, if the number returned will not be identical to that requested by the calling application, we will have to assume that something major has happened. Most probably the end of file has been reached, but we will not make a distinction if, instead, another type of failure has occurred—for example, the card has been removed during the process.

As usual, we will not trust the pointer passed in the argument, and we will check instead to see whether it is pointing to a valid, initialized, MFILE structure.

```
// 1. check if fp points to a valid open file structure
if (( fp->mode != 'r'))
{ // invalid file or not open in read mode
  FError = FE_INVALID_FILE;
  return 0;
}
```

Only then we will enter a loop to start transferring the data from the sector data buffer.

```
// 2. loop to transfer the data
while ( count>0)
{
```

Inside the loop, the first condition to check will be our current position with regard to the total file size.

```
// 2.1 check if EOF reached
if ( fp->seek >= fp->size)
{
  FError=FE_EOF;    // reached the end
  break;
}
```

Notice that this error will be generated only if the application calling the freadM() function will ignore the previous symptom: the last freadM() call returned with a number of data bytes inferior to what was requested or if the calling application has requested the exact number of bytes available in the file with the previous calls.

Otherwise we will verify whether the current buffer of data has already been used up completely.

```
// 2.2 load a new sector if necessary
if (fp->pos == fp->top)
{
```

If necessary we will reset our buffer pointers and attempt to load the next sector from the file.

```
fp->pos = 0;
fp->sec++;
```

If we already used up all the sectors in the current cluster, this might force us to step into the next cluster by peeking inside the FAT and following the chain of clusters.

```
// 2.2.1 get a new cluster if necessary
if ( fp->sec == mda->sxc)
{
  fp->sec = 0;
  if ( !nextFAT( fp, 1))
  {
    break;
  }
}
```

In either case we load the new sector of data in the buffer, paying attention to verify the possibility that it might be the last one of the file and it might be only partially filled.

```
// 2.2.2 load a sector of data
if ( !readDATA( fp))
{
  break;
}
// 2.2.3 determine how much data is inside buffer
if ( fp->size-fp->seek < 512)
  fp->top = fp->size - fp->seek;
else
  fp->top = 512;
} // load new sector
```

Now that we know we have data in the buffer, ready to be transferred, we can determine how much of it we can transfer in a single chunk.

```
// 2.3 copy as many bytes as possible in a single chunk
// take as much as fits in the current sector
if ( fp->pos+count < fp->top)
  // fits all in current sector
  len = count;
else
  // take a first chunk, there is more
  len = fp->top - fp->pos;

memcpy( dest, fp->buffer + fp->pos, len);
```

Using the memcpy() function from the standard C libraries (string.h) to move a block of data from the file buffer to the destination buffer, we get the best performance as these routines are optimized for speed of execution. The pointers and counters can be updated and the loop can be repeated until all the data requested has been transferred.

```
    // 2.4 update all counters and pointers
      count-= len;            // compute what is left
      dest += len;            // advance destination pointer
      fp->pos += len;         // advance pointer in sector
      fp->seek += len;        // advance the seek pointer

  } // while count
```

Finally, we can exit the function and return the number of actual bytes transferred in the loop.

```
  // 3. return number of bytes actually transferred
  return size-count;
} // freadM
```

The nextFAT() function helped us follow the cluster chain, hopping from the current cluster to the next one.

```
unsigned nextFAT( MFILE * fp, unsigned n)
// fp  file structure
// n   number of links in FAT cluster chain to jump through
//     n==1, next cluster in the chain
{
  unsigned c;
  MEDIA * mda=fp->mda;

  // loop n times
  do {
    // get the next cluster link from FAT
    c = readFAT( fp, fp->ccls);
    // compare against max value of a cluster in FATxx
    // return if eof
    if ( c >= FAT_MCLST)       // check against eof
    {
      FError=FE_FAT_EOF;
      return FAIL;   // seeking beyond EOF
    }
```

```
      // check if cluster value is valid
      if ( c >= mda->maxcls)
      {
        FError = FE_INVALID_CLUSTER;
        return FAIL;
      }

  } while (--n>0);// loop end

  // update the MFILE structure
  fp->ccls=c;

  return TRUE;
} // get next cluster
```

As you noticed, the nextFAT() function uses, in its turn, the services of the readFAT() function to perform the hard work of actually loading an entire segment (sector) of the FAT.

```
unsigned readFAT( MFILE *fp, unsigned ccls)
// mda        disk structure
// ccls       current cluster
// return     next cluster value,
//            0xffff if failed or last
{
  unsigned p, c;
  LBA l;

  // get page of current cluster in fat
  p = ccls >>8;                  // 256 clusters per sector

  // check if already cached
  if (fp->fpage != p)
  {
    // load the fat sector containing the cluster
    l = fp->mda->fat + p;

    if ( !readSECTOR( l, fp->buffer))
      return FAT_EOF;            // failed
```

```
      // note the sector contains a valid FAT page cache
      fp->fpage = ccls>>8;
   }
   // get the next cluster value
   // cluster = 0xabcd
   // packed as:      0  |   1   |   2   |   3   |
   // word p       0   1 |  2  3 |  4  5 |  6  7 |..
   //                 cd ab|  cd ab|  cd ab|  cd ab|
   c = ReadOddW( fp->buffer, ((ccls & 0xFF)<<1));

   return c;

} // readFAT
```

Since each sector of the FAT (we will call it a *page* from now on) contains 256 entries, it is very likely that when we follow a chain of clusters or, as soon will be the case when we look for an empty cluster, we will need to access the same page over and over. Instead of wasting time continuously reloading the same sector, the readFAT() function tries to keep track of the contents (cache) of the file buffer using the fpage element of the MFILE structure to maintain the index of the last FAT page loaded. This requires some cooperation from the readDATA() and readDIR() functions so that when they overwrite the buffer contents with their contents (file data and directory table entries, respectively), they update the fpage index, invalidating it, using the index value −1 to alert readFAT().

Closing a File

Since we can only open a file for reading with the fopenM() function as defined so far, there is not much work to perform upon closing the file.

```
unsigned fcloseM( MFILE *fp)
{
   unsigned e, r;
   r = TRUE;
   //  free up the buffer and the MFILE struct
   free( fp->buffer);
   free( fp);
   return( r);
} // fcloseM
```

The Fileio Module

We can save all the functions created so far in a file called **fileio.c**, the beginning of our
file input/output library. We will need to add the usual header and a few include files:

```
/*
** fileio.c
**
** FAT16 support
*/

// standard C libraries used
#include <stdlib.h>      // NULL, malloc, free...
#include <ctype.h>       // toupper...
#include <string.h>      // memcpy...

#include <sdmmc.h>       // sd/mmc card interface
#include "fileio.h"      // file I/O routines
```

And of course, we will need to create a **fileio.h** include file as well, with all the
definitions and prototypes that we want to publish for future applications to use.

```
/*
** fileio.h
**
** FAT16 support
*/

extern char FError;                      // mailbox for error reporting

// FILEIO ERROR CODES
#define FE_IDE_ERROR          1    // IDE command execution error
#define FE_NOT_PRESENT        2    // CARD not present
#define FE_PARTITION_TYPE     3    // WRONG partition type
#define FE_INVALID_MBR        4    // MBR sector invalid signtr
#define FE_INVALID_BR         5    // Boot Record invalid signtr
#define FE_MEDIA_NOT_MNTD     6    // Media not mounted
#define FE_FILE_NOT_FOUND     7    // File not found,open for read
#define FE_INVALID_FILE       8    // File not open
#define FE_FAT_EOF            9    // attempt to read beyond EOF
#define FE_EOF               10    // Reached the end of file
#define FE_INVALID_CLUSTER   11    // Invalid cluster>maxcls
#define FE_DIR_FULL          12    // All root dir entry are taken
```

```
#define FE_MEDIA_FULL          13      // All clusters taken
#define FE_FILE_OVERWRITE      14      // A file with same name exist
#define FE_CANNOT_INIT         15      // Cannot init the CARD
#define FE_CANNOT_READ_MBR     16      // Cannot read the MBR
#define FE_MALLOC_FAILED       17      // Could not allocate memory
#define FE_INVALID_MODE        18      // Mode was not r.w.
#define FE_FIND_ERROR          19      // Failure during FILE search

typedef struct {
  LBA    fat;                          // lba of FAT
  LBA    root;                         // lba of root directory
  LBA    data;                         // lba of the data area
  unsigned short maxroot;              // max entries in root dir
  unsigned short maxcls;               // max clusters in partition
  unsigned short fatsize;              // number of sectors
  unsigned char fatcopy;               // number of copies
  unsigned char sxc;                   // number sectors per cluster
  } MEDIA;

typedef struct {
  MEDIA * mda;                         // media structure pointer
  unsigned char * buffer;              // sector buffer
  unsigned short cluster;              // first cluster
  unsigned short ccls;                 // current cluster in file
  unsigned short sec;                  // sector in current cluster
  unsigned short pos;                  // position in current sector
  unsigned short top;                  // bytes in the buffer
  int    seek;                         // position in the file
  int    size;                         // file size
  unsigned short time;                 // last update time
  unsigned short date;                 // last update date
  char   name[11];                     // file name
  char   mode;                         // mode 'r', 'w'
  unsigned short fpage;                // FAT page currently loaded
  unsigned short entry;                // entry position in cur dir
  } MFILE;

// file attributes
#define ATT_RO          1              // attribute read only
#define ATT_HIDE        2              // attribute hidden
```

```
#define ATT_SYS      4                // "       system file
#define ATT_VOL      8                // "       volume label
#define ATT_DIR      0x10             // "       sub-directory
#define ATT_ARC      0x20             // "       (to) archive
#define ATT_LFN      0x0f             // mask for Long File Name

#define FOUND        2                // directory entry match
#define NOT_FOUND    1                // directory entry not found

// macros to extract words and longs from a byte array
// watch out, a processor trap will be generated if the address
//    is not word aligned
#define ReadW( a, f)  *(unsigned short*)(a+f)
#define ReadL( a, f)  *(unsigned short*)(a+f)+\
        (( *(unsigned short*)(a+f+2))<<16)

// this is a "safe" versions of ReadW
//    to be used on odd address fields
#define ReadOddW( a, f)  (*(a+f)+( *(a+f+1) << 8))

// prototypes
unsigned nextFAT( MFILE * fp, unsigned n);
unsigned newFAT( MFILE * fp);

unsigned readDIR( MFILE *fp, unsigned entry);
unsigned findDIR( MFILE *fp);
unsigned newDIR ( MFILE *fp);

MEDIA * mount( void);
void    unmount( void);

MFILE *  fopenM  ( const char *name, const char *mode);
unsigned freadM  ( void * dest, unsigned count, MFILE *);
unsigned fwriteM ( void * src, unsigned count, MFILE *);
unsigned fcloseM ( MFILE *fp);

unsigned listTYPE( char *list, int max, const char *ext );
```

Don't worry for now if we have not fleshed out all the functions yet; we will continue
working on them as we proceed through the rest of this chapter.

Testing `fopenM()` *and* `freadM()`

It might seem like a long time since we built the last project. To verify the code that we have developed so far, we had to reach a critical mass, a minimal core of routines without which no application could have worked. Now that we have this core functionality, we can develop for the first time a small test program to read from an SD/MMC card a file created in the FAT16 file system. We will call it **ReadTest**.

The idea is to copy a text file (any text file would work) on the SD/MMC card from your PC and then have the PIC32 read the file, count the number of lines, and display it on the LCD.

Here is the main module that you will save as **ReadTest.c**:

```c
/*
** ReadTest.c
**
** 07/18/07 v2.0 LDJ
** 11/23/07 v3.0 LDJ using the LCD display
*/

#include <p32xxxx.h>
#include <plib.h>
#include <explore.h>
#include <SDMMC.h>
#include <LCD.h>
#include "fileio.h"

#define B_SIZE 10
char data[ B_SIZE];

int main( void)
{
  MFILE *fs;
  unsigned r;
  int i, c;
  char s[16];

  //initializations
  initEX16();
  initLCD();                 // init LCD display
```

```
  // main loop
  while( 1)
  {
    putsLCD( "Insert card...");
    while( !getCD());        // wait for card to be inserted
    Delayms( 100);           // de-bounce
    clrLCD();

    if ( mount())
    {
      putsLCD( "mount\n");
      if ( (fs = fopenM( "Text.txt", "r")))
      {
        c = 0;
        putsLCD("Reading...");
        do{
          r = freadM( data, B_SIZE, fs);
          for( i = 0; i<r; i++)
          {
            if ( data[ i]=='\n')
            {
              c++;
              sprintf( s, "\n%d lines", c);
              putsLCD( s);
            }
          } // for i
        } while( r==B_SIZE);
        fcloseM( fs);
        homeLCD();
        putsLCD("File closed");
      }
      else
        putsLCD("File not found!");

      unmount();
    } // mounted
    else
        putsLCD("Mount Failed!");

    getKEY();
  } // loop
} // main
```

The sequence of operation is similar to the one we adopted when testing the basic SD/MMC access module, only this time instead of calling the initMedia() function and then starting to directly read and write sectors to and from the SD/MMC card, we called the mount() function to access the FAT16 file system on the card. We opened the data file using its "proper" name, and we read data from it in blocks of arbitrary length (B_SIZE), scanning them for new line characters to mark the end of each text line. Once we'd exhausted the content of the entire file, we closed it, deallocating all the memory used.

To build the project, you will need to remember to include all the following modules:

- SDMMC.c

- fileio.c

- LCDlib.c

- explore.c

- ReadTest.c

Remember to follow the checklist for your in-circuit debugger of choice, but also in the Project Build Options dialog box (**Project** | **Build Options** | **Project**), remember to reserve some space for the heap so that the fileio functions will be able to allocate memory dynamically for the file system structures and buffers. Even if 580 bytes should suffice, give the heap ample room to maneuver; I recommend you allocate at least 2 K bytes.

After building the project and programming the Explorer 16 board, we are ready to run the test. If all goes well you will be prompted to insert the SD card in the slot and you will quickly see a counter updating on the second line of the LCD, probably too fast for you to read anything but the last value.

Notice that you can recompile the project and run the test with different sizes for the data buffer from 1 byte to as large as the memory of the PIC32 will allow. The freadM() function will take care of reading as many sectors of data required to fulfill your request as long as there is data in the file.

Writing Data to a File

We are far from finished, though. The fileio.c module is not complete until we include the ability to create new files. This will require us to create an fwriteM() function but also to complete a piece of the `fopenM()` function and a considerable extension of the `fcloseM()` function. So far we had `fopenM()` return with an error code when a file could not be found in the root directory or the mode was not `r`. But this is exactly what we want when we open a new file for writing. When we check for the mode parameter value, we need to add a new option. This time, it is when the file is `NOT_FOUND` during the first scan of the directory that we want to proceed.

```
else // 11. open for 'write'
  {
    if ( r == NOT_FOUND)
    {
```

A new file needs a new cluster to be allocated to contain its data. The function `newFAT()` will be used to search in the FAT for an available spot, a cluster that is still marked (with `0x0000`) as available. This search could fail and the function could return an error that, among other things, could indicate that the storage device is full and all data clusters are taken. Should the search be successful, though, we will take note of the new cluster position and update the `MFILE` structure, making it the first cluster of our new file.

```
            // 11.1 allocate a first cluster to it
            fp->ccls = 0;                  // indicate brand new file
            if ( newFAT( fp) != TRUE)
            { // must be media full
              FError=FE_MEDIA_FULL;
              goto ExitOpen;
            }
            fp->cluster = fp->ccls;
```

Next, we need to find an available entry space in the directory for the new file. This will require a second pass through the root directory, this time looking for either the first entry that is marked as deleted (code `0xE5`) or for the end of the list where an empty entry is found (marked with the code `0x00`).

```
            // 11.2 create a new entry
            // search again, for an empty entry this time
            if ( (r = newDIR( fp)) == FAIL)
            {    // report any error
              FError = FE_IDE_ERROR;
              goto ExitOpen;
            }
```

The function newDIR() will take care of finding an available entry and, similarly to the findDIR() function used before, will return one of three possible codes:

- FAIL, indicating a major problem occurred (or the card was removed)

- NOT_FOUND, the root directory must be full

- FOUND, an available entry has been identified

```
// 11.3 new entry not found
if ( r == NOT_FOUND)
{
  FError=FE_DIR_FULL;
  goto ExitOpen;
}
```

In both the first two cases we have to report an error and we cannot continue. But if an entry is found, we have plenty of work to do to initialize it.

After calculating the offset of the entry in the current buffer, we will start filling some of its fields with data from the MFILE structure. The file size will be first.

```
else // 11.4 new entry identified fp->entry filled
{
    // 11.4.1
    fp->size = 0;

    // 11.4.2 determine offset in DIR sector
    e = (fp->entry & 0xf) * DIR_ESIZE;

    // 11.4.3 init all fields to 0
    for (i=0; i<32; i++)
            fp->buffer[ e +i ] = 0;
```

The time and date fields could be derived from the RTCC module registers or any other timekeeping mechanism available to the application, but a default value will be supplied here only for demonstration purposes.

```
// 11.4.4 set date and time
fp->date = 0x378A; // Dec 10th, 2007
fp->buffer[ e + DIR_CDATE] = fp->date;
```

```
fp->buffer[ e + DIR_CDATE+1] = fp->date>>8;
fp->buffer[ e + DIR_DATE] = fp->date;
fp->buffer[ e + DIR_DATE+1] = fp->date>>8;

fp->time = 0x6000; // 12:00:00 PM
fp->buffer[ e + DIR_CTIME] = fp->time;
fp->buffer[ e + DIR_CTIME+1] = fp->time>>8;
fp->buffer[ e + DIR_TIME] = fp->time+1;
fp->buffer[ e + DIR_TIME+1] = fp->time>>8;
```

The file's first cluster number, the filename, and the attributes (defaults) will complete the directory entry.

```
                // 11.4.5 set first cluster
                fp->buffer[ e + DIR_CLST] = fp->cluster;
                fp->buffer[ e + DIR_CLST+1] = (fp->cluster>>8);

                // 11.4.6 set name
                for ( i = 0; i<DIR_ATTRIB; i++)
                    fp->buffer[ e + i] = fp->name[i];

                // 11.4.7 set attrib
                fp->buffer[ e + DIR_ATTRIB] = ATT_ARC;

                // 11.4.8 update the directory sector;
                if ( !writeDIR( fp, fp->entry))
                {
                    FError=FE_IDE_ERROR;
                    goto ExitOpen;
                }
            } // new entry
    } // not found
```

Back to the results of our first search through the root directory. In case a file with the same name was indeed found, we will need to report an error.

```
    else // file exist already, report error
    {
        FError = FE_FILE_OVERWRITE;
        goto ExitOpen;
    }
```

Alternatively, we would have had to delete the current entry first, release all the clusters used, and then start from the beginning. After all, reporting the problem as an error is an easier way out for now.

So much for the changes required to the fopenM() function. We can now start writing the proper new fwriteM() function, once more modeled after a similarly named standard C library function.

```
unsigned fwriteM( void *src, unsigned count, MFILE * fp)
// src      points to source data (buffer)
// count    number of bytes to write
// returns  number of bytes actually written
{
    MEDIA *mda = fp->mda;
    unsigned len, size = count;

    // 1. check if file is open
    if ( fp->mode != 'w')
    {   // file not valid or not open for writing
        FError = FE_INVALID_FILE;
        return FAIL;
    }
```

The parameters passed to the function are identical to those used in the freadM() function. The first test we will perform on the integrity of the MFILE structure, passed as a parameter, is the same as well. It will help us determine if we can trust the contents of the MFILE structure having been successfully prepared for us by a call to fopenM().

The core of the function will be a loop as well:

```
// 2. loop writing count bytes
while ( count>0)
{
```

Our intention is that of transferring as many bytes of data as possible at a time, using the fast memcpy() function from the string.h libraries.

```
    // 2.1 copy as many bytes at a time as possible
    if ( fp->pos+count < 512)
      len = count;
    else
      len = 512- fp->pos ;

memcpy( fp->buffer+ fp->pos, src, len);
```

We need to update a number of pointers and counters to keep track of our position as we add data to the buffer and increase the size of the file.

```
// 2.2 update all pointers and counters
fp->pos+=len;        // advance buffer position
fp->seek+=len;       // count the added bytes
count-=len;          // update the counter
src+=len;            // advance the source pointer

// 2.3 update the file size too
if (fp->seek > fp->size)
  fp->size = fp->seek;
```

Once the buffer is full, we need to transfer the data to the media in a sector of the currently allocated cluster:

```
// 2.4 if buffer full, write current buffer to current
sector
if (fp->pos == 512)
{

  // 2.4.1 write buffer full of data
  if ( !writeDATA( fp))
     return FAIL;
```

Notice that an error at this point would be rather fatal. We will return the code FAIL, the value of which is 0, therefore indicating that not a single byte has been transferred. In fact, all the data written to the storage device thus far is now lost.

If all proceeds correctly, though, we can now increment the sector pointers, and if we have exhausted all the sectors in the current cluster, we must consider the need to allocate a new one, calling newFAT() once more.

```
// 2.4.2 advance to next sector in cluster
fp->pos = 0;
fp->sec++;

// 2.4.3 get a new cluster if necessary
if ( fp->sec == mda->sxc)
{
  fp->sec = 0;
  if ( newFAT( fp)== FAIL)
```

```
        return FAIL;
      }
    } // store sector
  } // while count
```

Shortly, when developing newFAT(), we will have to make sure that the function accurately maintains the chaining of the clusters in the FAT as they get added to a file.

```
  // 3. number of bytes actually written

  return size-count;

} // fwriteM
```

The function is now complete and we can report the number of bytes written upon exit from the loop.

Closing a File, Take Two

Closing a file opened for reading was a mere formality and a matter of releasing some memory from the heap, but when we close a file that has been opened for writing, there is an additional amount of housekeeping work that needs to be performed.

A new and improved fcloseM() function is needed, and it will start with a check of the mode field.

```
unsigned fcloseM( MFILE *fp)
{
  unsigned e, r;
  r = FAIL;

  // 1. check if it was open for write
  if ( fp->mode == 'w')
  {
```

In fact, when we close a file, there might still be some data in the buffer that needs to be written to the storage device, although it does not fill an entire sector.

```
    // 1.1 if the current buffer contains data, flush it
    if ( fp->pos >0)
    {
      if ( !writeDATA( fp))
          goto ExitClose;
    }
```

Once more, any error at this point is a rather fatal event and will mean that all the file data is lost, since the fcloseM() function will not properly complete.

The proper root directory sector must be retrieved and an offset for the directory entry must be calculated inside the buffer.

```
// 1.2      finally update the dir entry,
// 1.2.1    retrive the dir sector
if ( !readDIR( fp, fp->entry))
  goto ExitClose;

// 1.2.2 determine position in DIR sector
e = (fp->entry & 0xf) * DIR_ESIZE;
```

Next we need to update the file entry in the root directory with the actual file size (it was initially set to zero).

```
// 1.2.3 update file size
fp->buffer[ e + DIR_SIZE]   = fp->size;
fp->buffer[ e + DIR_SIZE+1] = fp->size>>8;
fp->buffer[ e + DIR_SIZE+2] = fp->size>>16;
fp->buffer[ e + DIR_SIZE+3] = fp->size>>24;
```

Finally, the entire root directory sector containing the entry is written back to the media.

```
// 1.2.4 update the directory sector;
if ( !writeDIR( fp, fp->entry))
  goto ExitClose;
} // write
```

If all went well, we will complete the fcloseM() function, deallocating the memory used.

```
// 2. exit with success
r = TRUE;

ExitClose:
  // 3. free up the buffer and the MFILE struct
  free( fp->buffer);
  free( fp);

return( r);

} // fcloseM
```

Accessory Functions

In completing fopenM(), fcloseM() and creating the new fwriteM() function,
we have used a number of lower-level functions to perform important repetitive
tasks.

We will start with newDIR(), used to find an available spot in the root directory to create
a new file. The similarity with findDIR() is obvious, yet the task performed is very
different.

```
unsigned newDIR( MFILE *fp)
// fp       file structure
// return  found/fail, fp->entry filled
{
  unsigned eCount;              // current entry counter
  unsigned e;                   // current entry offset
  int a;
  MEDIA *mda = fp->mda;

  // 1. start from the first entry
  eCount = 0;
  // load the first sector of root
  if ( !readDIR( fp, eCount))
    return FAIL;

  // 2. loop until you reach the end or find the file
  while ( 1)
  {
  // 2.0 determine the offset in current buffer
    e = (eCount&0xf) * DIR_ESIZE;

  // 2.1 read the first char of the file name
    a = fp->buffer[ e + DIR_NAME];

  // 2.2 terminate if it is empty (end of the list)or deleted
    if (( a == DIR_EMPTY) ||( a == DIR_DEL))
    {
      fp->entry = eCount;
      return FOUND;
    } // empty or deleted entry found
```

```
  // 2.3 get the next entry
    eCount++;
    if ( (eCount & 0xf) == 0)
    { // load a new sector from the root
      if ( !readDIR( fp, eCount))
          return FAIL;
    }

  // 2.4 exit the loop if reached the end or error
    if ( eCount > mda->maxroot)
      return NOT_FOUND;                    // last entry reached
  }// while

  return FAIL;
} // newDIR
```

The function newFAT() was used to find an available cluster to allocate for a new block of data/new file:

```
unsigned newFAT( MFILE * fp)
// fp      file structure
// fp->ccls    ==0 if first cluster to be allocated
//             !=0 if additional cluster
// return   TRUE/FAIL
// fp->ccls new cluster number
{
  unsigned i, c = fp->ccls;

  // sequentially scan through FAT
  do {
    c++;    // check next cluster in FAT
    // check if reached last cluster in FAT,
    // re-start from top
    if ( c >= fp->mda->maxcls)
      c = 0;

    // check if full circle done, media full
    if ( c == fp->ccls)
    {
      FError = FE_MEDIA_FULL;
      return FAIL;
    }
```

```
  // look at its value
  i = readFAT( fp, c);

} while ( i!=0);      // scanning for an empty cluster

// mark the cluster as taken, and last in chain
writeFAT( fp, c, FAT_EOF);

// if not first cluster, link current cluster to new one
if ( fp->ccls >0)
  writeFAT( fp, fp->ccls, c);

// update the MFILE structure
fp->ccls = c;

// invalidate the FAT cache
//    (since it will soon be overwritten with data)
fp->fpage = -1;

return TRUE;
} // newFAT
```

When allocating a new cluster beyond the first one, newFAT() keeps linking the clusters in a chain, and it marks every cluster as properly used. In its turn, the function uses one more accessory function. The writeFAT() function updates the contents of the FAT and keeps all its copies current.

```
unsigned writeFAT( MFILE *fp, unsigned cls, unsigned v)
// fp       MFILE structure
// cls      current cluster
// v        next value
// return   TRUE if successful, or FAIL
{
  unsigned p;
  LBA l;

  // get address of current cluster in fat
  p = cls * 2; // always even
  // cluster = 0xabcd
  // packed as:    0 | 1     | 2     | 3     |
  // word p        1 2 | 3 4 | 4 5 | 6 7 |..
  //                 cd ab| cd ab| cd ab| cd ab|
```

```
// load the fat sector containing the cluster
l = fp->mda->fat + (p >> 9 );
p &= 0x1fe;
if ( !readSECTOR( l, fp->buffer))
  return FAIL;
// get the next cluster value
fp->buffer[ p] = v;        // lsb
fp->buffer[ p+1] = (v>>8); // msb
// update all FAT copies
for ( i=0; i<fp->mda->fatcopy; i++, l += fp->mda->fatsize)
  if ( !writeSECTOR( l, fp->buffer))
      return FAIL;

  return TRUE;
} // writeFAT
```

Finally, writeDATA() was used by both fwriteM() and fcloseM() to write actual sectors of data to the storage device, computing the sector address based on the current cluster number.

```
unsigned writeDATA( MFILE *fp)
{
  LBA l;

  // calculate lba of cluster/sector
  l=fp->mda->data+(LBA)(fp->ccls-2) * fp->mda->sxc+fp->sec;

  return ( writeSECTOR( l, fp->buffer));

} // writeDATA
```

Testing the Complete Fileio Module

It is time to test the functionality of the entire fileio.c module we just completed. This time, after mounting the file system, we will open a source file (which could be any file) and copy its contents into a new "destination" file that we will create on the spot. Here is the code we will use for the **WriteTest.c** main file.

```
/*
**WriteTest.c
**
*/
```

```
#include <p32xxxx.h>
#include <explore.h>
#include <LCD.h>
#include <SDMMC.h>
#include "fileio.h"

#define B_SIZE 100

char data[B_SIZE];

int main( void)
{
  MFILE *fs, *fd;
  unsigned c, i, p, r;
  char s[32];

  //initializations
  initEX16();
  initLCD();                  //init LCD display

  putsLCD( "Insert card...\n");
  while( !getCD());           // wait for card to be inserted
  Delayms( 100);              // wait for card to power up

  if ( mount())
  {
    clrLCD();
    if ( (fs = fopenM( "source.txt", "r")))
    {
      if ( (fd = fopenM( "dest.txt", "w")))
      {
        c = 0;              // init byte counter
        p = 0;              // init progress index
        i = fs->size/16;    // progress bar increment

      putsLCD("Copying\n");
      do{
        // copy data
        r = freadM( data, B_SIZE, fs);
        r = fwriteM( data, r, fd);

        // update progress bar
        c += r;
        while (p < c/i)
```

```
        {
          p++;
          putLCD( 0xff);   // add one bar
        }
    } while( r == B_SIZE);

    r = fcloseM( fd);
    if ( r == TRUE)
    {
      clrLCD();
      sprintf( s, "Copied \n%d bytes", c);
      putsLCD( s);
    } // close dest
    else
      putsLCD("ER:closing dest");
    } // open dest
    else
      putsLCD("ER:creating file");

    fcloseM( fs);
  } // open source
  else
    putsLCD("ER:open source");

  unmount();
  } // mount
  else
    putsLCD("ER:mount failed");

  // main loop
  while( 1);
} // main
```

Make sure you replace the source filename (SOURCE.TXT) with the actual name of the file you copied on the card for the experiment.

After creating a new project (let's call it **WriteTest** this time), we will need to add all the necessary modules to the project window, including:

- SDMMC.c
- fileio.c

- explore.c
- LCDlib.c
- WriteTest.c

Once more, remember to follow the checklists for a new project and for the in-circuit debugger setup, but this time remember to add even more space for the heap so that we will be able to dynamically allocate two buffers for two MFILE structures.

> **Note**
>
> Once enough space is left for the global variables and the stack, there is no reason to withhold any memory from the heap. Allocate as large a heap as possible to allow malloc() and free() to make optimal use of all the memory available.

After building the project and programming the executable on the Explorer 16 board, we are ready to run the test. Insert the SD card in the slot when prompted, and if all goes well for a fraction of a second, dependent on the size of the source file chosen, you will be able to see a progress bar gradually filling the second line of the LCD. When the copy is completed, a message similar to the following will appear on the LCD:

```
Copied
1806 bytes
```

Once more the actual number of bytes should reflect the size of the source file used. At this point if you transfer the SD/MMC card back to your PC, you should be able to verify that a new file has been created (see Figure 15.10).

Its size and contents are identical to the source file, whereas the date and time reflect the values we set in the fopenM() function.

Notice that if you try to run the test program a second time, it is bound to fail now.

```
ER:creating file
```

This is because, as discussed during the development of the fopenM() function, we chose to report an error when trying to create a new file (open a file for writing) and we find a file with the same name already present.

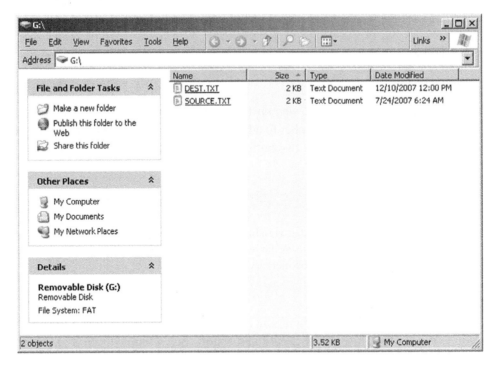

Figure 15.10: Windows Explorer Screen capture.

Notice that you can recompile the project and run the test with different sizes for the data buffer, from 1 byte to as large as the memory of the PIC32 will allow. Both the `freadM()` and `fwriteM()` functions will take care of reading and writing as many sectors of data as are required to fulfill your request. The time required to complete the operation will change slightly, though.

Code Size

The size of the code produced by the WriteTest project is considerably larger than the simple SDMMC.c module we tested in the previous chapter (see Figure 15.11).

Still, with all optimization options turned off, the code will add up to just 8,743 words. This represents only 6 percent of the total program memory space available on the PIC32MX360. I consider this a very small price to pay for a lot of functionality!

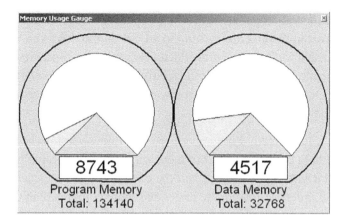

Figure 15.11: The memory usage gauge.

Debriefing

In this lesson we learned the basics of the FAT16 file system and developed a small interface module that allows a PIC32 microcontroller to read and write data files to and from a generic mass storage device. By using the SDMMC.c module, developed in the previous lesson for the low-level interface, we have created a basic file I/O interface for SD/MMC memory cards.

Now you can share data between a PIC32 application and almost any other computer system that is capable of accessing SD/MMC cards, from PDAs to laptops and desktop PCs; from DOS, Windows, and Linux machines to Apple computers running OS-X.

Tips & Tricks

A frequent question I am asked by embedded-control engineers is: "How can I interface to a 'thumb drive' (sometimes referred to as a *USB stick*), a USB mass storage device, to share/transport data between my application and a PC?"

The short answer is simple: "Don't, if you can help it!" The longer answer is: "Use an SD card instead!" and here is why. As you have seen in this lesson and the previous one, reading and writing to an SD card (miniSD and microSD included) is really simple and requires very little code and only one SPI port.

The USB interface, on the other side, has all the appeal and appearance of simplicity from the user perspective, but reading and writing to a USB thumb drive can be deceptively complex and expensive for a modest embedded-control application. First, the simplicity of the SPI interface must be replaced by the relatively greater complexity of the USB bus interface. What is required, then, is not just the standard USB interface but a host USB interface and corresponding software stack.

As of this writing, it has already been announced that future versions of the PIC32 will offer an integrated host USB interface, but there will be a considerable price to pay in terms of Flash and RAM required to support the complete software stack. This can be estimated at several orders of magnitude larger and more complex than the basic SD/MMC card solution we examined today.

Exercises

1. Review the FAT16 support libraries offered with the PIC32 tool suite. Now you have the tools to understand all that code and use the most advanced features with confidence.

2. Use the RTCC to provide the current time and date information when writing to a new file.

3. Evaluate the opportunity to use a separate buffer for more advanced FAT page caching, to further improve read/write performance.

4. Evaluate the modifications required to perform buffering of entire clusters and perform multiblock read/write operations to optimize the SD card low-level performance.

Books

Pate, Steve D., *Unix Filesystems: Evolution, Design, and Implementation* (John Wiley, 2003). Windows is our primary concern when we think of sharing files with a personal computer, but you have to look at Unix (and Linux) to find serious file systems for mission-critical data storage.

Links

www.tldp.org/LDP/tlk/tlk-title.html. *The Linux Kernel*, by David A Rusling, is an online book that describes the inner workings of Linux and its file system.

http://en.wikipedia.org/wiki/File_Allocation_Table. Once more, this is an excellent page of Wikipedia that describes the history and many ramifications of the FAT technology.

http://en.wikipedia.org/wiki/List_of_file_systems. An attempt to list and classify all major computer file systems in use.

http://en.wikipedia.org/wiki/ISO-9660. Want to know how files are written on a CD-ROM? The ISO-9660 file system is the answer.

Musica, Maestro!

The Plan

Gone is the time when music was an analog thing and the home stereo took up an entire rack of expensive electronics. Starting with music CDs almost 20 years ago and continuing today with iPods and MP3 players, music is now stored and consumed in digital form. For consumer and embedded applications, audio is an available and inexpensive option to delight but also to communicate with the user.

In this lesson we will explore the possibility to produce audio signals using the Output Compare modules of the PIC32. In Pulse Width Modulation (PWM) mode, and in combination with a more or less sophisticated low-pass filters, the Output Compare modules can be used effectively as DACs to produce an analog output signal. By modulating the analog signal with frequencies that fall into the range recognized by the human ear, between approximately 20 Hz and 20 kHz, we get sound!

Preparation

In addition to the usual software tools, including the MPLAB® IDE, the MPLAB C32 compiler, and the MPLAB SIM simulator, this lesson will require the use of the Explorer 16 demonstration board, an In-Circuit Debugger of your choice, and a soldering iron and a few components you'll need ready at hand to expand the board capabilities using the prototyping area or a small expansion board. You can check on the companion Web site (www.exploringPIC32.com) for the availability of expansion boards that will help you with the experiments that follow.

The Exploration

The way a PWM signal works is pretty simple. A pulse is produced at regular intervals (*T*), typically provided by a timer and its period register. The pulse width (*Ton*), though, is not fixed, but it is programmable and it could vary between 0 and 100 percent of the timer period. The ratio between the pulse width (*Ton*) and the signal period (*T*) is called the *duty cycle* (see Figure 16.1).

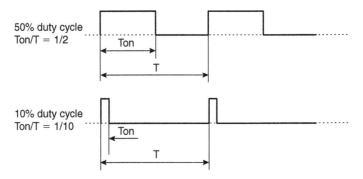

Figure 16.1: Examples of PWM signals of different duty cycles.

Two extreme cases are possible for the duty cycle: 0 percent and 100 percent. The first one would correspond to a signal that is always off. The second one would be the case when the output signal is always on. The number of possible cases in between, typically a relatively small finite number expressed as a logarithm in base 2, is commonly referred to as the *resolution* of the PWM. If, for example, there are 256 possible pulse widths, we say that we have a PWM signal with an 8-bit resolution.

If you could feed an ideal PWM signal with a fixed duty cycle to a spectrum analyzer to study its composition, you would discover that it contains three parts (see Figure 16.2):

- A DC component, with an amplitude directly proportional to the duty cycle

- A sinusoid at the fundamental frequency ($f = 1/T$)

- Followed by an infinite number of harmonics whose frequency is a multiple of the fundamental (*2f, 3f, 4f, 5f, 6f. . .*)

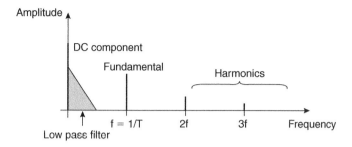

Figure 16.2: Frequency spectrum of a PWM signal.

Therefore, if we could attach an "ideal" low-pass filter to the output of a PWM signal generator to remove all frequencies from the fundamental and up, we could obtain just a clean DC analog signal whose amplitude would be directly proportional to the duty cycle.

Of course, such an ideal filter does not exist, but we can use more or less sophisticated approximations of it to remove as much of the unwanted frequency components as needed (see Figure 16.2). This filter could be as simple as a single passive R/C circuit (first-order low-pass filter) or could require several (*N*) active stages (*2xN*-order low-pass filter).

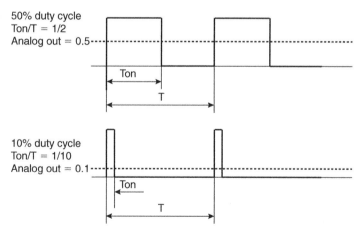

Figure 16.3: Analog output of PWM and ideal low-pass filter circuit.

If we aim to produce an audio signal and we choose the PWM frequency wisely, we can take advantage of the natural limitation of the human ear that will act as an additional filter, ignoring any signal whose frequency is outside the 20 Hz–20 kHz range. In addition, most of the audio amplifiers we might want to feed the output signal into will include a similar type of filter in their input stages. In other words, if we make sure that the PWM signal operates on a frequency at or above 20 kHz, both phenomena will contribute to help our cause and will allow us to use a simpler and more inexpensive filter circuit.

Intuitively enough, since we can change the duty cycle only once every PWM period (T), the higher the frequency of the PWM, the faster we will be able to change the output analog signal, and therefore the higher will be the frequency of the audio signal we will be able to generate.

In practical terms, this means that the highest audio signal a PWM can produce is only half of the PWM frequency. So, for example, a 20 kHz PWM circuit will be able to reproduce only audio signals up to 10 kHz, whereas to cover the entire audible frequency spectrum we need a base period of at least 40 kHz. Now you understand why it is not a coincidence that music CDs are digitally encoded at the rate of 44,100 samples per second.

OC PWM Mode

In a previous chapter we used the PIC32 Output Compare modules to produce precise timing intervals (to obtain the horizontal synchronization signal required to generate a composite video output). This time we will use the OC modules in PWM mode to generate a continuous stream of pulses with the desired duty cycle.

All we need to do to initialize the OC module to generate a PWM signal is set the three OCM bits in the OCxCON control register (see Figure 16.4) for the basic PWM

U-0	U-0	U-0	U-0	U-0	U-0	U-0	U-0
—	—	—	—	—	—	—	—
Bit 31							Bit 24

U-0	U-0	U-0	U-0	U-0	U-0	U-0	U-0
—	—	—	—	—	—	—	—
Bit 23							Bit 16

R/W-0	R/W-0	R/W-0	U-0	U-0	U-0	U-0	U-0
ON	FRZ	SIDL	—	—	—	—	—
Bit 15							Bit 8

U-0	U-0	R/W-0	R-0	R/W-0	R/W-0	R/W-0	R/W-0
—	—	OC32	OCFLT	OCTSEL	OCM<2:0>		
Bit 7							Bit 0

Figure 16.4: The Output Compare module main control register OCxCON.

configuration `0x110`. A second PWM mode is available (`0x111`), but we have no use for the fault input pins, commonly required by a different set of applications as a protection mechanism (motor control/power conversion). Next we need to select the timer on which to base the PWM period. The choice is limited to Timer2 or Timer3, but since we already used the latter for the video projects, this time we will give Timer2 our preference (see Figure 16.5).

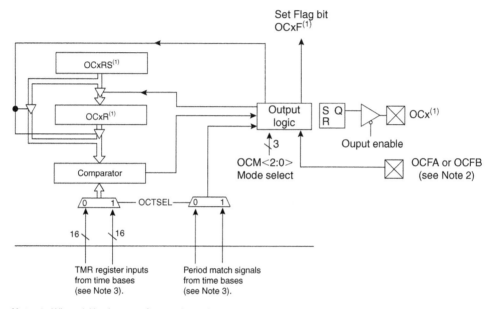

Note 1: Where 'x' is shown, reference is made to the registers associated with the respective output compare channels 1 through 5.
2: The OCFA pin controls OC1-OC3 channels. The OCFB pin controls OC4-OC5 channels.
3: Each output compare channel can use one of two selectable 16-bit time or a single 32-bit timer base.

Figure 16.5: Output Compare module block diagram.

Keeping in mind that we want to be able to produce at least a 44.1 kHz PWM period, and assuming a peripheral clock of 36 MHz, our standard configuration when using the Explorer 16 board, we can calculate the optimal configuration of the Timer2 (`T2CON`) and its period register (`PR2`). With a prescaler set to a 1:1 ratio, we obtain 816 clock ticks per period when generating an exact 44.1 kHz PWM period. This value dictates also the maximum resolution of the duty cycle for the Output Compare module.

Since we will have 816 possible values of the duty cycle, we could claim a resolution between 9 and 10 bits because we have more than 512 (2^8) but fewer than 1024 (2^9) steps.

Reducing the frequency to 20 kHz would give us 1 bit (literally) of additional resolution (taking us between 10 and 11), but that would also mean that we would be limiting the output frequency range to a maximum of 10 kHz, probably a small but noticeable difference to the human ear.

Once the chosen timer is configured and just before writing to the OCxCON register, we will need to set, for the first time, the value of the duty cycle writing to the register OCxR, and the register OCxRS. In PWM mode, the two registers will work in a master/slave configuration. Once the PWM module is started (writing the mode bits in the OCxCON register), we will be able to change the duty cycle by writing only to the OCxRS (slave) register. The OCxR register (master) will update, copying a new value from the slave OCxRS only and precisely at the beginning of each new period, to avoid glitches and to leave us with an entire period (*T*) of time to prepare the next duty cycle value.

Here is an example of a simple initialization routine for the OC1 module:

```
void initDA( int samplerate)
{
  // init OC1 module
  OpenOC1( OC_ON | OC_TIMER2_SRC | OC_PWM_FAULT_PIN_DISABLE, 0, 0);

  // init Timer2 mode and period (PR2)
  OpenTimer2( T2_ON | T2_PS_1_1 | T2_SOURCE_INT,
      FPB/samplerate);
  PR2 = FPB/samplerate-1;
  mT2SetIntPriority( 4);
  mT2ClearIntFlag();
  mT2IntEnable( 1);
} // initDA
```

Notice that we have also taken the opportunity to enable the timer interrupts so that we will be alerted each time a new period starts and we can decide how and whether to update the next duty cycle value writing to OC1RS (or using the SetDCOC1PWM() function).

Testing the PWM as a D/A Converter

To start experimenting on the Explorer 16, we will need to add just a couple of discrete components to the prototyping area. A resistor of 1 kOhm value and a capacitor of 100 nF value will produce the simplest low-pass filter (first order with a 1.5 kHz cutoff frequency). We can connect the two in series and wire them to the output pin of the OC1 module found on pin 0 of PORTD, as represented in the schematic in Figure 16.6.

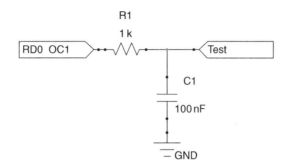

Figure 16.6: Using a PWM signal to produce an analog output.

A couple of more lines of code will complete our short test project:

```
void __ISR( _TIMER_2_VECTOR, ipl4) T2Interrupt( void)
{
  // clear interrupt flag and exit
  mT2ClearIntFlag();
} // T2 Interrupt

main( void)
{
  initEX16();        // init and enable vectored interrupts
  initDA( 44100);    // init the PWM for 44.1kHz
  SetDCOC1PWM( PR2/2);

  // main loop
  while( 1);

}// main
```

Add the usual header and include files, and save the code in a new file called **TestDA.c**. You can then create a quick test project that will contain this single file (I called it **Audio**), build it, and using the in-circuit debugger of your choice, program the Explorer 16 board. Connect a meter or an oscilloscope probe, if available, to the test point in Figure 16.6 and run the program to verify the output average DC level.

The needle of the meter (or the trace of the scope) will swing to indicate an average value of 1.5V—that is, 50 percent of the regular voltage output of a digital I/O pin on the Explorer 16 board. This is consistent with the value of the duty cycle set by the initialization routine to half of the PWM period (PR2/2). If you have an oscilloscope, you can also point the probe directly at the other end of the R1 resistor (directly to the output pin of the OC1 module) and verify that a square wave of the exact frequency of 44.1 kHz is present with a duty cycle of 50 percent (see Figure 16.7).

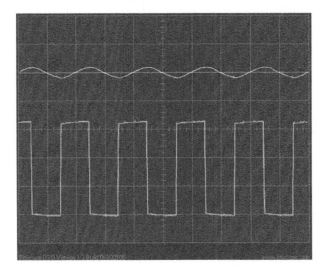

Figure 16.7: Snapshot of OC1 output (bottom) and filter (top).

You can now change the initialization routine to experiment with other values of the duty cycle between 0 and PR2 to verify the response of the circuit and the proportionality of the average output signal between 0 and 3V.

Producing Analog Waveforms

With help from the OC1 module, we have just crossed the boundary between the digital world, made of ones and zeros, and the analog world, where we have been capable of generating a multitude of values between 0 V and 3 V.

We can now start playing with the duty cycle, changing it from period to period to produce waveforms of any sort and shape. Let's start by modifying the project a little bit, adding some code to the interrupt routine that so far was left empty:

```
void __ISR( _TIMER_2_VECTOR, ipl4) T2Interrupt( void)
{
  OC1RS = (count < 22) ? PR2 : 0;
  count++;
  if ( count >= 44)
    count = 0;

  // clear interrupt flag and exit
  mT2ClearIntFlag();
} // T2 Interrupt
```

You will need to declare count as a global integer and remember to initialize it to 0.

Save the new code as **TestDA2.c**, and after replacing it as the main file in the project, rebuild the project and test it on the Explorer 16 board.

Every 20 PWM periods the filter output will alternate between the value 3 V (100 percent) and the value 0 V (0 percent), producing a square wave visible on the oscilloscope at a frequency of approximately 1 Khz (44.1 kHz/44), as shown in Figure 16.8.

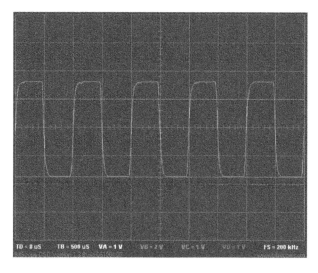

Figure 16.8: TestDA2 output, 1 kHz square wave.

A more interesting waveform could be generated by the following algorithm:

```
void __ISR( _TIMER_2_VECTOR, ipl4) T2Interrupt( void)
{
  OC1RS = count*PR2/44;
  count++;
  if ( count >= 44)
    count = 0;

  // clear interrupt flag and exit
  mT2ClearIntFlag();
} // T2 Interrupt
```

This will produce a triangular waveform (saw tooth) of approximately 3 V peak amplitude, with a gradual ramp of the duty cycle from 0 to 100 percent in 40 steps (2.5 percent each), followed by an abrupt fall back to 0, where it will repeat. This signal will repeat with a frequency of approximately 1 kHz as well (see Figure 16.9).

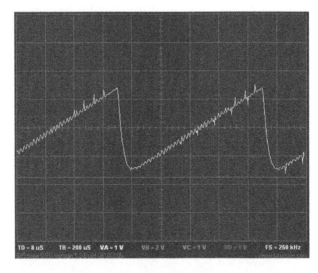

Figure 16.9: TestDA3 output, 1 kHz triangular wave.

Save the new code as "TestDA3.c", replace it as the main file of the project and rebuild.

None of the two examples will qualify as a İnice" sound if you try and feed them to an audio amplifier, although they will both have a recognizable (fundamental) high-pitched tone at about 1 kHz. Lots of harmonics will be present and audible in the audio spectrum and will give the sound an unpleasant buzz.

To generate a single clean tone, what we need is a pure sinusoid. The interrupt service routine that follows would serve the purpose, generating a perfect sinusoid at the frequency of 441 Hz; in musical terms that would be very close to an A4 (a *La* for those of "us" who have not studied music using the modern Boethian notation but rather the older *Do-Re-Mi-Fa-Sol-La-Si*).

```
void __ISR( _TIMER_2_VECTOR, ip14) T2Interrupt( void)
{
  // compute the new sample for the next cycle
  OC1RS = PR2/2 + PR2/2 * sin(count* 2*M_PI/100);
  count++;

  // clear interrupt flag and exit
  mT2ClearIntFlag();
} // T2 Interrupt
```

Unfortunately, as fast as the PIC32 and the math libraries of the MPLAB C32 compiler are, there are no chances for us to be able to use the (floating point) `sin()` function and perform the multiplications and additions required to calculate a new duty cycle value in time at the required rate of 440 Hz. The Timer2 interrupt hits every 22 us, too short a time for such a complex floating-point calculation. So, the interrupt service routine would end up "skipping" interrupts and producing a sinusoidal output that is only half (or less) than the required frequency (one octave lower). For real-time performance, we need to pretabulate the sinusoid values to perform the smallest number of calculations possible, preferably working on integers only. Here is an example that uses a constant table containing precomputed values stored in the Flash program memory of the PIC32:

```
const short Table[ 100]={
// insert comma separated values here...
};
```

To obtain the table values, let's use a spreadsheet program to compute the following formula:

```
= offset + INT( amplitude * SIN( ROW * 6.28/ PERIOD))
```

Substituting a period of 100 samples (441 Hz), an offset of 410, and an amplitude of 400, we obtain:

```
=410 + INT( 400*SIN(6.28*A1/100))
```

Let's fill the first column (**A**) of the spreadsheet with a counter and copy the formula over the first 100 rows of the second column (**B**), formatting the output for zero decimal digits (see Figure 16.10).

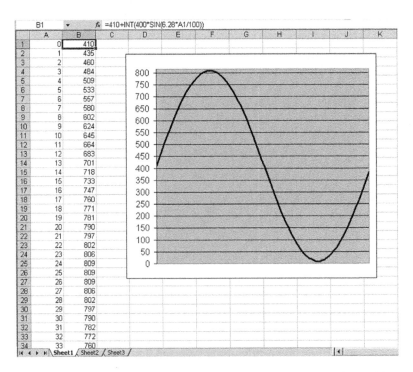

Figure 16.10: Spreadsheet to compute a 100-point sinusoid.

Select the first **100** cells of the B column and paste them directly into the MPLAB Editor. Add commas at the end of each line and close the curly brackets at the end of the table:

```
const short Table[ 100]={
// insert comma separated values here...
410,
435,
460,
484,
509,
533,
...
383};
```

The new interrupt routine will simply cycle through each element of the table:

```c
void __ISR( _TIMER_2_VECTOR, ipl4) T2Interrupt( void)
{
  OC1RS = Table[ count++];
  if ( count >= 100)
    count = 0;
  // clear interrupt flag and exit
  mT2ClearIntFlag();
} // T2 Interrupt
```

This time we will be able to easily produce the desired tone, and there will be plenty more time between the Timer2 interrupt calls to perform other tasks as well.

Save the new file as **TestDA4.c** and replace it as the main file of the project. Build and program the Explorer 16 demonstration board to check the resulting output (see Figure 16.11).

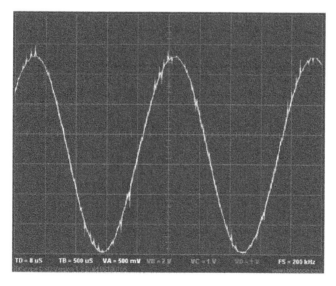

Figure 16.11: TestDA4 output, 440 Hz sinusoid.

Reproducing Voice Messages

Once we learn how to produce sound, there's no stopping us. There are infinite applications in embedded control in which we can put these capabilities to use. Any "human" interface can be greatly enhanced by using sound to provide feedback, to capture the attention of the user with alerts and error messages, or, if done properly, to simply enhance the user experience. But we don't have to limit ourselves to simple tones or basic melodies. We can reproduce any kind of sound, as long as we have a description of the required waveforms. Just like the table used for the sinusoid in the previous example, we could use a larger table to contain the unmistakable sound produced by a particular instrument or even a complete vocal message. The only limit is the room available in the Flash program memory of the PIC32 to store the data tables next to the application code.

If, in particular, we look at the possibility of storing voice messages, knowing that the energy of the human voice is mostly concentrated in the frequency range between 400 Hz and 4 kHz, we can considerably reduce our output frequency requirements and limit the PWM playback at the rate of only 8,000 samples per second. Notice that we should still maintain a high PWM frequency to keep the PWM signal harmonics outside the audio frequency range and the low-pass filter simple and inexpensive. It is only the rate at which we change the PWM duty cycle and we read new data from the table that will have to be reduced. For example, modifying the duty cycle only once every four interrupts would give us an 11,025 Hz sample rate. At this rate we would theoretically be able to play back as much as 40 seconds of voice messages (8-bit mono) stored inside the PIC32MX360 Flash memory. That is already a lot of talking for a single chip solution.

To further increase the capacity, potentially doubling it, we could start looking at simple compression techniques used for voice applications, such as ADPCM, for example. ADPCM stands for *Adaptive Differential Pulse-Coded Modulation*, and it is based on the assumption that the difference between two consecutive samples is smaller than the absolute value of each sample and can therefore be encoded using a smaller number of bits. The actual number of bits used is then optimized, and it changes dynamically to minimize signal distortion while providing a desired compression ratio. Hence the use of the term *adaptive*.

A Media Player

In the rest of this chapter, we will explore a much more ambitious project. Putting to use all the libraries and capabilities we have acquired in the last several chapters, we will attempt to create a basic multimedia application capable of playing stereo music files off an SD/MMC memory card.

The idea is to use two of the five OC modules available on the PIC32MX360, and since we care about the quality of the output, we will need a slightly more sophisticated filter than the single resistor and capacitor circuit (first-order low-pass filter) used so far in the TestDA project.

Using a low-cost dual operational amplifier like the MCP602, we can design a very simple Sallen Key (second-order) low-pass filter for the audio band that's perfectly capable of driving a small headset or to feed a more powerful stereo amplifier (see Figure 16.12).

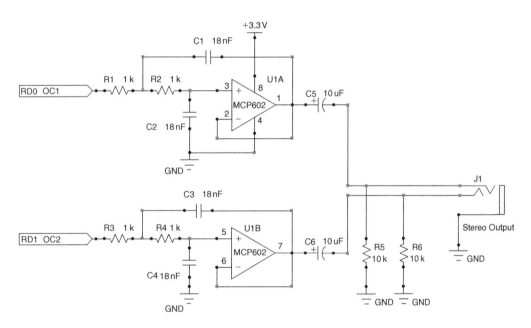

Figure 16.12: A simple audio PWM filter circuit.

As per the media format of choice, it will be the uncompressed *WAVE format* that is compatible with almost any audio application and is often the default "lossless" destination format for extracting files from a music CD.

We will start by creating a brand-new project that we will call **Wave**. We will immediately add to the project source files list the SD/MMC low-level interface (**SDMMC.c**) and the file I/O library (**fileio.c**) for access to a FAT16 file system.

The WAVE File Format

After opening a file for reading, we will need to understand the specific format used to encode the data. Files with the .wav extension, encoded in the WAVE format, are among the simplest and best documented. The WAVE format is a variant of the *RIFF file format*, a standard across multiple operating systems that uses a particular technique to store multiple pieces of information/data, dividing them into *chunks*. A chunk (see Table 16.1) is nothing more than a block of data prefixed by a header containing two 32-bit elements: the *chunk ID* and the *chunk size*.

Table 16.1: Format of a generic "chunk."

Offset	Size	Description	Value
0x00	4	Chunk ID	ASCII
0x04	4	Chunk size (size of the content)	Size
0x08	Size	Data content	
0x08 + size	0–1	Optional padding	0x00

Note also that the chunk total size must be a multiple of two so that all the data in a RIFF file ends up being nicely word aligned. If the data block size is not a multiple of two, an extra byte of padding is added to the chunk.

A chunk with the RIFF ID is always found at the beginning of a WAVE file, and its data block begins with a 4-byte *type* field. This *type* field must contain the string *WAVE*. Chunks can be nested like Russian dolls, but there can also be multiple subchunks inside a given type of chunk.

Table 16.2 illustrates a WAVE file RIFF chunk structure.

Table 16.2: RIFF chunk of type WAVE.

Offset	Size	Description	Value
0x00	4	This is the RIFF chunk ID	RIFF
0x04	4	Size of the data block + 4	Size
0x08	4	Type ID	WAVE
0x10	Size-4	Data block (subchunks)	

The data block in his turn contains a *fmt* chunk followed by a *data* chunk. As often is the case, one image is worth a million words (see Figure 16.13).

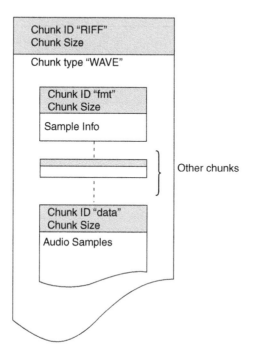

Figure 16.13: Basic WAVE file layout.

The *fmt* chunk contains a defined sequence of parameters that fully describes the stream of samples that follows in the *data* chunk, as represented in Table 16.3.

In between the *fmt* and *data* chunks, there could be other chunks containing additional information about the file, so we might have to scan the chunk IDs and skip through the list until we find the (*data*) chunk we are looking for.

The `play()` *Function*

Let's create a new `playWAV()` function that will take care of opening a WAVE file and, after capturing and decoding the information in the *fmt* chunk, will configure two PWM modules and will feed them with audio samples to reproduce a complete song in stereo. We will add the function to the **TestDA4.c** module, promptly renamed **AudioPWM.c**.

Table 16.3: The *fmt* chunk content.

Offset	Size	Description	Value
0x00	4	Chunk ID	*Fmt*
0x04	4	Chunk size	16 + extra format bytes
0x08	2	Compression code	Unsigned *int*
0x0a	2	Number of channels	Unsigned *int*
0x0c	4	Sample rate	Unsigned *long*
0x10	4	Average bytes per second	Unsigned *long*
0x14	2	Block align	Unsigned *int*
0x16	2	Significant bits per sample	Unsigned *int* (>1)
0x18	2	Extra format bytes	Unsigned *int*

```c
/*
** AudioPWM.c
**
*/
#include <p32xxxx.h>
#include <plib.h>
#include <stdlib.h>
#include <explore.h>
#include <sdmmc.h>
#include <fileio.h>
#include "AudioPWM.h"

#define B_SIZE 512      // audio buffer size

// audio configuration
typedef struct {
    char stereo;        // 0 - mono 1- stereo
    char fix;           // sign fix 0x00 8-bit, 0x80 16-bit
    char skip;          // advance pointer to next sample
    char size;          // sample size (8 or 16-bit)
} AudioCfg;
```

```
// chunk IDs
#define RIFF_DWORD      0x46464952UL
#define WAVE_DWORD      0x45564157UL
#define DATA_DWORD      0x61746164UL
#define FMT_DWORD       0x20746d66UL
#define WAV_DWORD       0x00564157UL


typedef struct {
  // data chunk
  unsigned int dlength;          // actual data size
  chardata[4];                   // "data"

  // format chunk
  unsigned   short    bitpsample; // bit per sample
  unsigned   short    bpsample;   // bytes per sample
                                  // (4=16bit stereo)

  unsigned   int      bps;        // bytes per second
  unsigned   int      srate;      // sample rate in Hz
  unsigned   short    channels;   // # of channels
                                  // (1= mono,2= stereo)


  unsigned   short    subtype;    // always 01
  unsigned   int      flength;    // size of this block (16)
  char       fmt_[4];             // "fmt_"

  char type [4];                  // file type name "WAVE"
  unsigned   int      tlength;    // size of encapsulated block
  char       riff[4];             // envelope "RIFF"
} WAVE;
```

The WAVE and AudioCfg structures will be useful to collect all the *fmt* parameters and organize the useful information in one place while the chunk ID macros will help us recognize the different unique IDs, treating them as 32-bit integers and allowing us a quick and efficient comparison.

Let's start coding the playWAV() function. It needs just one parameter: the filename.

```
int playWAV( char *name)
{
  WAVE    wav;
  MFILE   *f;
  unsigned int lc, r;
  int wi, pos, rate, period, last;
  char s[16];

  // 1. open the file
  if ( (f = fopenM( name, "r")) == NULL)
  { // failed to open
    return FALSE;
  }
```

After trying to open the file and reporting an error if unable, we will immediately
start looking inside the data buffer for the RIFF chunk ID and the WAVE type ID as a
signature. This will confirm that we have the right kind of file:

```
  // 2. verify it is a RIFF formatted file
  if ( ReadL( f->buffer, 0) != RIFF_DWORD)
  {
    fcloseM( f);
    return FALSE;
  }

  // 3. look for the WAVE chunk signature
  if ( (ReadL( f->buffer, 8)) != WAVE_DWORD)
  {
    fcloseM( f);
    return FALSE;
  }
```

If successful, we should verify that the *fmt* chunk is the first in line inside the data block.
Then we will harvest all the information needed to process the *data* block for the playback.

```
  // 4. look for the chunk containing the wave format data
    if ( ReadL( f->buffer, 12) != FMT_DWORD)
    {
      fcloseM( f);
      return FALSE;
    }
```

```
wav.channels    = ReadW( f->buffer, 22);
wav.bitpsample = ReadW( f->buffer, 34);
wav.srate       = ReadL( f->buffer, 24);
wav.bps         = ReadL( f->buffer, 28);
wav.bpsample    = ReadW( f->buffer, 32);
```

Next, we start looking for the *data* chunk, inspecting the chunk ID fields of the next block of data after the end of the *fmt* chunk and skipping the entire block if there's no matching.

```
// 5. search for the data chunk
wi = 20 + ReadW( f->buffer, 16);
while ( wi < 512)
{
  if (ReadL( f->buffer, wi) == DATA_DWORD)
    break;
  wi += 8 + ReadW( f->buffer, wi+4);
}
if ( wi >= 512) // could not find in current sector
{
  fcloseM( f);
  return FALSE;
}
```

If, in the process, we exhaust the content of the currently loaded buffer of data, we know we have a problem.

Note

Typical .wav files produced by extracting data from a music CD will have just the *data* chunk immediately following the *fmt* chunk. Other applications (MIDI interfaces, for example) can generate WAVE files with more complex structures, including multiple *data* chunks, playlists, cues, labels, and the like, but we aim at playing back only the plain-vanilla type of WAVE files.

Once it's found, the size of the *data* chunk will tell us the real number of samples contained in the file.

```
// 6. find the data size (actual wave content)
wav.dlength = ReadL( f->buffer, wi+4);
```

The playback sample rate must now be taken into consideration to determine whether we can play that "fast." It could happen that the requested sample rate exceeds our capabilities, and we might have to skip every other sample to reduce the data rate. We will consider 48k samples/sec our limit, although strictly speaking, at up to 96k samples/s, the PIC32 would still be able to produce a PWM output with 8 bits of resolution. Higher rates will be treated by gradually dividing the rate by a factor of two and doubling the skip factor.

```
// 7. if sample rate too high, skip
rate = wav.bps / wav.bpsample;     // rate = samples per second
ACfg.skip = wav.bpsample;          // skip to reduce bandwith
while ( rate > 48000)
{
  rate >>= 1;                      // divide sample rate by two
  ACfg.skip <<= 1;                 // multiply skip by two

}
```

We can then compute the required PWM period value (to be used to set the PR2 register). A problem could occur if the required period exceeds the available bits in the register (16), resulting in a period value greater than 65,536.

```
// 8. check if sample rate too low
period = (FPB/rate)-1;
if ( period > ( 65536L))           // max timer period 16 bit
{ // period too long
  fcloseM( f);
  return FALSE;
}
```

Next, the global structure ACfg is initialized with a few parameters that will help our interrupt service routine manage the audio playback:

```
// 9. init the Audio state machine
CurBuf = 0;
pos = wi+8;                        // data begin
ACfg.stereo = (wav.channels == 2);
ACfg.size = 1;                     // #bytes per channel
ACfg.fix = 0;                      // sign fix / 16 bit file
if ( wav.bitpsample == 16)
```

```
{                                  // if 16-bit
  pos++;                           // add 1 to get the MSB
  ACfg.size = 2;                   // two bytes per sample
  ACfg.fix = 0x80;                 // fix the sign
}
```

During the playback we will keep track of the number of samples extracted from the file, to determine when we have reached the end. The 32-bit integer variable lc will help us keep track of the number of samples left to play.

```
// 10 # of bytes composing the wav data chunk
lc = wav.dlength;
```

Notice that so far we have not used the freadM() function; we have been (cheating) peeking inside the file buffer, knowing fopenM() already had it loaded.

To make the playback smooth, we will use a double buffering scheme so that as the audio interrupt routines are fetching data from one buffer, we will take our time to refill the other buffer with new data from the file. The array ABuffer[] is defined as two blocks of B_SIZE bytes each (see Figure 16.14).

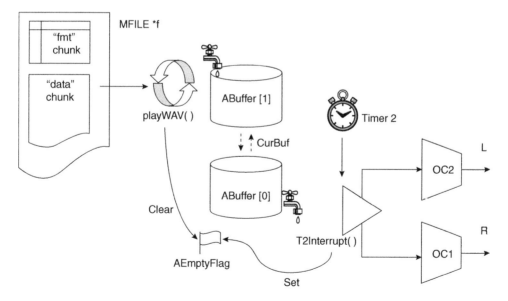

Figure 16.14: WAVE player dataflow.

For maximum performance, B_SIZE should be chosen as the size of a sector or an integer multiple of it so that the calls to the freadM() function will be able to transfer entire sectors of data at a time. We will have to verify that the time required for freadM() to fill one buffer will be shorter than the time required to play back (consume) all the data in the second buffer. When starting the double-buffering scheme, we can fill both buffers to get a head start:

```
// 11. pre-load both buffer
r = freadM( ABuffer[0], B_SIZE*2, f);
lc -= r;
AEmptyFlag = FALSE;
```

At this point we are ready to initialize the audio playback "machine," which will be simply our T2Interrupt() function modified to accommodate two channels for stereo playback using the OC1 and OC2 modules. We will initialize the OC modules first calling the initAudio() function and then we will start the Timer2 module and its interrupt to activate the playback.

```
// 12. configure Player state machine and start
initAudio();
startAudio( rate, pos, r-pos);
```

As the timer interrupt is activated, the service routine immediately starts consuming data from the first buffer, and as soon as its whole content is exhausted, it will set the AEmptyFlag flag to let us know that new data needs to be retrieved from the WAVE file and the second buffer will be selected as the active one. Therefore, to maintain the playback flowing smoothly, we will sit in a tight loop, constantly checking for the AEmptyFlag, ready to perform the refill, counting the bytes we read from the file until we use them all up.

```
// 13. keep feeding the buffers in the playing loop
// as long as entire buffers can be filled
while (lc > 0)
{ // 13.1 check user input to stop playback
  if ( readKEY())                    // if any button pressed
  {
    lc = 0;                          // playback completed
    break;
  }
```

```
  // 13.2 check if a buffer needs a refill
  if ( AEmptyFlag)
  {
    r = freadM( ABuffer[1-CurBuf], B_SIZE, f);
    lc-= r;                     // decrement byte count
    AEmptyFlag = FALSE;         // refilled

    // 13.3 <<put here additional tasks>>
    putsLCD("\n");              // on the second line
    sprintf( s, "%dKB", (wav.dlength-lc)/1024);
    putsLCD( s);                // byte count
  }
} // while wav data available
```

In the "feeding" loop we need to check for user input, reading the Explorer 16 buttons status, so that pressing a button we can stop the playback at any time. Immediately after loading a new buffer full of data, we will have a little time to spare, so this is the perfect place to put additional (short) tasks, such as updating a byte count on the LCD display, for example.

When the data left in the file is no longer sufficient to fill an entire buffer load, we can pad the buffer to size, repeating the last sample.

```
  // 14. pad the rest of the buffer
  last = ABuffer[1-CurBuf][r-1];
  while( r<B_SIZE)
    ABuffer[1-CurBuf][r++] = last;
  AEmptyFlag = FALSE;           // refilled
```

We wait then for the completion of the playback of the very last buffer, and we immediately terminate the audio playback.

```
  // 15.finish the last buffer
  AEmptyFlag = FALSE;
  while (!AEmptyFlag);

  // 16. stop playback
  haltAudio();
```

Closing the file, we will release the allocated memory and we will return to the calling application.

```
// 17. close the file
fcloseM( f);

// 18. return with success
return TRUE;
} // play
```

The Audio Routines

The `playWAV()` function we have just completed relied heavily on the lower-level audio functions to perform the actual Timer and OC peripheral initialization as well as the actual periodic update of the PWM duty cycle. The OC1 and OC2 modules are used simultaneously to produce the left and right channels. The timer interrupt routine will remain the real core of the playback functionality, just as in the previous TestDA project. A global pointer `BPtr` will keep track of our position inside each buffer as we will be using up the data to feed the PWM modules with new samples at every period.

```
void __ISR( _TIMER_2_VECTOR, ipl4) T2Interrupt( void)
{
  // 0. allow interrupt nesting
  asm( "ei");

  // 1. load the new samples for the next cycle
  OC1RS = 30+(*BPtr ^ ACfg.fix);
  if ( ACfg.stereo)
    OC2RS = 30 + (*(BPtr + ACfg.size) ^ ACfg.fix);
  else    // mono
    OC2RS = OC1RS;
```

Note

Although we can assign a medium priority to the Timer2 interrupt, we want to immediately reenable the interrupts so that a higher-priority interrupt can be immediately nested and serviced. After all, we have the luxury of an entire sampling period (22 us @ 44.1 kH) available to update the duty cycles of the two OC modules, whereas other higher-priority interrupts (for example, the composite video module, if we're to use it simultaneously . . . hint) might be less willing to wait for this interrupt to complete.

The pointer is advanced by a number of bytes that depends on both the size of the samples (16 or 8 bits each) as well as the need to skip samples to reduce the sample rate when the `playWAV()` function determines it is necessary.

```
// 2. skip samples to reduce the bitrate
BPtr += ACfg.skip;
```

As soon as a buffer-load of data is used up, we need to swap the active buffer.

```
// 3. check if buffer emptied
if ( --BCount == 0)
{
  // 3.1 swap buffers
  CurBuf = 1- CurBuf;

  // 3.2. place pointer on first sample
  BPtr = &ABuffer[ CurBuf] [ACfg.size-1];

  // 3.3 restart counter
  BCount = B_SIZE/ACfg.skip;

  // 3.4 flag a new buffer needs to be filled
  AEmptyFlag = 1;
}
```

We also reload the samples pointer, reset the samples counter, and set a flag to alert the `playWAV()` routine that we need a new buffer to be prepared before we run out of data again. Only then can we exit after clearing the interrupt flag.

```
// 4. clear interrupt flag and exit
mT2ClearIntFlag();
} // T2Interrupt
```

The initialization routine is also minimally changed from the original of the TestDA project.

```
void initAudio( void)
{ // configures peripherals for Audio playback
  // 1. activate the PWM modules
  // CH1 and CH2 in PWM mode, TMR2 based
  OpenOC1( OC_ON | OC_TIMER2_SRC | OC_PWM_FAULT_PIN_DISABLE,
          0, 0);
  OpenOC2( OC_ON | OC_TIMER2_SRC | OC_PWM_FAULT_PIN_DISABLE,
          0, 0);
```

```
// 2. init the timebase
// enable TMR2, prescale 1:1, internal clock, period
OpenTimer2(T2_ON | T2_PS_1_1 | T2_SOURCE_INT, 0);
mT2SetIntPriority( 4);    // set TMR2 interrupt priority

} // initAudio
```

But the actual audio playback is only started when we enable the Timer 2 interrupts, and that happens only after the playback state machine is properly initialized:

```
void startAudio( int bitrate, int position, int count)
{ // begins the audio playback

    // 1. init pointers and flags
    CurBuf = 0;                 // buffer 0 active first
    BPtr = ABuffer[ CurBuf] + position;
    AEmptyFlag = FALSE;

    // 2. number of actual samples to be played
    BCount = count/ACfg.skip;

    // 3. set the period for the given bitrate
    PR2 = FPB / bitrate-1;

    // 4. enable the interrupt state machine
    mT2ClearIntFlag();          // clear interrupt flag
    mT2IntEnable( 1);           // enable TMR2 interrupt
} // startAudio
```

Correspondingly, the haltAudio() function is just a matter of disabling the timer interrupts and therefore freezing the Output Compare module update, and with it the entire state machine.

```
void haltAudio( void)
{ // stops playback state machine
    mT2IntEnable( 0);
} // halt audio
```

To complete the audio module, we need just a simple header to publish the details of the playWAV() function and make it available to the project main module.

```
/*
** AudioPWM.h
*/
int playWAV( char *name);
```

A Simple WAVE File Player

Let's create a new main module that we will call **WavePlayer.c**. We will use the LCD display to prompt the user and to provide a little visual feedback in case of error as well as during the playback (see the notes 13.3 inside the core playWAV() function loop).

```
/*
** WavePlayer.c
*/
// configuration bit settings, Fcy=72 MHz, Fpb=36 MHz
#pragma config POSCMOD=XT, FNOSC=PRIPLL
#pragma config FPLLIDIV=DIV_2, FPLLMUL=MUL_18, FPLLODIV=DIV_1
#pragma config FPBDIV=DIV_2, FWDTEN=OFF, CP=OFF, BWP=OFF

#include <p32xxxx.h>
#include <plib.h>
#include <explore.h>
#include <SDMMC.h>
#include <fileio.h>
#include <LCD.h>
#include "AudioPWM.h"

main( void)
{
  initEX16();
  initLCD();
  putsLCD( "Insert card...\n");
  while ( !getCD());
  Delayms( 100);

  if ( !mount())
    putsLCD("Mount Failed");
```

```
  else
  {
    clrLCD();
    putsLCD("Playing...");
    if (!playWAV( "VOLARE.WAV"))
    {
      clrLCD();
      putsLCD("File not found");
    }
  }

  while( 1)
  {
  } // main loop

} //main
```

Build the project and program the code on the Explorer 16 board using your in-circuit debugger of choice, but don't forget to reserve room for the heap because the fileio module will use it to allocate buffers and data structures (remember to be generous . . .).

To proceed gradually, I recommend that you test the program with WAVE files of increasingly high sample rates and sizes. For example, you should run the first test with a WAVE file using 8-bit samples, mono, at 8k samples/second. Then proceed by gradually increasing the complexity of the format and the speed of playback, possibly aiming to reach with a last test the full capabilities of our application with a 16-bit per sample, stereo, 44,100 samples/second file. The reason for this gradual increase is that we need to verify whether the performance of the fileio.c module is up to the task. As the sample rate, number of channels, and size of the samples increase, so does the bandwidth required from the file system. We can quickly calculate the performance levels required by a few combinations of the above parameters.

Table 16.4 shows the byte rate required by each file format—that is, the number of bytes that get consumed by the playback function for every second (sample size × channels × sample rate). In particular, the last column shows how often a new buffer full of data will be required to be replenished (512 / byte rate), which gives us the time available for the playWAV() routine to read the next sector of data from the WAV file.

Table 16.4: WAVE file playback bandwidth requirements.

File	Sample Size	Channels	Sample Rate	Byte Rate	Reload Period (ms)
Voice mono	1	1	8,000	8,000	64.0
Voice stereo	1	2	8,000	16,000	32.0
Audio 8-bit mono	1	1	22,050	22,050	23.2
Audio 8-bit stereo	1	2	22,050	44,100	11.6
Audio 8-bit high bit-rate mono	1	1	44,100	44,100	11.6
Audio 8-bit high bit-rate stereo	1	2	44,100	88,200	5.8
Audio 16-bit mono	2	1	44,100	88,200	5.8
Audio 16-bit stereo	2	2	44,100	176,400	2.9

Now if you start experimenting gradually, as I suggested, moving down the table, you should be able to verify that you can obtain a smooth playback with any type of WAVE file all the way down to the latest row, where a sustained bit rate of more than 1.4 Mbit per second (8*Byterate) is required to keep the playback going uninterrupted.

Note

Since we decided for simplicity to use uniformly 8 bits of resolution for the PWM outputs, you shouldn't expect any increase in the quality of the audio output once you attempt to play back a WAVE file in one of the last two formats. All you will obtain at that point is a waste of the space on the SD/MMC memory card. If you want to maximize the use of the available storage space, make sure that when you copy a file onto the card, you reduce the sample size to 8 bits. That way you will be able to pack a double number of music files on the card.

Debriefing

This final lesson was perhaps the ideal conclusion for our long journey as we mixed the most advanced software and hardware capabilities in a project that covered both the digital and the analog domain. We started using the Output Compare peripherals to produce analog signals in the audio spectrum of frequencies. We used this new capability together with the fileio.c module, developed in the previous lesson, to play back uncompressed music files (WAVE file format) from a mass storage device (SD/MMC card). The basic media player application we obtained represents only a new starting point. There is no limit to the possible expansions of this project, and if I have managed to excite your curiosity and imagination, there is no limit to what you can do with the PIC32 and the MPLAB C32 compiler.

Tips & Tricks

The beginning and the end of the playback are two critical moments for the PWM modules. At rest the output filter capacitor is discharged and the output voltage is 0 V. But as soon as the playback begins, a 50-percent duty cycle will force it to ramp very quickly to approximately a 1.5 V level, producing a loud and unpleasant click. The opposite might happen at the end should we turn off the PWM modules instead of simply disabling the interrupts as we did in the demo project. The phenomenon is not dissimilar to what happens to analog amplifier circuits at power-on and -off. A simple workaround consists of adding just a couple of lines of code. Before the timer interrupt is enabled and the playback machine starts, add a small (timed) loop to gradually increase the output duty cycle from zero all the way up to the value of the first sample taken from the playback buffer.

Exercises

1. Investigate ADPCM decoding for use with voice messages (see application note AN643).

2. Search for all the .wav files on the card and build a playlist.

3. Implement a shuffle mode using the pseudo-random number generator and the playlist.

4. Perform a real-time signal spectrum analysis (FFT) and display the results with a video animation (graphic equalizer visualization).

Books

Mandrioli, D. and Ghezzi, C., *Theoretical Foundations of Computer Science* (John Wiley & Sons, NY, 1987). Not easy reading, but if you are curious about the deep mathematical theoretical foundations of computer science . . .

Links

http://en.wikipedia.org/wiki/RIFF. The RIFF file format explained.

http://en.wikipedia.org/wiki/WAV. The WAVE file format explained.

http://ccrma.stanford.edu/courses/422/projects/WaveFormat/. Another excellent description of the WAVE file format.

Disclaimer

Don't try this at home!

Final Note for the Experts

"Nel Blu Dipinto di Blu"
Italy, 1958; Domenico Modugno
Written by Franco Migliacci and Domenico Modugno

> *Penso che un sogno cosí non ritorni mai piú:*
> *Mi dipingevo le mani e la faccia di blu*
> *Poi d'improvviso venivo dal vento rapito*
> *E incominciavo a volare nel cielo infinito*
> *Volare, oh . . . cantare, oh . . .*

The lyrics are in Italian. The title translates to "In the Blue (Sky), Painted in Blue" (*volare* = to fly). Modugno sings about dreaming of painting his face and hands blue and, after being lifted by a sudden wind gust, flying away in the blue sky.

Dare to make your dreams came true!

Index

A

ABuffer [], 505
Adaptive differential pulse-coded
        modulation (ADPCM),
        496
ADC1BUF0 register, 253
AD1CHS register, 253, 265
ADC library, 255–256
AD1CON1 register, 250, 251
AD1CON2 register, 250
AD1CON3 register, 250, 252
AD1CSSL register, 250
addrLCD (), 230
ADON, 250
AD1PCFG register, 250
ADPCM; *see* Adaptive
        differential pulse-coded
        modulation (ADPCM)
AEmptyFlag flag, 506
AINPUTS, 250
Alphanumeric modules, 220
amask, 251
Analog-to-digital conversion,
        251–252
   demo for, 253–255
Analog-to-digital converter
        (ADC), 247
   basic conversion routine, 252
   block diagram of ten-bit
       high-speed, 248–249
   control registers, 249–251
   creating mini library,
       255–256
   and game program, 256–259
   initADC(), 250–251
   and potentiometer, 249–250
   sampling timing in, 252–253
   and temperature sensing,
       259–264
   voltage input in, 251

Analog waveforms, 490–495
   algorithm for, 492
   and Boethian notation, 493
   spreadsheet program for,
       493–494
Animate mode, 31–35
   delay loop for, 33–34
   Timer1 for, 31–33
ANSI C standard, integers in, 62
Arithmetic expressions; *see*
        Logic expression
Arithmetic libraries
   floating point, 69–70
   integers; *see* Integers
   measuring performance of; *see*
       StopWatch tool
Arrays, 49–50
   initialization for message, 50–51
ASCII characters, 220
   arrays, 116
   visual displaying of, 385–388
   visual printing of, 390–392
ASCII Setup dialog box, 206
Assignment statement, 8
Asynchronous serial application,
        classes of, 198
Asynchronous serial
        communication
        interface; *see* Universal
        asynchronous serial
        communication interface
        (UART)
AT (x, y), 392
AudioCfg file format, 501
AudioVideo32 board, 404

B

Baud rate, 200, 201, 202, 215
   in asynchronous serial
       interfaces, 176

   in SPI synchronous serial
       interfaces, 179, 180, 196
Baud Rate Generator (UxBREG),
        200
Binary operators, 38
BitBLT (bit block transfer), 389
Bit reversal array, 155, 156
Blocking function, 277
Block read command, 408
Block write command, 408
BMX; *see* Bus matrix (BMX)
Boolean logic values, 28
Boot record, 438
Break code, 323
Bresenham, Jack E., 368
Bresenham algorithm, 368–371
Buffer, 433
Build Project checklist, 356
Bus matrix (BMX), 131, 345
   in memory splitting, 132
busyLCD (), 230
Button inputs, 270–273
   debouncing, 275–278
   packing, 273–275
Byte command, 182

C

Cache, 345, 346
Cache memory module,
        163–164
   pre-fetched cached data in,
       164–165
Cartesian coordinate system, 362,
        389
CGRAM; *see* Character
        generator RAM buffer
        (CGRAM)
Change notification (CN) module,
        300–306
   cost evaluation, 306–307

Character generator RAM buffer
(CGRAM), 220
\n character (new line), 234
\t character (tab), 234
char integer, 46
Checklist
project build, 179, 184, 188,
190
project setup, 200, 204, 211
Chip On Glass (COG) technology,
220, 331
chunks, 498
C language
inner iteration, 379–380
pseudo-random number-
generator functions of,
364
clearHScreen (), 398
clearScreen (), 354
C library stdio.h, 442
Clipping, 363
Clock output (SCK), 406
Clock-polling state machine,
308–312
Clock system
configuration, 148–149
configuration bits; *see*
Configuration bits
oscillators, 142–143
peripheral bus clock, 147–148
primary oscillato clock chain,
146–147
Clrscr ( ), 392
clusters, 426–427
CNCON register, 301, 327
CNEN register, 301, 327
CNPUE register, 301, 327
COG technology; *see* Chip On
Glass (COG) technology
Communication device class
(CDC), 215
Communication protocol, PS/2,
287–288
Compiling, 9
Composite video interface, 348,
362
Composite video signal

defined, 332, 333
generating, 335–340
hardware interface, 335
interface, 348, 362
NTSC, 333
testing, 355–358
Configuration bits, 148
in codes, 150–152
Console library
building, 206–209
testing; *see* VT100 terminal
testing
Contact bouncing, 271
Cyclic redundancy check (CRC),
407

D

D/A converter
testing of PWM as, 488–490
data chunk, 499, 503
DDRAM; *see* Display Data RAM
buffer (DDRAM)
Debouncing, 275–278
contact, 271
Debugging, 12–13
Deinterlacing, 334
Delay loops, 33–34
Digital signal processing
coding, 152–153
FFT; *see* Fast Fourier
Transform
DIN connector, 286, 287
Direct memory access (DMA)
controller, 344
channel chaining, 349
functions, 346
library; *see* Dma.h
source pointer, 353
Display data RAM buffer
(DDRAM), 220
Division, of integers, 67–68
DMA; *see* Direct memory Access
(DMA)
DMA channel chaining, 349
DmaChnSetControl ( ),
346, 349

DmaChnSetEventControl
( ), 346
DmaChnSetTxfer ( ), 346,
354
DmaCHOpen ( ), 346
dma.h, 345
Do loops, 44–45
DONE control bit, 251, 253
Double buffering, 397–399; *see
also* Image buffers
Drawing, lines, 366–368
drawProgressBar(), 243
duty cycle, 484; *see also* Pulse
width modulation (PWM)
mode

E

EEPROM, serial, 179, 180, 181
32-bit, library, 187–191
testing, 191–193
sending commands to, 183
status register, 185–186
writing data to, 186–187
8-bit registers, 221
8088 processor; *see*
Microprocessors
Embedded-control applications,
173
communication in; *see*
Synchronous serial
communication interfaces;
Universal asynchronous
receiver and transmitters
(UART)
Embedded-control memory map,
134–135
kernel mode virtual map, 135
equal-to operator, 28
"escape sequences", 209, 210
Exceptions, 82–83
vectors table, 83
_exit () function, 27
Explorer 16 buttons
inputs, 270–278
layout, 270, 273
Explorer 16 demonstration board
interfacing, 404–405

message testing with, 54–55
for R6 potentiometer, 249
to SD/MMC memory
technology, 404
for TC1047A temperature
sensor, 260
External clock source (EC) mode,
143
External low-frequency and
low-power oscillator, 142
External primary oscillator
(POSC), 142

F
False logic value, 28, 29
Fast Fourier Transform (FFT),
153
algorithm, 154–158
arrays initialization in, 155–156
configuration bit settings,
157–158
initializations, 158
symbols used in, 156–157
FAT, 16
accessory functions, 457,
471–474
books, 480
closing a file, 457, 469–470
code size, 478–479
debriefing, 479
exploration, 426
file allocation table, 427–428
fileio module, 458–460
fundamental questions related
to, 431–442
links, 481
opening a file, 442–452
preparation, 425–426
reading data from a file,
452–457
root directory, 428–431
sectors and clusters, 426–427
testing fopenM() *and*
freadM(), 461–463
testing the complete fileio
module, 474–478
tips & tricks, 479–480

writing data to a file, 463–469
FAT file system, 425
fcloseM(), 469
FFT; *see* Fast Fourier Transform
(FFT)
File Allocation Table (FAT), 426,
427–428
fileio.c, 458, 463
Files, in project build
header files, 10
library files, 9
object files, 9
other files, 10
source files, 9
findDIR(), 446, 447, 449, 465
First-in/first-out (FIFO) buffer,
317–320
Fixed mapping translation (FMT),
130–131
Flash memory
bus offering access to, 118, 121
mapping, 132–133, 134, 135
memory space allocation, 118,
121
wait states configuration,
160–163
Flash memory, of PIC32, 388
Floating point, 69–70
measuring performance of; *see*
StopWatch tool
fmt chunk, 499, 502
Font8x8 [ ] array, 388
fopenM(), 443, 448, 457
For loops, 47–48
examples of, 48–49
Fractals, definition of, 378
Frame; *see* Video frame signals
Framed Slave mode, 347
freadM(), 452, 453
fwriteM(), 463–464

G
Gates, Bill, 425
"getC ()" function, 325–326
"getK ()" button encoding,
275–278, 319
getKey ( ), 371

getLCD ( ), 230
Graphic card, 383
greater-or-equal to operator, 29
greater-than operator, 29
Group priority level, 85–86

H
haltAudio (), 510
haltVideo ( ), 354
Hardware interface
for generation of composite
video signal, 335
HD44780 Controller
command bits, 224
compatibility of with LCD
display modules, 221–224
instruction set, 222–223
Header files, 10
Heap, 128–129
Hex dump format, 126
Home ( ), 392
Home computers; *see* ZX
Spectrum
Horizontal line signal, 334
Horizontal synchronization pulse,
334, 336
generating, 340
HRES (horizontal resolution); *see*
Resolution
HyperTerminal Properties dialog
box, 205

I
IBM PC XT, 383
ICSP/ICD interface, 16
I²C synchronous serial interfaces,
174, 175
block diagram, 174
vs. SPI synchronous serial
interfaces, 176–177
vs. UART, 176–177
ICW rotary encoder, 281
ICxC32 control bit, 327
ICxCON register, 327
ICxFEDGE control bit, 327
Image buffers, 343–344
VH pointer, 362

Image memory map, 354
#include, 235
include directory, creating,
    236–239
include search path, 237, 239
Incremental encoders, 278
initADC(), 250
initAudio(), 506
"initEX16 ()" function, 285
initialization, 30
initLCD(), 232
initMedia() function, 410
initVideo ( ), 398
Inner iteration, in C language,
    379–380
Input capture modules, 288–294
    cost evaluation, 306–307
Input/output (I/O) pins, 200, 202
    direction of, 14–15
    PortA in; *see* PortA
    PortB in, 17–19
Input/output (I/O) polling,
    307–312
    cost and efficiency evaluation,
        315–317
    testing, 312–315
Integer data type; *see* Integers
Integers
    in ANSI C standard, 62
    code generated by compiler, 63
    divisions, 67–68
    int integer, 62, 63
    long long integers, 62, 65–66
    measuring performance of; *see*
        StopWatch too
    multiplication, 63
    optimizations, 64
    testing, 64–65
Interlacing, 334
Internal low-frequency and low-
    power oscillator (LPRC),
    142
Internal oscillator (FRC), 142
Interrupt
    application of, 103–108
    handler, 82
        declaration, 88–89

latency, 82
library management, 90
managing multiple interrupt,
    95–98
multivectored management; *see*
    Multivectored interrupt
    management
priorities, 85–88
single vector management; *see*
    Single vector interrupt
    management
sources of, 84–85
Interrupt-driven rotary encoder
    input, 281–285
Interrupt Enable bit, 85
Interrupt Flag bit, 85
interrupt service routine (ISR),
    82; *see also* Interrupt
    handler
int integers, 62, 63
Isometric projection, 374

J
JTAG port, 16–17
    and PortA, 16, 17
    vs. ICSP/ICD, 16

K
Kata Kana characters, 220
Kernel mode virtual map, 135
Keyboards, 286
    interfacing to PS/2, 288–322
Keyboard-to-host communication
    waveform, 287
Key code decoding, 322–326

L
latency, interrupt, 82
LCD busy flag, 229, 230
LCD display modules
    busyLCD() function for,
        230–232
    and COG technology, 220
    Explorer 16 for, 219–221
    HD44780 compatibility with,
        221–224

initialization sequence,
    227–229
small library of functions to
    access, 226–232
for WAVE file player,
    511–512
25LC256 device datasheet,
    179, 180, 182; *see also*
    EEPROM, serial
LCDlib.c module, 274
LCD library, 232–236
LCD module control
    advanced, 339–340
    PMP configuration for,
        225–226
    LCD module controller RAM
        buffer, 220; *see also* LCD
        module control
LCD module Read Busy Flag,
    229
LCD status register, 229, 230
LED, 413, 416, 418, 419
    connected to PortA, 50
less-or-equal to operator, 29
less-than operator, 29
lib directory, creating, 236–239
Library files, 9
LINE_T, 338
Linker script, 9, 10–11, 125–126
Linking, 9
Logical block addresses (LBA),
    412
Logic analyzer, 35–37, 356–357
    measuring performance
        of video interface by,
        358–359
    message testing with, 53–54
    view, 316, 317
Logic expression, 28
Logic operators, 28–29
long integer, 45
long long integer, 46, 62, 65–66
Loops
    delay loop, 33–34
    do loops, 44–45
    for loops; *see* For loops
    main loop, 30, 33–34

for sending message; *see*
Message, loops for
while loops, 28–30, 43, 45
Low-frequency oscillator,
108–109
Low-pass filter circuit, 485
analog output of, 485
Luminance pulse, 335

M

main ( ) function, 7
infinite loop for, 44
Main loops, 30
delay loops in, 33–34
Make code, 323
malloc (), 433
Mandelbrot, Benoit, 378
Mandelbrot set
algorithm, 379–380
cardiod, 382
defined, 378
program, 380–382
Map files, 123–126
list of archives in, 124
memory configuration table,
124–125
memory sections, 125–126
Mass storage technologies,
401; *see also* Multi media
card (MMC); Secure
digital (SD)card
criteria, 402
master boot record (MBR), 434
Math functions, 371–374
McDonald, Marc, 425
Mechanical switch, 270
button inputs, 270–278
electrical response of,
270–271
rotary encoders, 278–285
MEDIA, 433, 440, 442, 443
Media player, 496–497
memcpy (), 454, 467
Memory allocation techniques,
118–123
Memory management unit
(MMU), 130

Memory mapping
embedded-control applications
in, 134–135
PIC32MX, 130–134
Memory Usage Gauge, 21
Message, loops for
initializing arrays for, 50–51
main program with variable
declarations, 51–52
testing
with Explorer 16
demonstration board,
54–55
with Logic analyzer,
53–54
with PIC32 Starter Kit,
55–57
timing constants, 50
Messages, voice, 495–496
MFILE, 442, 443, 444, 446, 447,
448, 451, 452, 457, 464,
465
Microchip TC1047A device,
259–264
Microprocessors, 383
MicroSD cards, 403
MiniSD cards, 403
MIPS core, 39
assembly programming
interface, 64
mount (), 441, 443
MPLAB C32 compiler, 344
MPLAB C32 linker, 433–434
MPLAB memory usage gauges,
422
MPLAB SIM, for debugging,
12–13
MPLAB SIM simulator, 355, 358
MPLAB SIM software simulator,
294, 299–300
mPMPMasterReadByte(),
232
mPMPOpen(), 232
MPSetAddress(), 232
MSb first, 362
Multi media card association
(MMCA), 402

Multi media card (MMC), 402
connectors pin-out, 403
Multiple interrupt, managing,
95–98
coding, 95–96
steps for new code, 97
Multivectored interrupt
management, 98–103
coding for, 101–102
Timer2 for, 103
vector table for, 99–100
MUXA, 250

N

newDIR(), 465, 471
newFAT(), 464, 468, 472, 473
nextFAT(), 455
NOT-equal to operator, 28
NTSC video standard, 333, 362

O

Object files, 9
OC32 control bit, 394
OCM bits, 486
OCxCON control register, 340,
341, 390
OCxCON register, 486, 488
OCxR register; *see* OCxCON
register
OCxRS register, 488
OLED; *see* Organic LED displays
(OLED)
OpenTimerXX ( ) function,
159
Optimizations, integers on, 64
testing, 64–65
Organic LED displays (OLED),
219
OR operation, binary, 363
Other files, 10
Output compare modules,
340–342, 486–488
initialization routine for, 488
media player and, 496–497
producing analog waveforms
with, 490–495
Output window, 299

P

Pac-Man game program, 256–259
Painted image, 332
PAL video standard, 333, 334
Parallel interfaces; *see* Parallel
  master port (PMP)
Parallel Master Port (PMP), 177,
  224–225
  configuration for LCD module
    control, 225–226
partition table, 434
Performance, 144, 145
peripheral bus clock, 147–148
Peripheral libraries, 40–41
Phase locked loops (PLL), 146
  multiplication factor of, 147
PIC24, 397; *see also*
  Microprocessors
PIC32, 327
  interfacing to PS/2, 288
PIC32 microcontroller
  amount of RAM to store video
    image in, 338, 343
  cache, 345, 346
  flash memory of, 388
PIC32MX bus, 129–130
PIC32MX memory mapping,
  130–134
PIC32 Starter Kit
  message testing with, 55–57
PICTail, 294
PICTail daughter board, 405
PICTail™, 359
Pixels, coordinate position of,
  362–363
play(); *see* PlayWAV()
playWAV(), 499–508
  and audio routines, 508–510
  coding of, 501–503
  and playback sample rate,
    504–508
plot ( ), 364
Plotting, of graphical objects,
  362–364
PMCON register, 225, 244
PMMODE register, 229, 245

PMP busy flag, 229, 230
PMP data buffer (PMPDIN),
  229
PMPDIN; *see* PMP data buffer
  (PMPDIN)
PMP library, 232–236
PMPMasterWrite(), 232
PMP mode; *see* Parallel Master
  Port (PMP) mode
PMP-to-LCD connection block
  diagram, 229
Pointers, 127–128
PortA, 7, 8
  direction of pins in, 15
  and JTAG port pins, 16, 17
  LEDs connected to, 50
PortB, 17–19
PORTD pins, 224
PORTE pins, 224
POSTEQ_N, 338
Potentiometer
  and ADC, 249–250
Power consumption, 144–145
PR4, 307, 308
PREEQ_N, 338
Preprocessor, 6
Primary oscillator clock chain,
  146–147
Printing text, on video screen, 389
Progress bar project, 240–244
  code for, 241–242
Progressive scanning, 334
Project build
  compiling, 9
  debugging, 12–13
  files in, 9–10
  linking, 9
Project Wizard, 4
PR1 registers, 31
PS/2
  communication protocol,
    287–288
  keyboard, interfacing methods
    buffering mechanism,
      317–322
    change notification (CN)

module, 300–306
  cost and efficiency
    evaluation of modules,
      306–307, 315–317
  input capture modules,
    288–294
  I/O polling, 307–317
  testing using stimulus
    scripts, 294–299
  physical interface, 286–287
  PIC32 interfacing to, 288
Pseudo-random number
  generators, 256, 258
  to test efficiency of Bresenham
    algorithm, 369
  to test video library project,
    364
Pulse width modulation (PWM)
  mode, 483
  audio routines, 508–510
  and low-pass filter, 485
  OC modules; *see* OC modules
  playWAV(), 499–508
  and reproduction of voice
    messages, 495–496
  resolution of, 484
  signals, 484–486
  testing as D/A converter,
    486–488
putcU ( ), 394
putcV ( ), 390, 394
putLCD(), 239
putsLCD(), 231, 233
PWM; *see* Pulse width
  modulation (PWM)
PWM filter circuit, audio, 497

Q

Quadrature encoders, 279

R

\r character (line end), 234
RAM, amount of
  on PIC32, 338, 343
RAM memory

bus offering access to, 129
map files, 126
mapping, 132–134
memory space allocation, 118, 121
placing heap in, 128–129
rand(), 256
RCA jack, 359
readDATA(), 447, 448
readFAT(), 456, 457
"readK ()" button encoding, 273–275
readSECTOR(), 412, 414, 434, 447
READ_SINGLE (CMD17) command, 411–412
Read status register command, testing, 182–186
ReadW() macro, 439
Real-Time Clock and Calendar (RTCC), 109–111
configuration of, 110–111
Resolution, horizontal and vertical, 338, 343, 355
RIFF chunk, 498
RIFF file format, 498; *see also* WAVE file format
root directory, 428–431
Rotary encoders, 278–281
interrupt-driven inputs, 281–285
state machine, 282
rotations array, 155
RS232 transceiver device, 198
RWTest program, 421
RX, 176

S

SAMP control bit, 251, 253, 265
Sampling timing, automating in ADC, 252–253
Scan codes, 322–323
Scanning
progressive, 334
video image, 332

SCK clock line, 195
SCK pin, 178, 180
SCL, 174, 175
SCL Generator timing example for basic, 296
SDA, 174, 175
SDI, 174, 175, 178, 193
SD/MMC cards, 402; *see also* SPI interface
to explorer 16 demo board, 404
project, 405–406
reading data from, 411–413
testing, 417–422
writing data to, 414–417
SDMMC.c module functions, 432
SDO, 174, 178, 193
SECAM video standard, 333, 334
Secondary oscillator; *see* Low-frequency oscillator
Sectors, 426
Secure Digital Card Association (SDCA), 402
Secure digital (SD) card, 402
command response code, 409
connectors pin-out, 403
initialization, 409–411
modes of communication, 403
specifications, 402, 409
writing data to, 412
SEE; *see* EEPROM, serial
Serial communication interfaces; *see* I²C synchronous serial communication interfaces; SPI synchronous serial communication interfaces; Universal asynchronous receiver and transmitters (UART)
Serial interface engine (SIE), USB, 215
Serialization, 344–351
Shadow registers, 101
short integer, 46
Simulator profiler, 299–300
sin (), 373, 493

singleV ( ), 399
Single vector interrupt management, 90–95
coding, 91, 92
testing, 93
Timer2 for, 90–91, 92, 94
Sinusoidal function graph, 373
Slave select (SS), 175
Software simulator, 10–11
Source files, 9
SPI baud rate generator (SPI2BRG), 406
SPI2CON register, 405
SPI interface, 403
selecting, 406
sending commands in, 406–409
SPI module, 344
testing, 355–358
SPI peripheral module (SPI1), 404
SPI synchronous serial interfaces, 174; *see also* SPIxCON control register
advantage of, 175
baud rate in, 179, 180, 196
block diagram, 175
clock frequency of, 180
communication using, 179–182
module block diagram, 178
PIC32, 175
vs. I²C, 176–177
vs. UART, 176–177
SPIxCON control register, 179, 180, 194
Spreadsheet
to compute 100-point sinusoid, 494
Startup code, 7
Stimulus scripts, 294–299
StopWatch tool, 70–73
coding, 70–71
StepOver command execution, 71–72
String declaration, 116–117
Subpriority level, 86
S-Video, 362

SV_LINE, 353
SV_POSTEQ, 356
swapV ( ), 398
Synchronization, 344–351
Synchronization pulses
  horizontal, 334, 336
  vertical, 334, 337
Synchronous serial
    communication interfaces
  I²C; *see* I²C synchronous serial
      communication interfaces
  SPI; *see* SPI synchronous serial
      communication interfaces
  versus UART, 174

T

T1CON, 32–33
Temperature sensing
  in ADC, 259–264
Temperature sensors; *see*
    Microchip TC1047A
    device
Text; *see* ASCII character set
Text Test project, 393
Timer1, 31–33, 227
  application of, 103–108
  low-frequency oscillator for,
    108–109
Timer2
  for multivectored interrupt
    management, 103
  for single vector interrupt
    management, 90–91, 92,
    94
Timers
  combining, 159–160
  OpenTimerXX ( ) for, 159
  Timer1; *see* Timer1
  Timer2; *see* Timer2
  WriteTimerXX ( ) for,
    159
T2Interrupt(), 506
TM162JCAWG1, Tianma, 220
TMR1, 31
Tracing function, 35–36
Triangular waveform, 492

TRISA register, 15
True logic value, 28, 29
TV broadcasting, 334
Two-dimensional function, graph
    of
  visualization, 374–378
TX, 176

U

UART; *see* Universal
    asynchronous receiver and
    transmitters (UART)
U2MODE, 201; *see also*
    UxMODE control
    registers
  initialization value for, 201
Universal asynchronous receiver
    and transmitters (UART);
    see also Console library
  basic functionality of, 199
  baud rate, 200, 201, 202,
    215
  baud rate in, 176
  block diagram, 176
  configuration, 200–202
  control registers; *see* UxMODE
    control registers
  as debugging tool, 211
  demo project, matrix,
    211–214
  modules block diagram, 199
  receiving data from, 203
  sending data to, 202–203
  testing, 204–206
  vs. I²C, 176–177
  vs. SPI synchronous serial
    communication interfaces,
    176–177
  vs. synchronous serial
    communication interfaces,
    174
USB bus, 198, 286, 288
  serial interface engine (SIE),
    215
User-defined symbols, 239
U2STA, 201

initialization value for, 202
UxMODE control registers, 201

V

Variable declarations, 45–46
Vectored interrupts, 98–103
Vertical synchronization pulses,
    334, 337
VGA, 362
VH pointer, 362
Video frame signals, 334
Video image
  buffering, 343–344
  drawing lines, 366–368
  memory map, 344
  scanning, 332
Video interfaces, 362
Video library, 351–353
Video memory
  direct memory access
    controller, 344
  image map, 344
  writing text on, 385–388
Video pins, 335, 343, 344
Video project, 354–355
Video standards, international,
    333
VirtToPhys ( ), 354
Voice messages
  PWM and reproduction of,
    495–496
Voltage
  input in ADC, 251
  output in ADC, 259
VRES (vertical resolution); *see*
    Resolution
VT100 terminal, 206, 394
  testing, 209–211

W

Wait states configuration, for flash
    memory, 160–163
WAVE file format, 498–499
  uncompressed, 497
WAVE file player, 511–513
  bandwidth require for, 513

dataflow, 505
Waveforms, analog, 490–495
    algorithm for, 492
    and Boethian notation, 493
    spreadsheet program for,
        493–494
While loop, 28–30, 43, 45
    logic expression, 28–29
window array, 155

`writeFAT()`, 473
`writeLCD()`, 239
`writeSECTOR()` function,
    414
`WriteSPI2()`, 181, 182
`WriteTimerXX ( )` function,
    159
Writing text, on video memory,
    385–388

X
xxCON registers, 225

Z
ZX80 processor; *see*
    Microprocessors
ZX Spectrum, 382–383